Laser-Based Optical Detection of Explosives

Devices, Circuits, and Systems

Series Editor
Krzysztof Iniewski
CMOS Emerging Technologies Research Inc.,
Vancouver, British Columbia, Canada

PUBLISHED TITLES:

Atomic Nanoscale Technology in the Nuclear Industry
Taeho Woo

Biological and Medical Sensor Technologies
Krzysztof Iniewski

Building Sensor Networks: From Design to Applications
Ioanis Nikolaidis and Krzysztof Iniewski

Circuits at the Nanoscale: Communications, Imaging, and Sensing
Krzysztof Iniewski

Design of 3D Integrated Circuits and Systems
Rohit Sharma

Electrical Solitons: Theory, Design, and Applications
David Ricketts and Donhee Ham

Electronics for Radiation Detection
Krzysztof Iniewski

**Embedded and Networking Systems:
Design, Software, and Implementation**
Gul N. Khan and Krzysztof Iniewski

Energy Harvesting with Functional Materials and Microsystems
Madhu Bhaskaran, Sharath Sriram, and Krzysztof Iniewski

**Graphene, Carbon Nanotubes, and Nanostuctures:
Techniques and Applications**
James E. Morris and Krzysztof Iniewski

High-Speed Devices and Circuits with THz Applications
Jung Han Choi

High-Speed Photonics Interconnects
Lukas Chrostowski and Krzysztof Iniewski

**High Frequency Communication and Sensing:
Traveling-Wave Techniques**
Ahmet Tekin and Ahmed Emira

Integrated Microsystems: Electronics, Photonics, and Biotechnology
Krzysztof Iniewski

PUBLISHED TITLES:

Optical Fiber Sensors: Advanced Techniques and Applications
Ginu Rajan

Organic Solar Cells: Materials, Devices, Interfaces, and Modeling
Qiquan Qiao

Radiation Effects in Semiconductors
Krzysztof Iniewski

Semiconductor Radiation Detection Systems
Krzysztof Iniewski

Smart Grids: Clouds, Communications, Open Source, and Automation
David Bakken

Smart Sensors for Industrial Applications
Krzysztof Iniewski

Soft Errors: From Particles to Circuits
Jean-Luc Autran and Daniela Munteanu

Solid-State Radiation Detectors: Technology and Applications
Salah Awadalla

Technologies for Smart Sensors and Sensor Fusion
Kevin Yallup and Krzysztof Iniewski

Telecommunication Networks
Eugenio Iannone

Testing for Small-Delay Defects in Nanoscale CMOS Integrated Circuits
Sandeep K. Goel and Krishnendu Chakrabarty

VLSI: Circuits for Emerging Applications
Tomasz Wojcicki

Wireless Technologies: Circuits, Systems, and Devices
Krzysztof Iniewski

Wireless Transceiver Circuits: System Perspectives and Design Aspects
Woogeun Rhee

FORTHCOMING TITLES:

Advances in Imaging and Sensing
Shuo Tang, Dileepan Joseph, and Krzysztof Iniewski

Analog Electronics for Radiation Detection
Renato Turchetta

Cell and Material Interface: Advances in Tissue Engineering, Biosensor, Implant, and Imaging Technologies
Nihal Engin Vrana

Circuits and Systems for Security and Privacy
Farhana Sheikh and Leonel Sousa

CMOS: Front-End Electronics for Radiation Sensors
Angelo Rivetti

Laser-Based Optical Detection of Explosives

Paul M. Pellegrino
UNITED STATES ARMY RESEARCH LAB
ADELPHI, MARYLAND, USA

Ellen L. Holthoff
UNITED STATES ARMY RESEARCH LAB
ADELPHI, MARYLAND, USA

Mikella E. Farrell
UNITED STATES ARMY RESEARCH LAB
ADELPHI, MARYLAND, USA

CRC Press
Taylor & Francis Group
Boca Raton London New York

CRC Press is an imprint of the
Taylor & Francis Group, an **informa** business

CRC Press
Taylor & Francis Group
6000 Broken Sound Parkway NW, Suite 300
Boca Raton, FL 33487-2742

First issued in paperback 2017

© 2015 by Taylor & Francis Group, LLC
CRC Press is an imprint of Taylor & Francis Group, an Informa business

No claim to original U.S. Government works

ISBN-13: 978-1-4822-3328-5 (hbk)
ISBN-13: 978-1-138-74805-7 (pbk)

Visit the Taylor & Francis Web site at
http://www.taylorandfrancis.com

and the CRC Press Web site at
http://www.crcpress.com

Contents

Preface

Detecting explosive hazards is an ongoing challenge of high priority for defense, academics, and industry groups, both nationally and internationally. The effort to detect materials at both point of contact and standoff distances has generated significant technological advances in hazard detection systems. In particular, the necessity to detect and identify hazardous materials at standoff distances has led to the development of many laser-based optical explosive hazard detection systems with immediate application to military, national security, and law enforcement agencies, as well as emergency response teams. The unique properties of lasers make them a versatile component for explosives detection in that they offer multiple detection approaches that are not possible with other techniques, such as wavelength specificity that allows for molecular and atomic spectroscopy, as well as the capability for long-distance propagation of intense energy, enabling standoff detection. After a thorough literature review, it was found that there is no current comprehensive literature volume that compiles state-of-the-art laser-based explosive detection technologies. Therefore, this project provides readers with a comprehensive review of past, current, and emerging laser-based methods for the detection of a variety of explosives in which we focus on the advantages, disadvantages, and areas where future research is being directed.

With the extensive current and emerging technologies in the field of laser-based optical detection of explosives, we are fortunate to work with many distinguished contributors who are leading experts in a broad spectrum of topics. We feel our contributing authors have provided the reader with a significant amount of information necessary to not only understand the technology but also appreciate its significant contribution and potential for standoff hazard detection. This volume begins with a brief introduction by the editors in Chapter 1, followed by a discussion of important considerations and evaluation methods for laser-based explosive detection systems that must be addressed before the technologies presented herein can be transitioned to a real-world environment, and considerations for sample preparation and system training. In Chapter 2, William G. Holthoff discusses laser propagation considerations and safety, while J. Greg Gillen and Jennifer R. Verkouteren, from the National Institute of Standards and Technology (NIST), discuss the preparation of standard test materials for the evaluation of standoff optical-based detection systems in Chapter 3. These introductory chapters are followed by detailed discussions on a variety of laser-based technologies applied to the detection of explosives. In Chapter 4, Rohit Bhartia, William F. Hug, Ray D. Reid, and Luther W. Beegle present a chapter on explosives detection using deep ultraviolet native fluorescence and researchers from the United States Army Edgewood Chemical Biological Center discuss Raman spectroscopy in Chapter 5. Charles M. Wynn, from Massachusetts Institute of Technology Lincoln Laboratory, discusses the capabilities of photodissociation followed by laser-induced fluorescence in Chapter 6. This is followed by chapters by David A. Cremers on laser-induced breakdown spectroscopy (Chapter 7) and by Anish K. Goyal and Travis R. Myers on reflectometry and hyperspectral imaging (Chapter 8). Mark C. Phillips and Bruce E. Bernacki, from Pacific Northwest National Laboratory, present additional reflectance measurement techniques and analysis of explosives residues in Chapter 9. In Chapter 10, researchers from the United States Naval Research Laboratory present photothermal methods for laser-based detection of explosives, while J. Brian Leen, Manish Gupta, and Douglas S. Baer discuss cavity-enhanced absorption spectrometry in Chapter 11, and Henry O. Everitt and Frank C. De Lucia describe the detection and recognition of explosives using terahertz-frequency spectroscopic techniques in Chapter 12. This volume concludes with Robert J. Levis and John J. Brady discussing short-pulse laser-based techniques in Chapter 13.

Although the editors believe that each chapter in this volume can stand alone and be read without having to refer to additional chapters, we would like to suggest some chapter pairings so that the reader may gain a complete understanding of the complexities and capabilities associated with some of the technologies presented herein. Chapters 4 and 5 both provide discussions pertaining to Raman spectroscopy; however, Chapter 4 focuses on deep ultraviolet resonance Raman and Chapter 5 discusses Raman cross sections in more detail. Chapters 8, 9, and 10 describe a variety of applications of reflectance spectroscopy. Chapter 8 focuses on reflection–absorption infrared spectroscopy, while Chapter 9 discusses hyperspectral imaging based on reflectance spectroscopy, and Chapter 10 explains reflectance as it relates to photothermal methodology used to investigate particles on surfaces.

While we have touched on several technologies in this book, obviously there are numerous examples of emerging techniques that are still in early developmental stages, and with additional research and maturity these could be included in follow-up editions. As the reader will find, laser-based explosives detection is a work in progress, with a variety of new capabilities emerging in recent years, due to the availability of more exotic laser sources. As these sources become more common, rugged, and less expensive, surely more applications will be discovered in the near future. We have enjoyed and appreciated working with all the talented authors to compile this book; we hope that you find it is informative gives an accurate portrayal of the technologies currently available.

Mikella E. Farrell, Ellen L. Holthoff, and Paul M. Pellegrino
United States Army Research Laboratory, Adelphi, MD

MATLAB® is a registered trademark of The MathWorks, Inc. For product information, please contact:

The MathWorks, Inc.
3 Apple Hill Drive
Natick, MA 01760-2098 USA
Tel: 508 647 7000
Fax: 508-647-7001
E-mail: info@mathworks.com
Web: www.mathworks.com

Editors

Editors Paul M. Pellegrino, Ellen L. Holthoff, and Mikella E. Farrell standing in a crater created with a small directed and supervised C4 explosion at Fort A.P. Hill, Virginia, August, 2011.

Paul M. Pellegrino is currently chief of the Optics and Photonics Integration Branch in the Sensors and Electron Devices Directorate at the US Army Research Laboratory (ARL), Adelphi, Maryland. He has been with the ARL as a physicist for approximately 17 years. In addition to his branch chief duties, he actively participates in numerous spectroscopic efforts for hazardous material sensing. He has more than 20 years' experience in the areas of optics, physics, and computational physics, with a strong emphasis in the last 15 years on the application of novel spectroscopy and optical transduction for chemical and biological sensing. Current research interests include surface-enhanced Raman scattering, quantum control, and photo-acoustic spectroscopy. He has authored and coauthored over 40 open literature publications on hazardous material detection. Dr. Pellegrino is a member of the Optical Society of America (OSA), the International Society for Optics and Photonics (SPIE), and the Society for Applied Spectroscopy and currently acts as a reviewer for the OSA and SPIE. He can be reached at paul.m.pellegrino.civ@mail.mil.

Ellen L. Holthoff (nee Shughart) is a research chemist at the US Army Research Laboratory (ARL), the army's central laboratory for combat material, located in Adelphi, Maryland. Dr. Holthoff works in the Sensors and Electron Devices Directorate of ARL, where her experimental work includes the development of microelectromechanical systems (MEMS)-scale photo-acoustic sensor platforms for gas detection, molecularly imprinted polymers for chemical and biological sensing applications, and drop-on-demand ink-jet printing for sample standardization. Other research interests include sol-gel chemistry and fluorescence spectroscopy. Dr. Holthoff held an Oak Ridge Associated Universities Postdoctoral Fellowship at ARL. She has authored and coauthored over 30 research papers and conference proceedings as well as three book chapters and numerous internal army reports. She can be reached at ellen.l.holthoff.civ@mail.mil.

Mikella E. Farrell (nee Hankus) is a research chemist at the US Army Research Laboratory (ARL) of the US Army Research Development and Engineering Command (RDECOM), in Adelphi, Maryland. ARL is the army's central laboratory, sporting unique facilities and a dedicated workforce of government and private sector partners who make up the largest source of world-class integrated research and analysis in the army. From ARL, technology and analysis products are moved into RDECOM Research, Development, and Engineering Centers (REDECs) and industry customers. Mikella Farrell works in the Sensors and Electron Devices Directorate of ARL, where she performs both basic and applied research in a wide variety of areas. Some of her work has included developing surface-enhanced Raman scattering (SERS) substrates for army-specific biological and hazard sensing, biomimetic hazard sensing employing designed peptides, the fabrication of a nanoscale SERS imaging probe, and transitioning a standardized technique for the fabrication of drop-on-demand hazard test evaluation coupons. She also has been involved with supporting Defense Advanced Research Projects Agency SERS Fundamentals programs, several university SERS-based research programs, and the evaluation of numerous fielded standoff hazard detection systems. She holds a US patent, has coauthored a book chapter, and has authored over 30 research papers in well-respected journals, proceedings, and internal Army reports. She can be reached at mikella.e.farrell.civ@ mail.mil.

Contributors

Douglas S. Baer
Los Gatos Research
Mountain View, California

Luther W. Beegle
Jet Propulsion Laboratory
California Institute of Technology
Pasadena, California

Bruce E. Bernacki
Optics and Infrared Sensing
Pacific Northwest National Laboratory
Richland, Washington

Rohit Bhartia
Jet Propulsion Laboratory
California Institute of Technology
Pasadena, California

John J. Brady
Signature Science, LLC
Egg Harbor Township, New Jersey

Jeff M. Byers
Naval Research Laboratory
Washington, DC

Steven D. Christesen
Research and Technology Directorate
Edgewood Chemical and Biological
  Center
Aberdeen, Maryland

David A. Cremers
Applied Research Associates, Inc.
Albuquerque, New Mexico

Frank C. De Lucia
Department of Physics
Ohio State University
Columbus, Ohio

Erik D. Emmons
Leidos
Chemical and Biological Solutions
  Directorate
Abingdon, Maryland

Henry O. Everitt
Army Aviation and Missile Research
  Development and Engineering Center
Redstone Arsenal, Alabama

Mikella E. Farrell
Sensors and Electron Devices Directorate
United States Army Research
  Laboratory
Adelphi, Maryland

Augustus W. Fountain III
Research and Technology Directorate
Edgewood Chemical and Biological
  Center
Aberdeen, Maryland

Robert Furstenberg
Naval Research Laboratory
Washington, DC

Greg Gillen
Materials Measurement Laboratory
National Institute of Standards and
  Technology
Gaithersburg, Maryland

Anish K. Goyal
Block Engineering, LLC
Marlborough, Massachusetts
and
Block MEMS, LLC
Marlborough, Massachusetts

Jason A. Guicheteau
Research and Technology Directorate
Edgewood Chemical and Biological
  Center
Aberdeen, Maryland

Manish Gupta
Los Gatos Research
Mountain View, California

Ellen L. Holthoff
Sensors and Electron Devices Directorate
United States Army Research
  Laboratory
Adelphi, Maryland

William G. Holthoff
Explosives Ordnance Disposal
  Technology Division
Naval Surface Warfare Center Indian Head
Indian Head, Maryland

William F. Hug
Photon Systems, Inc.
Covina, California

Christopher A. Kendziora
Naval Research Laboratory
Washington, DC

J. Brian Leen
Los Gatos Research
Mountain View, California

Robert J. Levis
Center for Advanced Photonics
  Research
Department of Chemistry
Temple University
Philadelphia, Pennsylvania

R. Andrew McGill
Naval Research Laboratory
Washington, DC

Travis R. Myers
Lincoln Laboratory
Massachusetts Institute of
  Technology
Lexington, Massachusetts

Viet K. Nguyen
Naval Research Laboratory
Washington, DC

Michael R. Papantonakis
Naval Research Laboratory
Washington, DC

Paul M. Pellegrino
Sensors and Electron Devices
  Directorate
United States Army Research
  Laboratory
Adelphi, Maryland

Mark C. Phillips
Optics and Infrared Sensing
Pacific Northwest National
  Laboratory
Richland, Washington

Ray D. Reid
Photon Systems, Inc.
Covina, California

Jennifer R. Verkouteren
Materials Measurement Science
  Division
Materials Measurement Laboratory
National Institute of Standards and
  Technology
Gaithersburg, Maryland

Charles M. Wynn
Lincoln Laboratory
Massachusetts Institute of
  Technology
Lexington, Massachusetts

1 Laser-Based Optical Detection of Explosives

Mikella E. Farrell, Ellen L. Holthoff, and Paul M. Pellegrino

CONTENTS

1.1 INTRODUCTION

Explosives and energetic precursor materials are becoming commonly used to purposefully harm people and property. Energetic attacks occur in foreign and domestic sites throughout the world. Increasingly, the Department of Defense, coalition military, and security forces are seeking out means to detect and identify explosive hazards before the detonation event occurs, preventing warfighters and innocent bystanders from being impacted. Foreign and national governments are pouring funds into industry-, academic-, and government-sponsored research initiatives to develop sensitive and accurate systems for the detection of many classes of hazardous materials. Of the many hazards commonly encountered by warfighters, energetics in the form of improvised explosive devices (IEDs) and homemade explosives (HMEs) remain the primary threat to military and civilian personnel. Unfortunately, the energetic materials used to construct these devices can be extremely difficult to detect. This is in large part due to the extensive variety of materials that can be used as explosives or explosive precursors. Therefore, to answer this need, various hazard detection technologies have emerged. For example, ion mobility spectrometry (IMS)[1,2] has been employed as an explosive screening technology in most airports since September 11, 2001, and the technology has been adopted for military use in the portable Joint Chemical Agent Detector (JCAD) for the detection of chemical warfare agents and toxic industrial chemicals. The development of specific polymers (e.g., amplified fluorescent polymers) for detection of explosives has been successful for a handful of materials,[3–5] including 2,4,6-trinitrotoluene (TNT), resulting in commercialization by FLIR®,[6] and various colorimetric kits,[7] based on chemical reactions that produce colored unambiguous products when exposed to a specific class of analyte, are currently being used by law enforcement and the military. Although these analysis methods have proven effective for certain classes of materials, they all require direct or close contact with the explosive hazards, which is not always ideal or safe for the operator. The need for noncontact, ranged detection of hazards has focused attention on the development of standoff capabilities. Most of the emerging technologies being developed to meet these requirements are primarily standoff laser-based optical detection systems, and each offers a multitude of benefits as well as some challenges to address the complexity of hazard detection.

1.2 THE CHALLENGE

There are many challenges that must be overcome for the successful detection of hazardous materials. The challenges include but are not limited to detection of a material, accurate differentiation of hazardous from nonhazardous material, overcoming interferents, and determination of various quantities of material. Generally, the detection of explosive or explosive precursor material is based on matching a particular physical or chemical property of the sample to an ever-expanding threat library. There are a variety of potential explosives and precursor materials encountered by the military, national security agencies, law enforcement agencies, and emergency response teams. The most common hazards, including their reaction chemistries, have been discussed thoroughly in the literature.[8,9] Example explosive materials and precursors, and their chemical compositions, are listed in Table 1.1.

TABLE 1.1
Examples of Common Explosives

Material	Composition	Application
EGDN	$C_2H_4N_2O_6$	Explosive
RDX	$C_3H_6N_6O_6$	Explosive
TNT	$C_7H_5N_3O_6$	Explosive
DNT	$C_7H_6N_2O_4$	Explosive
PETN	$C_6H_8N_4O_{12}$	Explosive
HMX	$C_4H_8N_8O_8$	Explosive
TATP	$C_9H_{18}O_6$	Explosive
AN	NH_4NO_3	Precursor material
Sucrose	$C_{12}H_{22}O_{11}$	Precursor material
Chlorate	$KClO_3, NH_4ClO_4$	Primer, propellant

EGDN, ethylene glycol dinitrate; RDX, 1,3,5-trinitroperhydro-1,3,5-triazine; TNT, 2,4,6-trinitrotoulene; DNT, 2,4-dinitrotoluene; PETN, pentaerythritol tetranitrate; HMX, octahydro-1,3,5,7-tetranitro-1,3,5,7-tetrazocine; TATP, triacetone triperoxide); precursor materials, sucrose and ammonium nitrate (AN); explosive primers and propellants, potassium chlorate ($KClO_3$) and ammonium perchlorate (NH_4ClO_4).

Explosives are organic compounds primarily composed of carbon (C), hydrogen (H), nitrogen (N), and oxygen (O) in various ratios. As compared to other classes of organic compounds, explosives are characterized as being both rich in nitrogen and oxygen and poor in carbon and hydrogen. For example, the high-velocity military explosive HMX (octahydro-1,3,5,7-tetranitro-1,3,5,7-tetrazocine) is composed of approximately 37.8% nitrogen, 43.2% oxygen, 16.2% carbon, and 2.7% hydrogen by weight. The similarities in chemical composition among explosive materials and other subtle nuances of data collection can cause various challenges associated with their accurate detection and identification.

It can be difficult to distinguish explosives from nonhazardous, benign materials based on physical properties alone; this is in part due to the lack of specificity that can occur in some detection techniques (e.g., colorimetric,[7] fluorescence[4]) and libraries. Common explosives and both HME and IED materials have motivated research efforts to focus on detecting both bulk and trace quantities on multiple surfaces in order to better safeguard military and civilian personnel. It has been previously shown that trace explosive residues can persist on surfaces[10] like car door handles, clothing,[11] and even human hair.[12,13] Being able to detect these trace materials may lead forensic investigators to the hazard material before the detonation event ("left of the boom" idea) and potential harm to the public occur.

Additionally, environmental conditions can impact the nature of the sample residue remaining.[14,15] In order to assess the capabilities of hazard detection systems that rely on chemical signatures for detection and identification, the persistence and fate of explosive residues in the operational environment must be understood. Specifically, some materials are known to experience photodegradation when exposed to the uncontrolled environment for some length of time, and in solution some materials can break down into constituent components or chemically react with the solvent. Even some mechanical disturbances can impact the way the residue materials present for detection and identification. Therefore, the capability of sensor systems to accurately detect and identify an explosive hazard is based on the ability to distinguish the explosives as well as any degradation products. The hazard detection systems must be evaluated with appropriate test material concentrations under controlled conditions in order to correctly identify and quantify unknown residues commonly utilized in theater.

For the detection of hazards, a system that embodies as many of the following attributes as possible is needed. Ideally, the system should be able to detect many types of explosives and be adaptable or expandable to new targets; it should demonstrate low analyte detection limits (lowest quantity detectable); it should have a high probability of positive detection and a low probability of false alarm; it should demonstrate good throughput or areal coverage; it should have limited vulnerabilities to countermeasures or interferents; it should be rugged and robust and have limited maintenance needs; it should be operationally easy to use; it should have low cost, space, and power requirements; and it should have standoff detection capabilities to minimize risk to the warfighter or security personnel. While technology does not yet exist that combines all these attributes, using several of the technologies that are described in this book in concert may bring us closer to preventing acts of terror before they occur, thus safeguarding warfighters and civilians.

1.3   A SOLUTION: LASER-BASED TECHNIQUES

Although broadband sources, such as lamps, are a common component of numerous optical spectroscopy techniques, modern research done in the area of explosives detection has been mainly performed using laser sources. For standoff detection techniques, the long-distance propagation of laser radiation provides capabilities not possible with other light sources. Pure, laser-based methods could enable the rapid, sensitive, and accurate point, remote, or standoff detection of energetic materials. Laser-based approaches typically have less complex experimental designs enabling such analysis as compared to broadband-based approaches. Commonly, a laser beam is focused and irradiates a gas-phase or condensed-phase sample. Light that is emitted, reflected, or scattered is collected and directed into a detector for analysis. The light is analyzed, allowing for determination of the electronic structure, vibrational structure, or atomic constituents of the adsorbed species (i.e., a chemical signature). Applications of laser-based spectroscopy to explosives detection have been widely studied and reported in the literature.

A continued and aggressive evolution of laser sources has changed the prospects of laser-based explosives detection. In particular, great progress has been made in semiconductor lasers operating throughout the infrared (IR) spectrum, and in the ability of the laser architecture to have a cascade effect and produce numerous photons per electron.[16,17] For example, the quantum cascade laser (QCL) has matured to a level at which numerous companies can produce gain material for laser systems. Along with this production, several companies have produced laser systems that are suitable for spectroscopic purposes. The current state of the art for external-cavity (EC) grating-tunable QCLs has demonstrated up to 350 cm^{-1} of continuous tunability from a single-gain element, allowing for collection of vibrational spectra in the IR fingerprint region for various materials.[18] Spectroscopically, these sources could have a dramatic impact on explosives detection due to their wide and continuous tuning over pertinent regions in the IR and a nominal resolution of ~1 cm^{-1}. Many of the chapters in this book discuss QCLs for explosives detection techniques, including hyperspectral imaging and reflectometry (Chapters 8 and 9), photothermal methods (Chapter 10), and cavity-enhanced absorption spectroscopy (Chapter 11).

In general, recent advances in the development of solid-state lasers have allowed for smaller, more cost-effective sources with extended wavelength emission capabilities throughout the IR, visible, and ultraviolet (UV) regions. Furthermore, the pulse duration and peak powers achievable with solid-state lasers make them a particularly attractive source for long-distance detection. Phenomenal advances in the area of femtosecond lasers have been achieved in the last decade. Commonplace limits of approximately 5 fs in solid-state sources and 100 fs in fiber sources have been achieved. These advances can be directly linked to the of advent of using a Kerr-lens mode-locked broadband source (e.g., titanium–sapphire laser) followed by chirped pulse amplification.[19,20] Since the length of the laser pulse is inversely proportional to the bandwidth or number of colors contained in the pulse, these sources have effectively provided users with a broadband coherent laser source without the need for nonlinear conversion. When focused, these ultrashort pulses are very intense, allow-ing for multiphoton excitation, resulting in translational, vibrational, rotational, and electronic excitation of molecules. The unique capabilities of ultrafast lasers make them an attractive source for a variety of spectroscopic applications, as discussed herein in a chapter dedicated to short-pulse laser-based techniques for explosives detection (Chapter 13).

Innovative research in deep-UV lasers with emission wavelengths below 250 nm has demonstrated resonance enhancements for organic compounds and fluorescence-free Raman spectroscopic measurements.[21] There are several different types of lasers (e.g., gas, solid state, semiconductors) that emit in the deep UV. While steady progress has been achieved in the development of UV semiconductor sources including light-emitting diodes and lasers, these technologies have not matured enough to reach desired deep-UV wavelengths, so significant use in energetic detec-tion has not been seen. These sources remain focused on biological hazard detection using fluorescence spectroscopy. Only two laser technologies provide fundamental emission in this region: excimer and metal vapor. One example, transverse excited hollow cathode (TEHC) glow discharge lasers, provides emission in the deep UV at 224 and 248 nm, depending on the gain medium and pump gas.[22] Recently, it has been demonstrated that the Raman spectral enhancements and fluorescence features enabled by the deep UV are applicable to the detection of explosives.[23,24] A number of chapters in this book describe the use of deep-UV lasers for explosives detection methods, including deep-UV native fluorescence (Chapter 4), Raman spectroscopy (Chapters 4 and 5), and photodissociation followed by laser-induced fluorescence (PD-LIF, Chapter 6).

Terahertz (THz) radiation, which occurs between the microwave and IR regions of the electromagnetic spectrum, provides the ability to create images and transmit infor-mation. Until recently, suitable coherent sources for THz generation were not available, resulting in what is commonly referred to as the THz gap. Gas lasers, free-electron lasers, and, more recently, QCLs and laser-driven THz emitters have all proven capable of generating THz radiation.[25,26] Again, given the enormous research activity in QCL and ultrashort pulse sources, these techniques have seen a large increase in their use to gener-ate THz radiation. Generally speaking, the most widely used optical sources of pulsed THz radiation are laser-driven THz emitters based on frequency down-conversion from the optical region. The two main techniques that have been developed to produce THz radiation rely on either a driven photoconductor system or an optical mixing scheme. Specifically, THz-driven photomixer sources consist of materials with a large second-order nonlinear coefficient (e.g., semiconductors, organic crystals) and depend on the production of current pulses that are proportional to the reciprocal of the frequency being generated; for example, the reciprocal of 1 THz is 1 picosecond.[26,27] Therefore, femtosecond lasers have also become attractive sources to enable the generation of THz radiation. The generated THz waves are capable of penetrating various materials, such as cloth, plastic, and other nonmetallic, dry items. This capability provides the possibility to detect explosives concealed in optically opaque materials, but as we will see in Chapter 12, fundamental limitations exist in the capability of THz radiation for identification of hazards in solid phase.[26] The collection of THz spectra for compounds such as TNT, RDX, and HMX has been demonstrated using THz wave time domain spectroscopy.[28,29] THz frequency spectroscopic techniques for the detection and recognition of explosives are discussed in more detail in Chapter 12.

Continued advances in laser development have propelled and will continue to propel laser-based detection technologies into the forefront of explosives detection and identification. However, in order to meet the requirements for standoff operation in a nonlaboratory setting, additional variables must be considered. For example, atmospheric absorption, scattering, and turbulence will potentially impact a laser beam propagating through the atmosphere. Furthermore, accidental targeting and/or reflection may cause injury to the operator and bystanders. These considerations are discussed herein in a chapter dedicated to the operation of laser-based detection systems at standoff ranges (Chapter 2). Additionally, standard test materials for standoff optical detection are necessary to appropriately evaluate detection systems. Recently, a variety of approaches for preparing these optical test standards have been developed. The factors relevant to their preparation and use are described in Chapter 3.

The detection of energetics (IEDs and HMEs) is complicated by the availability and use of a diverse range of materials. Currently, no technology exists that is capable of rapidly detecting this large variety of targets. Therefore, there is an urgent need to continue our investigations for alternative methodologies to detect explosives. Although technical hurdles associated with laser-based standoff detection will continue to be a challenge both in the laboratory and in the transition to field operation, the necessity to provide reliable, real-time detection of hazards at range remains a much sought-after goal by both the civilian and defense communities. The path forward for laser-based optical detection techniques is sure to contain bumps, but the future capabilities that may emerge along the way are very exciting.

REFERENCES

1. Ewing, R. D., Atkinson, D. A., Eiceman, G. A., and Ewing, G. J., A critical review of ion mobility spectrometry for the detection of explosives and explosive related compounds. *Talanta* **2001**, *54*(3), 515–529.
2. Makinen, M., Nousiainen, M., and Sillanpaa, M., Ion spectrometric detection technologies for ultra-traces of explosives: A review. *Mass Spectrom. Rev.* **2011**, *30*(5), 940–973.
3. McQuade, D. T., Pullen, A. E., and Swager, T. M., Conjugated polymer-based chemical sensors. *Chem. Rev.* **2000**, *100*, 2537–2574.
4. Thomas, S. W., Amara, J. P., Bjork, R. E., and Swager, T. M., Amplifying fluorescent polymer sensors for the explosives taggant 2,3-dimethyl-2,3-dinitrobutane (DMNB). *Chem. Commun.* **2005**, *36*, 4572–4574.
5. Yang, J.-S. and Swager, T. M., Fluorescent porous polymer films as TNT chemosensors: Electronic and structural effects. *J. Am. Chem. Soc.* **1998**, *120*, 11864–11873.
6. FLIR Systems, Inc., Fido XT Explosives Detector. gs.flir.com/detection/explosives/fido (accessed May 15, 2014).
7. Forzani, E. S., Lu, D. L., Leright, M. J., Aguilar, A. D., Tsow, F., Iglesias, R. A., Zhang, Q., Lu, J., Li, J. H., and Tao, N. J., A hybrid electrochemical-colorimetric sensing platform for detection of explosives. *J. Am. Chem. Soc.* **2009**, *131*(4), 1390–1391.
8. Marshall, M. and Oxley, J. C., *Aspects of Explosives Detection*. Elsevier: Oxford, UK, **2009**.
9. Woodfin, R. L., *Trace Chemical Sensing of Explosives*. Wiley: Hoboken, NJ, **2007**.

10. Cullum, H. E., McGavigan, C., Uttley, C. Z., Stroud, M. A. M., and Warren, D. C., A second survey of high explosives traces in public places. *J. Forensic Sci.* **2004**, *49*(4), 684–690.
11. Chirico, R., Almaviva, S., Botti, S., Cantarini, L., Colao, F., Fiorani, L., Nuvoli, M., and Palucci, A., Stand-off detection of traces of explosives and precursors on fabrics by UV Raman spectroscopy. In Lewis, C. and Burgess, D. (eds), *Proceedings of SPIE Conference on Optics and Photonics for Counterterrorism, Crime Fighting, and Defence VIII*, pp. 85460W-1–85460W-5. Edinburgh, Scotland, 26 September, **2012**.
12. Oxley, J. C., Smith, J. L., Kirschenbaum, L. J., and Marimganti, S., Accumulation of explosives in hair—Part II: Factors affecting sorption. *J. Forensic Sci.* **2007**, *52*(6), 1291–1296.
13. Oxley, J. C., Smith, J. L., Kirschenbaum, L. J., Marimganti, S., Efremenko, I., Zach, R., and Zeiri, Y., Accumulation of explosives in hair—Part 3: Binding site study. *J. Forensic Sci.* **2012**, *57*(3), 623–635.
14. Kunz, R. R., Gregory, K. E., Aernecke, M. J., Clark, M. L., Ostrinskaya, A., and Fountain, A. W., Fate dynamics of environmentally exposed explosive traces. *J. Phys. Chem. A* **2012**, *116*(14), 3611–3624.
15. Kunz, R. R., Gregory, K. E., Hardy, D., Oyler, J., Ostazeski, S. A., and Fountain, A. W., Measurement of trace explosive residues in a surrogate operational environment: Implications for tactical use of chemical sensing in C-IED operations. *Anal. Bioanal. Chem.* **2009**, *395*(2), 357–369.
16. Curl, R. F., Capasso, F., Gmachl, C., Kosterev, A. A., McManus, B., Lewicki, R., Pushkarsky, M., Wysocki, G., and Tittel, F. K., Quantum cascade lasers in chemical physics. *Chem. Phys. Lett.* **2010**, *487*(1–3), 1–18.
17. Hecht, J., Quantum cascade lasers prepare to compete for terahertz applications. *Laser Focus World* **2010**, *46*(10), 45–49.
18. Holthoff, E. L. and Pellegrino, P. M., Sensing applications using photoacoustic spectroscopy. In Iniewski, K. (ed.), *Optical, Acoustic, Magnetic, and Mechanical Sensor Technologies*, pp. 139–174. CRC Press: Boca Raton, FL, **2012**.
19. Spence, D. E., Kean, P. N., and Sibbett, W., 60-fsec pulse generation from a self-mode-locked Ti: Sapphire laser. *Opt. Lett.* **1991**, *16*(1), 42–44.
20. Kaiser, W., *Ultrashort Laser Pulses: Generation and Applications*. Springer: Berlin, **1993**.
21. Asher, S. A. and Johnson, C. R., Raman spectroscopy of coal liquid shows that fluorescence interference is minimized with ultraviolet excitation. *Science* **1984**, *225*(4659), 311–313.
22. Storrie-Lombardi, M. C., Hug, W. F., McDonald, G. D., Tsapin, A. I., and Nealson, K. H., Hollow cathode ion lasers for deep ultraviolet Raman spectroscopy and fluorescence imaging. *Rev. Sci. Instrum.* **2001**, *72*, 4452–4459.
23. Tuschel, D. D., Mikhonin, A. V., Lemoff, B. E., and Asher, S. A., Deep ultraviolet resonance Raman excitation enables explosives detection. *Appl. Spectrosc.* **2010**, *64*(4), 425–432.
24. Bhartia, R., Hug, W. F., and Reid, R. D. Improved sensing using simultaneous deep UV Raman and fluorescence detection. In Fountain, A. W. (ed.), *Proceedings of SPIE Conference on Chemical, Biological, Radiological, Nuclear, and Explosives (CBRNE) Sensing XIII*, pp. 83581A-1–83581A-9. Baltimore, MD, **2012**.
25. Gallerano, G. P. and Biedron, S., Overview of terahertz radiation sources. In *Proceedings of the 2004 FEL Conference*, pp. 216–221, **2004**.
26. Zhang, X.-C. and Xu, J., Terahertz radiation. In *Introduction to THz Wave Photonics*, pp. 1–26. Springer: New York, **2010**.
27. De Lucia, F. C., Spectroscopy in the terahertz spectral region. In Mittleman, D. (ed.), *Sensing with Terahertz Radiation*, pp. 39–115. Springer: Berlin, **2003**.

28. Kemp, M. C., Taday, P. F., Cole, B. E., Cluff, J. A., Fitzgerald, A. J., and Tribe, W. R., Security applications of terahertz technology. *Proceedings of SPIE* **2003**, *5070*, 44–52.

29. Chen, Y., Liu, H., Deng, Y., Veksler, D., Shur, M., Zhang, X.-C., Schauki, D., Fitch, M. J., and Osiander, R., Spectroscopic characterization of explosives in the far-infrared region. *Proceedings of SPIE* **2004**, *5411*, 1–8.

2 Additional Considerations for Laser-Based Detection Systems at Standoff Ranges

William G. Holthoff

CONTENTS

2.1 INTRODUCTION

Most of the technologies discussed in this book show great promise for laser-based explosives detection in a laboratory environment. However, transitioning these technologies outside the laboratory requires a careful and practical analysis of the engineering complexity, cost, and potential safety/environmental considerations, as well as the public perception of the safe use of lasers. This chapter is intended to provide the reader with a general overview of these considerations in order to provide a better understanding of the complexities involved with developing and fielding a laser-based system for standoff explosives detection.

2.2 OVERVIEW OF A LASER-BASED SYSTEM FOR STANDOFF EXPLOSIVES DETECTION

A laser-based standoff explosives detection system will likely require five main components: a laser, a detector/receiver, a beam director, a targeting (i.e., fire control) system, and support equipment. Figure 2.1 provides an overview of the major systems and potential subsystems required. The overall system cost, size, and complexity will be driven by the operational requirements (i.e., standoff range, limit of detection, etc.) and the optical detection technique selected.

The attributes of the laser system will vary depending on the optical technique selected. The wavelength necessary for the optical technique will drive the laser system type (i.e., solid-state, fiber, gas, semiconductor, etc.), as the properties of the active medium in the laser determine its output wavelength [1]. Some optical techniques offer greater flexibility in wavelength selection, such as laser-induced breakdown spectroscopy (LIBS) and Raman spectroscopy (refer to Chapters 7 and 5, respectively), which are not limited to a specific wavelength region. In contrast, optical techniques based on fluorescence and absorption will require lasers emitting radiation in the wavelength region where analytes of interest have absorption bands.

Optical properties of the laser system such as wavelength and beam quality impact the overall system design due to diffraction and atmospheric effects such as absorption, scattering, and turbulence-induced beam spreading and beam wander. The required beam quality and average power, coupled with the laser system type, electrical-to-optical conversion efficiency, and so on will be the primary drivers of laser system size, weight, and support equipment requirements.

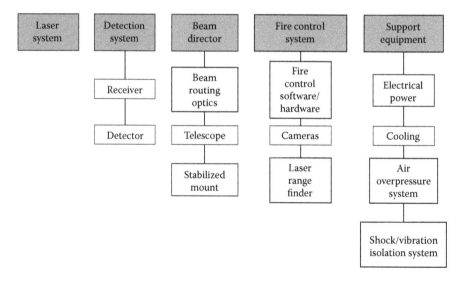

FIGURE 2.1   Schematic of the major systems for a laser-based standoff explosives detection system. The subsystems shown are not required for all applications but represent a general overview of subsystems that might be needed.

The design and system requirements of the detector/receiver will be unique to the optical technique selected and are not discussed in detail in this chapter. Detectors can range from thermal imagers for photothermal spectroscopy (refer to Chapter 10) to spectrometers for LIBS or Raman spectroscopy. The beam director may serve as a common aperture for both the laser and the detection system receiver for some techniques such as LIBS and Raman spectroscopy. In other cases, such as photo-acoustic spectroscopy (refer to Chapter 10), a different receiver is necessary.

The third major component, the beam director, consists of the optics and optical-mechanical systems required to project and focus the laser beam onto the target. The major beam director subsystems include the beam delivery system to route the laser beam from the laser to the telescope, the telescope to focus and/or collimate the beam to a spot size capable of inducing the intended effect at the target, and a steering mirror or stabilized mount to point and hold the beam on the target. In some cases, the beam director can be used as the receiver for the detection system as well.

Depending on the system design and environment in which it will operate, the beam delivery optical path and telescope may require an enclosure and purging with clean dry air to avoid contaminants on the optical surfaces, which can lead to damage. Computer-controlled optical-mechanical systems will be required to (1) control the telescope focus, via the primary–secondary mirror spacing, for example; (2) maintain the alignment of the beam delivery and telescope optics; and (3) control the steering mirror or stabilized mount used to direct the beam onto the target.

A fire control system will be necessary to provide targeting information, such as range and aim point, and then to control the system subcomponents necessary to focus the beam on the desired aim point. Fire control inputs would likely include target ranging information and video feeds to provide the system operator with aim-point selection and situational awareness to avoid accidental injury to bystanders. Depending on the laser system operating parameters, the system may be capable of causing damage to the skin or eyes. To mitigate this potential, the fire control software may need to be designed to shutter or shut off the laser if bystanders enter the beam path. If the intended target is moving, the software will also need to track the aim point while the target is interrogated. Ideally, the aim-point selection will be accomplished via a camera using the same telescope aperture as the laser. This approach minimizes the chances of misalignment between the laser and camera. A wide-field-of-view camera would also likely be necessary to provide the fire control system with the images necessary to act to prevent injury to bystanders prior to their entry into the beam path.

Support equipment includes all the ancillary equipment required to keep a laser-based system operational in the field. This category includes electrical power, cooling, purge air, and shock/vibration isolation systems. The overall weight and size of electrical power and cooling systems will in general scale with the average power and efficiency of the laser system. Because the conversion process from electrical power to optical power is inefficient (generally less than 30% efficiency), the majority of the energy is converted inside the laser head into heat that must be removed [1–3]. Failure to remove this heat can result in damage to the laser gain medium, reduction in output power, or distortion of the laser beam by processes such as thermal lensing [4]. The air overpressure system will vary based on the size of the laser and beam director, and the optical system design. The laser head and beam director will also need to be mounted on shock/vibration isolation mounts to reduce misalignment and prevent damage to the components. The weight and complexity of the shock/vibration isolation mounts will depend on the laser and beam director size and weight and on the desired ruggedness of the overall system.

Prior to embarking on system development, there must be a thorough understanding of the trade-offs associated with specific system requirements for a laser-based standoff explosives detection system. For example, a laser-based explosives detection technique may demonstrate good performance in a laboratory environment or at short ranges (i.e., several meters) in a simulated operational environment; however, it may prove cost prohibitive or technically infeasible to significantly extend the operational range without increasing the overall system size. The increase in system size and weight adds to the cost and technical complexity of the system. For longer propagation distances, certain atmospheric conditions may limit the range or operational effectiveness of the system. In addition, long-distance propagation for some wavelengths may increase the potential for injury to bystanders. These considerations, and others not discussed, need to be taken into account prior to selecting a specific detection technique and laser wavelength and other system attributes.

In the following sections, other considerations are discussed. Section 2.3 gives a synopsis of atmospheric propagation considerations. The first part provides the reader with a general understanding of how the laser system attributes (i.e., beam quality and wavelength), in addition to the operational requirements (i.e., range, required beam diameter at the focus, etc.), drive the overall size of the beam director. The second part provides an overview of atmospheric absorption, scattering, and turbulence and their potential impact on the system.

Section 2.4 is intended to educate the reader on the hazards associated with laser radiation. A common misperception, that laser radiation outside the retinal hazard region, which encompasses all wavelengths between 0.400 and 1.400 μm, is "eye safe," is discussed. Under the proper exposure conditions, laser radiation at any wavelength is capable of damaging the eye. Section 2.4.1 is a brief overview of the anatomy of the eye and ocular components most susceptible to damage. Section 2.4.2 reviews how tissue type, wavelength, and exposure conditions influence the damage mechanism and tissue(s) affected.

2.3   ATMOSPHERIC PROPAGATION CONSIDERATIONS

One of the fundamental barriers to the employment of laser-based detection systems at significant standoff ranges (hundreds of meters) is atmospheric propagation. Depending on the wavelength, output power, and atmospheric conditions, a laser beam will be attenuated or distorted or both. The same factors can affect the return signal, depending on the detection technique selected. This section is an overview of the difficulties imposed by propagation.

2.3.1 DIFFRACTION-LIMITED PROPAGATION

Although the output of a laser is considered collimated, even if the rays of a laser beam are initially parallel, the beam diverges as it propagates away from the source [5]. Diffraction effects are wavelength dependent, with the divergence angle for a beam with a Gaussian intensity profile directly proportional to the ratio between the wavelength, λ, and beam-waist radius, w_0 [6,7]. Diffraction directly defines the smallest spot size to which a particular wavelength can be focused or the level of collimation achievable under ideal conditions for a given aperture diameter. Depending on the optical technique a system uses, the spot size at the target will likely be a critical parameter. For example, a LIBS system requires a small spot size because the irradiance of the laser beam must exceed the plasma threshold of the surface targeted. As a result, the exit aperture (i.e., primary mirror) diameter of the telescope can be one of the fundamental drivers of overall system size.

As illustrated in Figure 2.2, an ideal Gaussian beam has an intensity that decreases radially from the axis of propagation. In this section, the radius at which the intensity reaches $1/e^2$ (about 13.5%) of the maximum intensity is defined as the beam radius, $w(z)$. Eighty-six percent of the total power in the laser beam is carried within the beam radius [7]. The beam radius varies along the propagation path, generally decreasing to a minimum at the beam waist before increasing again beyond the waist as depicted in Figure 2.3.

$$w(z) = w_0 \left[1 + \left(\frac{z}{z_R} \right)^2 \right]^{1/2} \tag{2.1}$$

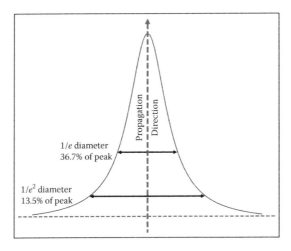

FIGURE 2.2 Example of Gaussian beam. The $1/e$ and $1/e^2$ beam diameters correspond to the circular area, which contain 68% and 86% of the total power in the laser beam, respectively.

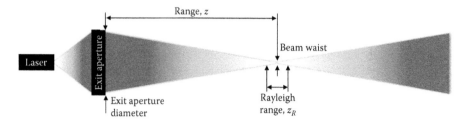

FIGURE 2.3   Conceptual drawing illustrating laser beam propagation from the laser aperture to the beam waist.

Equation 2.1 describes the variation of $w(z)$ along the propagation path, at distance z from the waist in terms of the Rayleigh range, z_R [7]. The Rayleigh range is defined in Equation 2.2:

$$z_R = \frac{\pi w_0^2}{\lambda} \tag{2.2}$$

where:
   w_0 = ideal beam-waist radius
   λ = wavelength [7]

The Rayleigh range is the distance over which the area of the beam increases to twice the area at the beam waist and describes the degree of focus. A short Rayleigh range means the laser beam is sharply focused, and the intensity decreases rapidly with distance in front of or behind the beam waist (i.e., focal point).

Equations 2.1 and 2.2 can be used to estimate the exit aperture required to achieve the necessary beam-waist radius, w_0, for a given application. The Rayleigh range can be determined for the desired waist size and laser wavelength using Equation 2.2. Equation 2.1 can then be used to determine the radius of the beam at the exit aperture of the beam director at distance z from the beam waist. Nominally, the spot size desired is the beam-waist size, and the distance from aperture to target is the beam waist to aperture distance.

To illustrate the effect of diffraction as a function of wavelength, Figure 2.4 presents the aperture diameter required to focus an ideal Gaussian laser beam to a 1 cm beam waist as a function of range. Longer wavelengths and longer ranges require larger aperture sizes to achieve a given beam-waist diameter.

Real laser beam propagation is more complex because the output from real-world lasers is not truly Gaussian. The quality factor, M^2 (pronounced "M squared"), has been defined to describe the variance between the output of a real laser and an ideal Gaussian beam. An ideal Gaussian laser beam has an M^2 value of 1, while the M^2 value of a real laser beam is always greater than 1. For a real-world laser beam, Equation 2.2 can now be rewritten as in Equation 2.3 [8]:

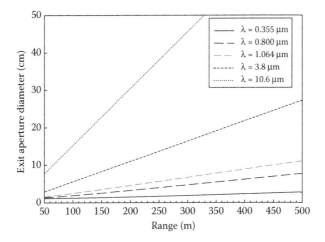

FIGURE 2.4 Exit aperture diameter required to focus a beam to a 1 cm diameter as a function of range. The exit aperture diameter required to achieve a 1 cm beam waist at a given range increases as the wavelength increases. The exit aperture diameters in this figure are equal to 2.25 times the $w(z)$ radius at the aperture.

$$z_R = \frac{\pi w_0^2}{M^2 \lambda} \tag{2.3}$$

The addition of the M^2 term to the Rayleigh range equation (Equation 2.3) for a real beam with an $M^2 > 1$ has several consequences for system development. Maintaining the same Rayleigh range as an ideal Gaussian beam results in a beam radius that is larger by M (i.e., $\sqrt{M^2}$) everywhere along the propagation path. In this case, a laser with an M^2 value of 4 produces a beam waist that is a factor of two larger than an ideal Gaussian beam and requires an exit aperture twice the size. To produce the same irradiance at the beam waist as the ideal Gaussian beam in this case requires a laser with four times the output power. Conversely, to maintain the same beam-waist radius, w_0, as the ideal Gaussian beam at a given distance requires the exit aperture for the real beam to be M^2 larger. In this case, a laser with an M^2 value of 4 produces a beam-waist radius equivalent to an ideal Gaussian beam; however, the exit aperture is four times the size and the Rayleigh range is reduced by a factor of four.

The final practical consideration is that of diffraction effects resulting from truncation of the ideal Gaussian beam by the exit aperture. An ideal Gaussian beam presumes an infinite aperture because the Gaussian distribution is infinite. The radius at which the Gaussian beam is truncated affects the total amount of power transmitted through the aperture and the peak intensity at the focus [6]. An aperture diameter π times the $1/e^2$ radius transmits 99% of the Gaussian beam power; however, diffraction effects result in a peak intensity reduction of approximately 17% at the focus [6]. For many applications where size, weight, and cost are a concern, apertures three times the $1/e^2$ radius or greater are impractical. Practical experience has shown that a ratio of approximately 2.25 between the aperture and the Gaussian beam $1/e^2$ radius provides the best compromise between reduction in peak intensity at the waist due to truncation and the aperture size required [6]. This compromise may be better understood as choosing the optimum beam size for a given aperture diameter. Reducing the beam radius at the aperture results in a larger beam size and reduced peak intensity at the waist. Increasing the beam radius at the aperture results in a smaller beam size and reduced peak intensity at the waist resulting from the increased truncation of the laser beam.

2.3.2 EFFECT OF ATMOSPHERIC ABSORPTION, SCATTERING, AND TURBULENCE

In addition to diffraction, a laser beam propagating through the atmosphere is subject to a variety of effects that can distort the beam. Gebhardt provides an excellent overview of the difficulties involved with high-power laser propagation [9]. The most important atmospheric effects are (1) absorption and scattering due to molecular and aerosol components of the atmosphere; (2) random wander, spreading, and distortion of the beam due to atmospheric turbulence; (3) self-induced thermal blooming resulting from absorption by the atmosphere of a small amount of the laser beam energy; and (4) the strong attenuating effects of a plasma resulting from air breakdown at high optical intensities. These effects depend on both the atmospheric conditions and the laser characteristics (wavelength [λ], pulse or continuous wave [CW] operation, power, etc.). For example, while carbon dioxide (CO_2) laser ($\lambda = 10.6 \ \mu m$) propagation is dominated by thermal blooming due to strong molecular absorption, propagation of shorter wavelength laser radiation (e.g., erbium [$\lambda = 1.54 \ \mu m$] and neodymium [Nd] lasers [$\lambda = 1.06 \ \mu m$]) is dominated by atmospheric turbulence and scattering effects [9].

Extinction (i.e., attenuation) of a laser beam in the atmosphere is the gradual loss of intensity as it is absorbed, scattered, and transmitted. Extinction results from scattering and absorption by molecules and aerosols and can have a significant effect on the beam parameters at the target plane. Typically, extinction is quantified using an extinction coefficient measured in units of inverse length [10]. Scattering and absorption attenuate the beam, with the extinction losses increasing as a function of pathlength. In addition, absorption can lead to beam distortion due to thermal blooming.

Molecular atmospheric absorption is dependent upon the temperature, pressure, and concentration of the absorbing species along the propagation path. For a horizontal path up to several kilometers in length at a given altitude they can be considered a constant [11]. As Figure 2.5 illustrates, atmospheric absorption is very irregular and complicated as a function of wavelength. In many regions of the spectrum, the absorption is due principally to the presence of CO_2 and water vapor; however, other atmospheric constituents, such as methane and nitrous oxide, can also contribute in certain wavelength regions [9]. The molecular (Rayleigh) scattering coefficient varies as λ^{-4} and only becomes comparable in importance to aerosol scattering for wavelengths less than 0.450 μm [9]. At 20°C and 100% relative humidity, the total molecular extinction at 1.54 μm was calculated to be 0.02/km [11]. This value of molecular extinction is considered insignificant compared with typical values of aerosol extinction. In

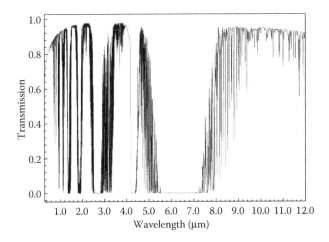

FIGURE 2.5   Transmission calculated with MODTRAN for a low-altitude (3 m above ground) horizontal 1 km path using the 1976 U.S. Standard Atmosphere model and rural aerosol model.

comparison, the molecular extinction due solely to molecular absorption at 10.6 μm is approximately an order of magnitude larger [12]. The absorption is due primarily to water vapor, with a small contribution from weak CO_2 lines [11].

In contrast with molecular absorption, the extinction of a laser pulse due to scattering and absorption by atmospheric aerosols is less dependent on wavelength [9]. Mie scattering theory relates the wavelength dependence to the size distribution and the aerosol complex refractive index, with the general trend showing an increase in the importance of aerosol effects at shorter wavelengths [9,11]. Shettle et al. developed models for aerosols and the effect of humidity on their optical properties [13]. Rural, urban, and maritime models were developed for wavelengths between 0.200 and 40.000 μm. Figure 2.6 presents the total aerosol extinction (scattering and absorption) for 0.337, 1.536, 2.000, and 10.591 μm as a function of relative humidity for all three models prior to adjustment for variations in visibility. These plots are suitable for comparing relative effects of the same atmosphere for different wavelengths but may not be suitable for comparing different models without accounting for specific visibility conditions. Scattering dominates the aerosol extinction coefficient for the 0.337, 1.536, and 2.000 μm wavelengths, accounting for on average 92%, 85%, and 85% of the coefficient, respectively. At 10.591 μm, the aerosol extinction coefficient is, on average, evenly split between scattering and absorption components.

2.3.2.1   Thermal Blooming

Thermal blooming results from the partial absorption of laser radiation by the atmosphere, causing localized heating and resulting in refractive index gradients that act as a distributed nonlinear lens distorting the beam intensity profile [12]. In addition to distorting the beam intensity profile, thermal blooming can limit the maximum irradiance that can be propagated to the target plane independent of the available transmitted power [9,14]. Typically, the most likely case of thermal

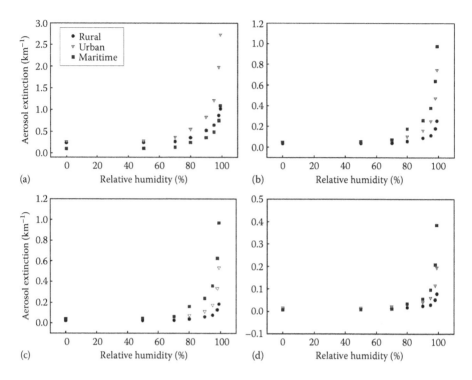

FIGURE 2.6   Aerosol extinction as a function of propagation environment for (a) 0.337 μm, (b) 1.536 μm, (c) 2.000 μm, and (d) 10.591 μm. Note that the vertical scale varies as a function of wavelength. (Data from Shettle, E.P. and R.W. Fenn, *Models for the Aerosols of the Lower Atmosphere and the Effects of Humidity Variations on their Optical Properties*, AFGL-TL-79-0214, Air Force Geophysics Laboratory. 1979.)

blooming for CW laser propagation is a wind-dominated condition where beam distortion takes on a characteristic crescent shape with the narrow portion of the crescent in the wind direction and more spreading in the direction transverse to the wind [12]. Figure 2.7 illustrates how the intensity pattern is bent into the wind. As the wind moves across the beam, the temperature increases because of absorption; as the temperature increases, the air density decreases along with the refractive index. This explains why the beam bends toward the wind as light is refracted toward a denser medium. Absorption is the most important parameter because the distortion is critically dependent on how much laser radiation is absorbed; even 1 part per million per centimeter (ppm/cm) can cause severe distortions at kilowatt power levels [12].

One way to avoid thermal blooming is to use a short-duration laser pulse [12]. While absorption is essentially an instantaneous process, it takes a finite amount of time for this energy to be converted into translational energy and change the density of the atmosphere in the propagation path. This time is associated with the acoustic transit time across the beam radius, a, which is given by Equation 2.4:

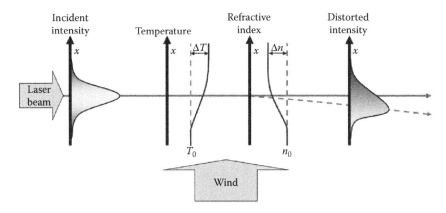

FIGURE 2.7 Illustration of thermal blooming effect. (Adapted from Gebhardt, F. G., *Proceedings of SPIE*, 1221: 2–25, 1990.)

$$t_{acoustic} = \frac{a}{v_a} \qquad (2.4)$$

where v_a is the velocity of the acoustic wave [12]. Laser pulses shorter than the acoustic transit time suffer less distortion than longer laser pulses with the same amount of energy.

Laser pulses shorter than the acoustic transit time can reduce the influence of thermal blooming. However, the problem can return when successive pulses propagate through the same air. While each pulse suffers minimal alteration due to the atmospheric distortion it causes, the residual distortion from previous pulses can cause blooming of ensuing pulses. The ideal repetition rate is thus determined by the limits of thermal blooming caused by overlap or residual heating of previous pulses. The ideal repetition rate for propagation is a function of the beam size, wind velocity, and also the energy absorbed by the air from individual pulses [12]. An important parameter in multiple-pulse blooming is the number of pulses per flow (gas transit) time (PPFT) across the beam and is given by Equation 2.5:

$$PPFT = \frac{2a}{v_w} PRF \qquad (2.5)$$

where:

v_w = wind speed
a = beam radius
PRF = pulse repetition rate [12]

2.3.2.2 Atmospheric Turbulence

Atmospheric turbulence results when sections of the atmosphere with different temperatures are mixed by wind, convection, or both [5]. The mixing process results in smaller turbulent cells of air with small variations in temperature (and, consequently, small variations in refractive index), creating an inhomogeneous medium. As the laser beam propagates through the turbulent cells, the random variations in refractive index distort the laser beam, causing fluctuations in beam size, beam position, and the intensity distribution within the beam [15]. The magnitude of these effects is dependent on the atmospheric index of refraction structure parameter, C_n^2, and the aperture size [5].

C_n^2 is the single most important parameter in equations describing the effects of atmospheric turbulence and is given by Equation 2.6:

$$C_n^2 = \frac{(n_1 - n_2)^2}{r_{12}^{2/3}} \tag{2.6}$$

for a homogeneous and isotropic turbulence where n is the index of refraction at points r_1 and r_2, which are separated by the distance r_{12} [16]. Because n varies as a function of wavelength, C_n^2 will as well. Values of C_n^2 vary from approximately 10^{-17} m$^{-2/3}$ for extremely weak turbulence to 10^{-13} m$^{-2/3}$ or greater when turbulence is strong [17]. The latter value is usually observed near the ground in direct sunlight [5]. The atmospheric turbulence strength varies as a function of the time of day and geographic location [18]. For example, a C_n^2 of 10^{-12} is very strong, most likely encountered just above the pavement under direct solar illumination. A C_n^2 of 10^{-14} is likely what one would expect near the surface over land, while 10^{-16} is probably what one could expect over water under many conditions. Low C_n^2 levels can occur at night and any time there is cloud cover, even temporary cover resulting from partly cloudy conditions during the day.

The aperture size has an effect on the magnitude of the atmospheric-turbulence-induced effects as a function of the lateral coherence parameter, r_0 or ρ_0 ($r_0 = 2.1\rho_0$), which is a measure of the spatial coherence of an electromagnetic wave as restricted by the turbulent atmosphere, and is given by Equation 2.7 [18,19]:

$$r_0 = 1.68 \left[\frac{2\pi^2}{\lambda} \int_0^R C_n^2(r)(1 - r/R)^{5/3} \, dr \right]^{-3/5} \tag{2.7}$$

where:
   R = pathlength
   $C_n^2(r) = C_n^2$ at a distance r along the path

The coherence length is important because it limits the average resolution of an optical system. For apertures smaller than the coherence length, the amplitudes across the aperture are correlated and added together; thus, the beam waist depends on the aperture in a predictable manner, although phase variations across the aperture will produce beam wander [5,18]. For apertures larger than the coherence length, the correlation across the aperture is limited to regions on the order of the coherence length. As a result, diffraction and refraction take place, distorting the beam's spatial profile and diameter. When the coherence length is less than or equal to the aperture diameter, increasing the pathlength has the effect of decreasing the effective aperture. Thus, the relative effectiveness of a laser system employing a focused beam may be greatly reduced by atmospheric turbulence. In practice, the observed spot sizes are often twice as large as spot sizes predicted by diffraction theory alone [5]. Under most conditions, this beam spreading is wavelength dependent, with the wavelength dependence proportional to $\lambda^{-1/5}$. This result is significant; it implies an optimum wavelength for a given aperture diameter, atmospheric turbulence level, and range, because beam spreading due to diffraction is proportional to λ [5,9].

Beam wander is a random variation in the position of the laser beam centroid on the target. Statistically, beam wander can be characterized by the standard deviation of the beam centroid displacement along an axis (σ_W), and for a focused beam it is given by Equation 2.8:

$$\sigma_W = \left(2.72 C_n^2 R^3 W_0^{-1/3}\right)^{0.5} \tag{2.8}$$

where W_0 is the beam radius [5,20]. Beam wander will be more pronounced for a beam propagating just above and parallel to the ground.

Scintillation is a phenomenon where the laser beam propagates through turbulent cells resulting in intensity fluctuations across the central beam position. The standard deviation in the irradiance (s_1) can be estimated for a horizontal beam path by Equation 2.9:

$$s_1 = \left(1.24 C_n^2 k^{7/6} r^{11/6}\right)^{0.5} \tag{2.9}$$

where $k = 2\pi/\lambda$ [20]. Based on this equation, we can conclude that the irradiance fluctuation should increase with range, and the fluctuation should decrease with increasing wavelength [20].

2.3.2.3 Atmospheric Propagation Examples

Hanson et al. compared the propagation of CW 1.565 and 3.603 μm laser radiation in maritime environments [21]. The experiments were conducted at low altitude in a variety of atmospheric conditions to quantify the relative effects of turbulence under realistic conditions. The spatial profiles were recorded for lasers emitting at 1.565 and 3.603 μm under identical atmospheric conditions using a common aperture. The long-term (10 s average) spot sizes (normalized to the diffraction-limited spot size) were observed to be 1.5–2.5 times larger for the 1.565 μm beam than for the 3.603 μm beam (see Figure 2.8). Aberrations in the optical components likely led to significantly greater degradation of the 1.565 μm beam [21]. The authors mention that in weak turbulence conditions, these aberrations were the limiting factor in the quality of 1.565 μm images while only having a minor effect on the 3.603 μm images At 1.565 μm, the observed scintillation ranged from approximately 0.25 to approximately 2.5. In comparison, at 3.603 μm, the observed scintillation was between approximately 30% and approximately 70% of the 1.565 μm scintillation value.

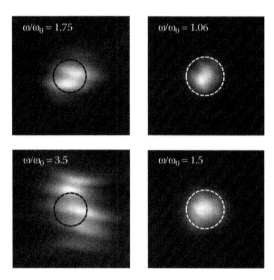

FIGURE 2.8 Typical images, averaged over 10 s of the focused 1.565 μm (*left*) and 3.603 μm (*right*) beam after propagating 2.82 km over water. The top images were collected in low turbulence and the bottom images in moderate turbulence. The turbulence-broadened average spot size ω was calculated by fitting the data to a one-dimensional Gaussian function. The calculated diffraction-limited spot with radius ω_0 is overlaid in the center of each image. The normalized spot size ω/ω_0 for each wavelength and turbulence level is given. (Adapted from Hanson, F. et al., *Applied Optics*, 48, 4149–4157, 2009.)

Laserna et al. studied the effect of atmospheric turbulence on LIBS measurements up to a distance of 120 m [20]. For LIBS measurements, a telescope is used to focus a laser pulse onto the target with sufficient irradiance to generate a plasma and then to collect the subsequent elemental emission lines from the plasma and focus them into a spectrometer for analysis. Thus atmospheric turbulence affects both optical paths. The experiments were performed using a LIBS-based sensor system consisting of dual neodymium-doped yttrium aluminum garnet (Nd:YAG) lasers ($\lambda = 1.06$ μm, 750 mJ/pulse, 10 Hz, 5.5 ns pulse width) whose beams were overlapped in time and space and propagated to the target plane using a 40 cm aperture Cassegrain-type telescope. Testing was performed in an indoor environment with a pathlength of 50 m and outdoors at ranges of 30, 50, and 120 m. The beam was propagated 1.5 m above the ground during daytime hours with a reported visibility of 10 km, windy conditions (maximum wind speed of 30 km/h with frequent wind gusts), and a temperature of 18°C. Figure 2.9 presents (a) the laser beam cross section prior to entering the telescope and (b) the focused beam cross section after propagating 50 m along a horizontal path outdoors. The focused beam cross section illustrates the effect of atmospheric turbulence and may include diffraction effects induced by the focusing optics.

The atmospheric-turbulence-induced effects observed included beam wander, spread, and scintillation [20]. To characterize the beam spread, 50 laser pulses were fired at an aluminum plate and the footprint left by their impact measured in one direction at a range of 30, 50, and 120 m. The measured footprint on the aluminum target increased from 2.18 to 2.93 to 3.74 mm at a range of 30, 50, and 120 m,

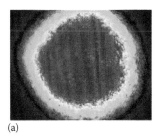
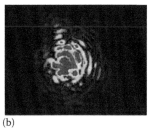

(a)   (b)

FIGURE 2.9   Laser beam cross sections obtained with a beam analyzer: (a) in the near field just before entering the telescope system and (b) when tightly focused with the telescope after propagating 50 m along a horizontal path at 1.5 m from the ground. (Adapted from Laserna, J. J. et al., *Optics Express*, 48, 10265–10276, 2009.)

respectively. As the range increased from 30 to 120 m, the LIBS signal decreased by 91%; however, the signal-to-noise ratio (SNR) only decreased by 29%.

Mukherjee et al. used a tunable infrared (IR) laser to study standoff detection of TNT using a photothermal technique up to a distance of 150 m [22]. A tunable CW CO_2 laser beam, which was mechanically chopped, was expanded through an approximately 30.5 cm Cassegrain telescope to collimate the beam for target illumination. The receiver consisted of an approximately 61 cm Ritchey-Chrétien telescope with a cooled mercury cadmium telluride (MCT) focal plane array (FPA) detector. Measurements were performed at 20, 25, 50, 75, and 150 m. Spectra for TNT were obtained at all standoff distances with the signal decreasing with increasing range. Mukherjee et al. state that the limitation in performing measurements at greater standoff distances resulted from the need for adaptive optics to minimize the effects of beam wander and intensity fluctuations during the measurement time, which was several seconds in duration.

2.4   INTERACTION OF LASER RADIATION WITH BIOLOGICAL TISSUE

A primary concern in the development and deployment of any laser-based system is the potential for injury to the warfighter, targeted individuals, and bystanders resulting from accidental targeting or reflection hazards. Article II, Protocol IV (Protocol on Blinding Laser Weapons) of the Convention on Certain Conventional Weapons (CCW) requires signatories to "take all feasible precautions to avoid the incidence of permanent blindness to unenhanced vision" from laser systems employed in military conflicts for uses such as communications, target discrimination, and range finding [23,24]. In addition to treaties limiting the employment of laser systems on the battlefield, international and national laser safety regulations and standards govern the manufacture and use of laser systems. At the international level, laser product and safety standards are defined in the International Electrotechnical Commission (IEC) 60825-1 standard. Exposure limits (EL) in the IEC 60825-1 standard are adopted from the International Commission of Non-Ionizing Radiation Protection (ICNIRP) guidelines [25,26]. Laser products sold in the United States must comply with the US Food and Drug Administration Center for Devices and Radiological

Health (FDA-CDRH) regulation 21 CFR1040.10. In addition, laser safety and ELs are based on the American National Standards Institute (ANSI) Z136.1 standard. As such, careful consideration must be given to safety when designing a laser-based standoff explosives detection system.

The eyes are most vulnerable to laser radiation in the retinal hazard region, which encompasses all wavelengths between 0.400 and 1.400 µm. This vulnerability is due to anatomical and optical characteristics that focus and transmit the optical radiation from the cornea to the retina, significantly increasing the potential of permanent injury because small exposures at these wavelengths can cause permanent loss of vision owing to damage of the retina. Wavelengths outside the retinal hazard region (<0.400, >1.400 µm) are generally referred to as being "eye safe"; however, this terminology is misleading. While significantly higher radiant exposures are required to cause permanent damage, lasers emitting radiation outside the retinal hazard region are still capable of causing damage to the anterior components of the eye and skin. Because the consequences of overexposure of the eyes are generally more serious than those associated with overexposure of the skin, safety standards such as ANSI Z136.1 emphasize protection of the eyes [26]. The ELs, also referred to as maximum permissible exposure (MPE) in some safety standards (ANSI Z136.1), vary as a function of wavelength and the potential exposure conditions (i.e., exposure duration, single or multiple exposure, etc.).

The ELs, or MPE, are derived from experimentally determined damage threshold values with an added safety factor. The safety factor chosen by the safety committees is based on the level of uncertainty in the experimental data. A safety factor of 10 is commonly stated for the ELs or MPE; however, if the threshold values are determined with small experimental uncertainty, the safety factor could be as little as 2.5–3 [27]. The experimental damage threshold studies are typically performed on animals with similar physiological attributes to humans [28]. Damage thresholds for a given set of exposure conditions are typically reported as the ED_{50} as determined by probit statistics [27]. ED_{50} is the dose that results in a 50% probability of causing a detectable lesion at some time post exposure (typically 1–48 h) [27,29]. For the cornea and lens, damage threshold values are typically reported as irradiance (W/cm^2) or radiant exposure (J/cm^2) incident on the cornea [30]. These values are typically computed by measuring the $1/e$ beam diameter in addition to the power or energy per pulse incident on the cornea. The $1/e$ beam diameter corresponds to the circular area that contains 68% of the total power in the laser beam. For the retina, damage threshold values are typically reported in terms of the power (W) or energy (J) that is incident on the cornea and enters the pupil, typically referred to as the total intraocular energy (TIE) or intraocular energy (IOE) [30]. To fully characterize the level of exposure to the retina, losses in power (W) or energy (J) due to reflection and absorption in the ocular media prior to the retina and the spot size at the retina must be calculated.

2.4.1 ANATOMY OF THE HUMAN EYE

To better understand the complexities of laser-induced eye damage requires a basic understanding of the anatomy of the human eye. As illustrated in Figure 2.10, the human eye is a complex structure comprising multiple types of biological tissue. The eye can be separated into two chambers: the anterior and the posterior.

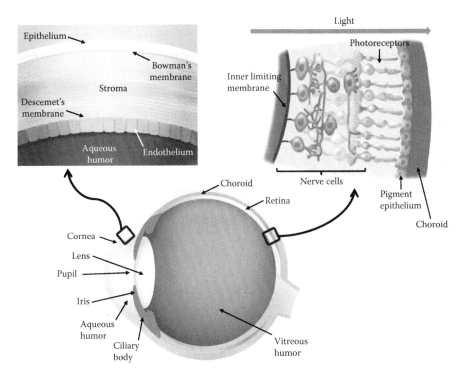

FIGURE 2.10 Anatomy of the human eye.

The anterior chamber is bounded by the cornea, lens, and iris and is filled with aqueous humor. The aqueous humor is a solution that provides nutrients to the cornea and is part of the optical pathway [1]. This chamber contains the majority of the structures responsible for focusing images onto the retina of the eye. The posterior chamber is where images are formed. It is bounded by the retina, lens, and ciliary body and is filled with vitreous humor, a gel-like structure. The retina, cornea, and lens are the structures of the eye that are most likely to be permanently damaged by laser radiation. Additional detail is provided on the retina and cornea because they are multilayer structures, with each layer having distinct properties (e.g., optical absorption and ability to regenerate).

The retina is a multilayer structure and is the most photosensitive structure in the eye. Prior to impinging on the photosensitive cells (i.e., rods and cones), light must pass through the inner limiting membrane, which separates the vitreous humor from the retina, and several nerve cell layers [1]. The photosensitive retinal cells are adjacent to the retinal pigment epithelium (RPE) layer, which is the innermost layer of the retina. The RPE serves as a final light sink for incoming photons that reduces intraocular glare. The black appearance of the pupil results from the light-absorbing pigmentation in the RPE layer. Underneath the retina is the choroid, which is a vascular structure that supplies nutrition to the outer third of the retina, which includes the rod and cone photoreceptors. Damage to tissue in the retina and choroid is typically permanent.

As illustrated in Figure 2.10, the cornea consists of five layers (epithelium, Bowman's membrane, stroma, Descemet's membrane, and endothelium) and is covered by tear film with an overall thickness of approximately 650 μm [31]. The tear film consists of three distinct layers: an oily outer layer, a watery layer, and a mucin layer. The epithelial layer is approximately five cells thick, with an overall thickness of 50–60 μm. This layer generally heals rapidly with a cellular turnover rate of approximately 2–5 days [29]. The Bowman's membrane is located beneath the epithelium and is 8–12 μm in thickness. Changes to this layer could be considered permanent because the membrane remains retracted and does not rejoin when disrupted. The stroma is the thickest layer of the cornea (approximately 0.5 mm thick) and does have an ability to regenerate. However, permanent vision impairment may occur due to edema and collagen shrinkage [29,31]. Descemet's membrane is 10 μm thick and has a low regenerative capacity. The endothelium consists of a single cell layer approximately 5 μm in thickness that does not regenerate after injury in humans. In humans, endothelial injuries heal primarily by undamaged cells adjacent to the wound enlarging and spreading to cover the injured area. As a consequence, the capability of the endothelium to repair damage is limited [32].

2.4.2   EFFECT OF WAVELENGTH AND PULSE WIDTH

Laser radiation absorbed by biological tissue results in one or more competing biophysical interaction mechanisms: thermal, photochemical, photomechanical, and optielectric breakdown [26,33]. Both the total energy absorbed into a finite volume of tissue and the time in which it is delivered affect the damage level and damage mechanism.

In the case of thermal injury, the absorption of laser radiation results in heating of the irradiated tissue. Temperature increases of 10°C–50°C above ambient can result in tissue coagulation via protein denaturation, while larger amounts of deposited energy can vaporize the water and carbonize tissue solids [29]. The threshold for thermal damage is generally greater for longer exposure durations because conduction and vascular convection transport some of the absorbed energy away from the irradiated tissue while energy is still being deposited [29]. Photochemical injury mechanisms occur when the photons absorbed by a molecule are of sufficient energy to break chemical bonds or induce chemical reactions or both. The wavelength dependence of photochemical interactions and the optical absorption properties of the irradiated tissue play a role in determining the dominant damage mechanism [33].

Photomechanical injury mechanisms occur at short pulse durations (0.1 ms or less), given the pulse is of sufficient energy, and consist of acoustic or pressure waves that cause mechanical damage to the irradiated tissue and tissue surrounding it. Optielectric breakdown mechanisms have been observed for picosecond and femtosecond-duration pulses, with injury resulting from nonlinear optical effects such as self-focusing, optical breakdown, and plasma formation [26]. The particular combination of exposure parameters (i.e., wavelength, peak power, beam diameter, pulse width, pulse repetition rate, and total energy delivered) influences the interaction mechanism and primary tissue affected [29].

As illustrated in Figure 2.11, the transmission of optical radiation varies as a function of wavelength and tissue type. Boettner measured the spectral transmittance for

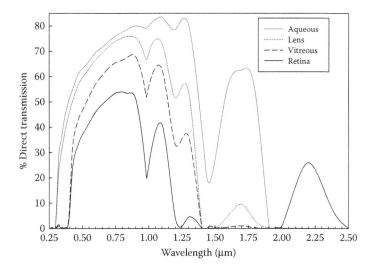

FIGURE 2.11   Calculated direct transmittance of the human eye. As radiation incident on the cornea propagates through the eye it is attenuated due to absorption and/or scattering. The magnitude of the attenuation is dependent on the optical properties of each ocular component. Each plot illustrates the calculated percentage of radiation incident on the anterior (front) surface of the aqueous humor, lens, vitreous humor, and retina; conversely, each plot illustrates the magnitude of radiation attenuated by all the ocular components preceding each surface. (Adapted from Boettner, E.A., *Spectral Transmission of the Eye*, The University of Michigan: Ann Arbor, MI. p. 43, 1967.)

each component of the human eye (i.e., cornea, aqueous humor, lens, vitreous humor, retina, and choroid) in the 0.22–2.80 μm wavelength range [34]. Boettner used this data to compute the successive transmittance as radiation incident on the cornea propagates through the eye. Figure 2.11 presents the calculated direct transmittance at the various anterior surfaces of the human eye as a function of wavelength. As Figure 2.11 illustrates, the cornea and lens are the primary absorbers of radiation below 0.400 μm with the exception of a small window centered at 0.320 μm. Between 0.400 and 1.400 μm, a significant portion of the radiation incident on the cornea is transmitted to the retina. While radiation between 0.700 and 1.400 μm is transmitted to the retina, it is not perceptible to the human eye. This, in addition to the lack of pain sensory nerves in the retina, increases the risks associated with this wavelength region because exposure may initially go unnoticed [1]. Beyond 1.400 μm, transmission to the retina ceases, with the anteior components of the eye becoming the primary absorber of incident radiation. Beyond 2.5 μm, water becomes the primary absorber of incident radiation, with the high water content of most tissue limiting the penetration depth to tens of microns [29].

2.4.2.1   Ultraviolet Radiation

Damage to the cornea, lens, and retina have resulted from exposure to ultraviolet (UV) laser radiation. Photon energies available from common UV laser sources range from approximately 3.5 to approximately 7.9 eV, which are sufficient to induce electron transitions and, in some cases, exceed the dissociation energies of many organic molecular bonds (approximately 3.0–7.5 eV) [29,35]. As a result, UV radiation is capable of causing photochemical damage to tissue. For example, in cellular tissue, shorter wavelength UV photons have sufficient energy to potentially induce photochemical reactions involving protein and nucleic acid constituents of the cell. Production of photoproducts at critical sites within the cell can lead to disruption of normal cell function and eventually cell death [36]. Longer-wavelength UV photons can also induce photochemical reactions in other cellular constituents with similar effects. There is typically a time delay on the order of hours between the onset of photochemical damage and ability to observe the damage on a macroscale level. Table 2.1 presents a summary of the experimentally determined ocular damage thresholds induced by CW and pulsed lasers.

Below approximately 0.300 μm, UV radiation is strongly absorbed by the cellular layers of the cornea, with the epithelial layer as the primary absorber. The reported threshold dose in this region is on the order of tens of millijoules per square centimeter [36]. The cornea is most sensitive in the 0.260–0.280 μm wavelength region, which corresponds to the first absorption bands of the aromatic amino acids of proteins and common nucleic acid bases [36]. Corneal transmission increases as the wavelength increases from 0.280 to 0.400 μm, with the lens becoming the primary

TABLE 2.1

Experimentally Determined Ocular Damage Thresholds Induced by CW and Pulsed Lasers

Laser Source	Wavelength (μm)	Threshold Dose (J/cm²)		
		Cornea	Lens	Retina
Argon fluoride (ArF) [33] 20 ns pulse duration	0.193	0.015[a]–0.025[b]	–	–
Krypton fluoride (KrF) [33] 20 ns pulse duration	0.248	0.059	–	–
Xenon chloride (XeCl) [33] 20 ns pulse duration	0.308	0.021	–	–
Helium cadmium (HeCd) [37] Mode-locked operation	0.325	14	–	0.361
Nitrogen (N₂) [38] 10 ns pulse duration	0.337	8.4 ± 3.3	1	–
Ruby (frequency-doubled) [33] 30 ns pulse duration	0.347	–	14	–
Krypton-ion [36] Continuous wave	0.351, 0.356[c]	67	–	–
Xenon fluoride (XeF) [33] 25 ns pulse duration	351	–	15	–

[a] Threshold dose for superficial clouding.

[b] Threshold dose for ablation.

[c] 3:1 intensity ratio.

absorber in this wavelength region. In this wavelength region, the corneal damage threshold is several orders of magnitude greater, on the order of ten to hundreds of joules per square centimeter [36].

Zuclich et al. investigated the ocular effects induced in Rhesus monkeys by exposure to UV laser radiation in the 350–360 nm region using krypton-ion and argon-ion CW lasers [36,38]. The krypton-ion laser simultaneously output 0.351 and 0.356 μm radiation with a 3:1 intensity ratio. Single-pulse exposures, ranging from 18 to 120 s, and multiple-pulse exposures (pulse width: 250 μs to 1 s; duty cycle: 50%; pulse train length: 30 s) were investigated. The best fit to the experimental data was reported as a threshold dose of 67 J/cm^2 [36]. Corneal damage resulting from exposures two to three times greater than the threshold dose were observed to be limited to the epithelium and repaired within 48–72 h post exposure. The argon-ion laser simultaneously output 0.351 and 0.364 μm radiation with a 1:1 intensity ratio. The threshold doses necessary to create corneal and lenticular lesions for a 4 s exposure were reported as 96 ± 14 and 76 J/cm^2, respectively [38]. For these exposure parameters, damage induced in the cornea is postulated to result from a photochemical damage mechanism. The lenticular threshold data indicates that the predominant damage mechanism for the lens is related to the peak power incident. At short exposure durations (1 ns to at least 1 μs), the damage mechanism appears to be thermal [33].

While the anterior components of the eye effectively shield the retina from most UV radiation, there is a small window centered at approximately 0.320 μm where the percentage of radiation transmitted to the retina approximates that at 0.400 μm [36]. Zuclich and Toboada used a mode-locked helium cadmium (HeCd) laser ($\lambda = 0.325$ μm) to investigate the ocular damage threshold in Rhesus monkeys in this region [37]. The laser was reported to have a pulse repetition rate of approximately 10.9 ns and a pulse duration of approximately 900 ps. The corneal damage threshold was reported as 14 J/cm^2; however, at exposure levels approximately two orders of magnitude lower, damage to the retina was observed. The retinal damage threshold was reported as 0.361 J/cm^2. The retinal damage threshold was found to be dependent on the total energy deposited, not the peak power, which suggests a photochemical damage mechanism. The retinal damage was permanent [36,37].

Research has been performed into the ocular damage threshold for nanosecond-duration pulses across the UV region [33,36,38]. The corneal threshold dose was determined for exposures to pulsed argon fluoride (ArF) ($\lambda = 0.193$ μm), krypton fluoride (KrF) ($\lambda = 0.248$ μm), and xenon chloride (XeCl) ($\lambda = 0.308$ μm) laser radiation. While the KrF and XeCl threshold doses were in reasonable agreement with threshold doses obtained from a low-power conventional lamp source, the ArF threshold dose was two or three orders of magnitude lower and believed to result from a photomechanical mechanism [36].

Zulich et al. investigated ocular damage thresholds in Rhesus monkeys using a pulsed nitrogen (N$_2$) laser ($\lambda = 0.337$ μm) with a 10 ns pulse width, a pulse repetition rate variable to 50 Hz, and a peak power of approximately 1 MW [38]. The threshold dose for corneal damage from exposure to trains of 10 ns pulses was 8.4 ± 3.3 J/cm^2. The visual appearance of the corneal lesions resulting from the pulsed N$_2$ laser radiation was different from that of those caused by a CW krypton-ion laser, with signs of mechanical tearing or fracture of the epithelial layer observed. Lenticular opacities (clouding of the lens) were also induced by the N_2 laser. A single 10 ns pulse (1.1 J/cm^2) was sufficient to cause a barely noticeable lenticular clouding. Trains of two or more pulses induced distinct cataracts visible immediately after exposure [38]. In this case, the damage mechanism postulated for the lenticular opacities is thermal in nature. The lens has an absorption peak centered at 0.365 μm, resulting in deposit of the incident energy in the anterior portion of the lens. The high peak power results in an irradiance sufficient to cause immediate thermal damage.

2.4.2.2   Visible and Near-Infrared Radiation (0.400–1.400 μm)

Damage thresholds for wavelengths inside the retinal hazard region are significantly lower, owing to minimal attenuation within the eye of wavelengths in this region and the focusing action of the corneal curvature and the lens. For example, the MPE to the eye for a 10 ns laser pulse as defined in ANSI Z136.1 for wavelengths in the retinal hazard region is approximately four to six orders of magnitude lower than for IR wavelengths greater than 1.4 μm. The cornea and lens focus the parallel rays of a collimated laser beam to a very small retinal image, increasing the irradiance at the retina by as much as 10^5 in comparison with the irradiance incident on the cornea [29].

The majority of radiation incident on the retina is absorbed by the pigments in the RPE layer, with less than 15% of the incident radiation absorbed by chromophores in the photoreceptor layer [1]. Because the RPE layer acts as a radiation sink, thermal damage typically originates there but can extend into the other layers of the retina and into the choroid.

The exposure duration and wavelength dictate the dominant damage mechanism at the damage threshold level [26,39,40]. Research has also demonstrated that the damage threshold is a function of retinal image size [41–44]. The damage threshold dependence on retinal image size varies depending on the specific damage mechanism [26,27].

For laser radiation in the 0.400 μm to approximately 0.550 μm wavelength region and exposure durations greater than 10 s, retinal injury predominantly results from a photochemical damage mechanism [1,26,29,40,45]. The damage threshold is on the order of tens to hundreds of joules per square centimeter, increasing as the wavelength increases [45].

For exposures ranging from several seconds to approximately 0.1 ms, the predominant damage mechanism is thermal [26,29,39,46,47]. This is also true for exposure durations greater than 10 s in the near-IR wavelength region [26]. Lund et al. experimentally determined retinal injury thresholds for laser radiation in the 0.400–1.00 μm wavelength region for 100 ms exposure durations using Rhesus monkeys as experimental subjects [39,46,47]. As illustrated in Figure 2.12, the general trend for ED_{50} damage threshold values is an increase as the wavelength increases.

For exposure durations shorter than approximately 10 μs, the predominant damage mechanism at the threshold level changes from thermal in nature to photomechanical. Melanin particles contained in melanosome spheroids located in the RPE layer of the retina strongly absorb laser energy, resulting in formation of a steam bubble, termed microcavitation [26,27,48]. A study conducted by Lee et al., showed that microcavitation damage thresholds are lower than thermally

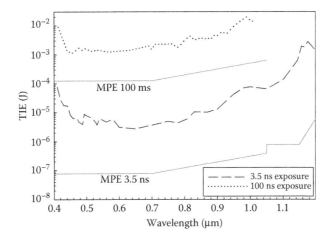

FIGURE 2.12 Experimentally determined minimum visible lesion threshold (ED_{50}) values as a function of wavelength for retinal exposures in the Rhesus monkey: (a) 100 ms exposure durations (dotted line) and (b) 3.5 ns exposure durations. The visible lesion determination was made 1 h post exposure. The data are presented as the TIE, which is defined as the energy incident on the cornea within the ocular pupil. (ED_{50} values from Lund, D. J., P. Edsall, and B. E. Stuck, *Proceedings of SPIE*, 5688, 383–393, 2005; exposure limits [MPE] from ICNIRP, *Health Physics*, 105, 271–295, 2013.)

induced damage thresholds [26,27]. Numerous studies have been conducted to determine the retinal damage threshold for short (ns to μs) and ultrashort (fs to ps) pulse lasers in the visible and near-IR region [39,45,49–59]. Lund et al. experimentally determined retinal injury thresholds for laser radiation in the 0.400–1.20 μm wavelength region for 3.5 ns exposure durations using Rhesus monkeys as experimental subjects [39,45,57]. As illustrated in Figure 2.12, the 100 ms and 3.5 ns data sets exhibit the same general trend as a function of wavelength. However, the 3.5 ns ED_{50} values are two to three orders of magnitude lower than the 100 ms ED_{50} values.

Cain et al. experimentally determined the single-pulse ED_{50} threshold dose necessary to create a minimally visible retinal lesion in Rhesus monkeys due to exposure to visible and near-IR laser radiation with exposure durations ranging from nanoseconds to femtoseconds [50,58]. As Figure 2.13 illustrates, for both the visible and near-IR data sets, the ED_{50} threshold dose decreases by approximately an order of magnitude as the pulse width decreases. At the shorter exposure durations (fs to ps), nonlinear optical effects, such as self-focusing and laser-induced breakdown, become significant [48,59].

The near-IR region between 1.100 and 1.400 μm is unique from a laser safety perspective [60–62]. As illustrated in Figure 2.11, transmission through the various ocular components decreases rapidly for wavelengths greater than 1.100 μm owing to water absorption. Between approximately 1.150 and 1.400 μm, the absorption is more evenly distributed across the various ocular components [60]. Depending on the exposure parameters, damage to the retina, lens, or cornea can occur.

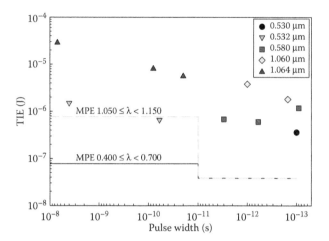

FIGURE 2.13 Experimentally determined minimum visible lesion thresholds (ED$_{50}$) values for retinal exposure in the Rhesus monkey as a function of exposure duration and wavelength. The visible lesion determination was made 1 h postexposure. The data are presented as the TIE, which is defined as the energy incident on the cornea within the ocular pupil. (ED$_{50}$ values from Cain, C. P., et al., *Ultrashort Pulse Laser Effects in the Primate Eye*, AL/OE-TR-1994-0141 PN. 1994; Cain, C. P., et al., *Near-Infrared Ultrashort Pulse Laser Bioeffects Studies*, AFRL-HE-BR-TR-2003-0029, US Air Force Research Laboratory: Brooks, TX. 2003; exposure limits [MPE] from ICNIRP, *Health Physics*, 105, 271–295, 2013.)

2.4.2.3   IR Radiation: 1.400 µm and Beyond

For wavelengths greater than 1.4 µm, absorption of laser radiation by the water component of the ocular media shields the retina from exposure. Exposure to laser radiation in this region primarily results in damage to the cornea; however, damage to the lens has been attributed to wavelengths below 3 µm [28]. The primary damage mechanism for exposure durations greater than 1 µs is thermal; at shorter exposure durations, damage can result from thermomechanical mechanisms [26].

At wavelengths less than 3 µm, the radiation penetrates more deeply into the cornea. For example, the $1/e$ absorption depth ranges from a low value of 10 µm at a wavelength of 10.6 µm to 0.124 mm at 1.93 µm to 1.1 mm at 1.54 µm [32,63]. The absorption depth is the depth at which irradiation is diminished to $1/e$ (0.37) of its incident level. Thus, approximately 63% of the incident radiation is absorbed in one absorption depth, and approximately 95% is absorbed in three absorption depths [63]. The increase in absorption depth results in a greater absorbing volume, which distributes the dose over a larger thermal mass, raising the damage threshold [33]. The increased penetration depth is also a liability once the injury threshold is reached because of the regenerative properties of each layer of the cornea.

Research has shown a correlation between epithelial and endothelial injury thresholds and absorption depth. Bargeron et al. found that the threshold dose for endothelial damage was approximately 10 times greater than the epithelial injury threshold dose for 10.6 μm laser radiation [64]. Rabbit eyes were exposed to CW CO_2 laser radiation at three different peak irradiances (24.5, 10, and 3.6 W/cm^2), and the exposure duration was varied until endothelial damage occurred. The reported ED_{50} threshold doses were 24.5, 52.0, and 864 J/cm^2, respectively. McCally et al. found that for 1.54 μm laser radiation the injury threshold dose for the endothelium was approximately 15% greater than the injury threshold dose for epithelial damage [32]. The eyes of New Zealand white rabbits were exposed to CW erbium fiber laser radiation with the ED_{50} threshold dose reported as 36.6 J/cm^2 (epithelium) and 44.0 J/cm^2 (endothelium) for an 11 s exposure duration. The convergence of the endothelial and epithelial damage thresholds at 1.54 μm results from the larger absorption depth at this wavelength, which results in a more uniform temperature distribution in the different corneal layers.

Numerous studies have been performed to determine the corneal damage threshold due to exposure to pulsed IR laser radiation (see Table 2.2) [65–78]. The variation in corneal damage threshold at each wavelength listed in Table 2.2 is the result of

TABLE 2.2
Reported ED_{50} Threshold Dose Necessary to Create a Visible Lesion on the Cornea

Laser Source	Wavelength (μm)	Pulse Width (s)	1/e Spot Size (mm)	ED_{50} Dose (J/cm)
Nd:YAG pumped OPO [78]	1.573	3×10^{-9}	0.4	26.6
			1	4.3[a]
Erbium glass [65]	1.54	5×10^{-8}	1	21
Erbium glass [68]	1.54	4×10^{-8}	1–2	4.7
		1×10^{-3}	1–2	7.2
Erbium glass [76]	1.54	8×10^{-4}	0.4×0.6	56
Erbium glass [77]	1.54	8×10^{-4}	0.58×0.57	30.9[b]
			0.66×0.61	21.2[c]
Erbium glass [69]	1.54	9.3×10^{-4}	1	9.6
Holmium LiYF4 [69]	2.06	4.2×10^{-8}	0.032	5.2
Holmium LiYF4 [69]	2.06	1×10^{-4}	0.18	2.9
Carbon dioxide [70]	10.6	1.7×10^{-9}	2.12	0.66
		2.5×10^{-8}	2.12	1.08
		2.5×10^{-7}	2.12	0.36
Carbon dioxide [72]	10.6	8×10^{-8}	3.88	0.36
Carbon dioxide [74]	10.6	8×10^{-8}	3.72	0.307
Carbon dioxide [71]	10.6	9.6×10^{-4}	1.02	0.599
Carbon dioxide [66]	10.6	1×10^{-3}	2.8	0.8

[a] Extrapolated value.
[b] *Ex vivo* tissue.
[c] *In vitro* tissue.

differences in experimental conditions (i.e., differences in pulse duration, laser beam spot size, animal type tested, etc.) [77,79].

Exposure to short-duration pulses (less than 1 µs) from 10.6 µm radiation can result in thermal damage to all the corneal layers, depending on the exposure conditions. However, radiation at this wavelength is strongly absorbed by the tear film and epithelial layers, with 99% being absorbed in the first 50 µm [74]. This energy is rapidly converted to heat, which is initially concentrated in the volume of absorption and subsequently is conducted to deeper layers of the cornea [73]. Zuclich et al. experimentally determined that nanosecond-duration exposures to 10.6 µm laser radiation at a dose greater than or equal to the threshold dose resulted in an immediate opacification of the corneal epithelium [69,70]. The observed epithelium damage was no longer observable after 48 h, with the exception of severe exposures (greater than or equal to twice the threshold dose).

In contrast with exposures at 10.6 µm, exposure to doses greater than or equal to the threshold dose from 1.54 or 2.06 µm laser radiation can result in permanent stromal scars [69]. Lund et al. investigated the ocular hazard posed by a Q-switched erbium glass laser using owl monkeys [65]. Near-threshold exposures resulted in corneal lesions extending into the Bowman's membrane with more severe exposures reaching the Descemet's membrane. While the epithelial defects healed over the course of 24–72 h, the stromal opacification remained essentially the same over a 3 week period. Stuck et al. investigated the ocular injury threshold for pulsed holmium (2.06 µm) and erbium (1.54 µm) laser radiation [69]. Rhesus monkeys were subjected to long-pulse and short-pulse laser radiation. The depth and diameter of the corneal lesions were found to be both dose and wavelength dependent, with the depth of the lesions varying from half to full corneal thickness. For both erbium and holmium laser exposures, the corneal opacity (stromal scarring) persisted for the duration of the study (10 months). Lesions produced by erbium laser exposures were deeper, with little change over the 10-month period. While some holmium laser-induced lesions visible 1 h postexposure were not observable 24–48 h postexposure, most persisted.

2.5 CONCLUSIONS

As discussed in Section 2.3, certain atmospheric conditions may limit the range or operational effectiveness of a laser-based explosives detection system. Multiple modeling and simulation (M&S) programs for atmospheric propagation of laser radiation have been developed to evaluate system performance in a range of different atmospheric conditions [80]. Performing M&S during system development can provide significant insight into how different design considerations will affect system performance.

Irrespective of how well a specific optical technique performs in a laboratory or developmental test environment, prior to developing or fielding a laser-based explosives detection system, it would be prudent to consider both the engineering and safety aspects of such a system. From a cost and engineering standpoint, an analysis should be performed to assess the trade-offs associated with specific system capabilities and attributes (i.e., limit of detection, standoff range, size and weight, etc.).

The end-user perspective should also be considered during system development. In addition to being effective, an ideal system to the end user (1) is easy to operate, (2) is compact and lightweight, and (3) requires minimal maintenance and logistical support.

As Section 2.4 illustrates, from a safety aspect, significant advantages are gained from using a laser that emits radiation outside the retinal hazard region. The damage thresholds for the anterior components of the eye are in general several orders of magnitude higher than the retinal damage threshold. While laser radiation in the UV or IR regions ($\lambda \geq 1.4$ μm) is still capable of causing injury under some exposure conditions, the higher damage threshold provides researchers with greater latitude in developing techniques or procedures to reduce the laser radiation hazard. For example, Alakai Defense Systems has submitted a patent application that uses an aversion response (i.e., the blink reflex) from a visible laser to reduce the ocular exposure resulting from a laser emitting radiation that is invisible to the human eye [81]. In this case, two lasers propagate to the target simultaneously. The first laser is used to interrogate the target, and the second laser produces an aversion response in the same manner as a "laser dazzler" device [82].

Regardless of the optical detection technique selected, consideration should be given at each development phase to reducing the laser safety hazards associated with the system. For example, selection of the laser parameters (wavelength, pulsed or CW operation, etc.) and telescope design (collimated or focused beam, beam diameter, etc.) can have a profound effect on the Nominal Ocular Hazard Distance (NOHD). The NOHD is the distance between the source and the point along the propagation path beyond which the radiant exposure, or irradiance, no longer exceeds the MPE [83]. Ideally, the parameters can be selected to minimize the NOHD and the distance at which diffuse and specular reflections are a hazard.

The safety aspect should be considered from two points of view: that of experts knowledgeable in the hazards posed by laser radiation and that of everyone else. For example, in many cases, a system can be engineered to mitigate the potential laser radiation hazard under a specific set of operating conditions. However, in the author's experience, while such a system has the potential to gain the approval of the experts who sit on safety committees, obtaining authorization for fielding or employment of such a system from policy makers is a much harder sell. While scientists and engineers tend to focus on the practical aspects of such a system, policy makers recognize that the system's capabilities need to outweigh the potential negative publicity associated with such a system potentially causing injury to the warfighter, targeted individuals, and bystanders. The scientific community needs to do a better job of educating policy makers regarding the hazards associated with lasers so that they have the knowledge required to make informed decisions.

ACKNOWLEDGMENTS

I would like to express my appreciation to my colleague, Dr. Barton Billard, and his wife Linda for their assistance reviewing and editing this book chapter.

REFERENCES

1. Träger, F., *Springer Handbook of Lasers and Optics*. 2007, New York: Springer.
2. Koechner, W. and M. Bass, *Solid-State Lasers: A Graduate Text*. Advanced Texts in Physics. 2003, New York: Springer.
3. Patel, C. K. N., High power infrared QCLs: Advances and applications. *Proceedings of SPIE*, 2012. **8268**: 826802-1–826802-13.
4. Koechner, W., *Solid-State Laser Engineering*. Springer Series in Optical Sciences, Vol. 1. 2010, New York: Springer.
5. Weichel, H., *Laser Beam Propagation in the Atmosphere*. Tutorial Texts in Optical Engineering v. TT 3. 1990, Bellingham, WA: SPIE Optical Engineering Press.
6. Siegman, A. E., *Lasers*. 1986, Sausalito, CA: University Science Books.
7. Saleh, B. E. and M. C. Teich, *Fundamentals of Photonics*. 1991, New York: John Wiley & Sons.
8. Melles Griot, Gaussian beam optics, in *The Practical Application of Light*. 1999, Carlsbad, CA: Melles Griot.
9. Gebhardt, F. G., High power laser propagation. *Applied Optics*, 1976. **15**(6): 1479–1493.
10. Burley, J. L., Comparison of high energy laser expected dwell times and probability of kill for misson planning scenarios in actual and standard atmospheres. Thesis. 2012, Wright-Patterson Air Force Base, OH: Air Force Institute of Technology.
11. Hutt, D. L., et al., Estimating atmospheric extinction for eyesafe laser rangefinders. *Optical Engineering*, 1994. **33**(11): 3762.
12. Smith, D. C., High-power laser propagation: Thermal blooming. *Proceedings of the IEEE*, 1977. **65**(12): 1679–1714.
13. Shettle, E. P. and R. W. Fenn, Models for the aerosols of the lower atmosphere and the effects of humidity variations on their optical properties, AFGL-TL-79-0214. 1979, Hanscom AFB, MA: Air Force Geophysics Laboratory.
14. Gebhardt, F. G., Twenty-five years of thermal blooming: An overview. *Proceedings of SPIE*, 1990. **1221**: 2–25.
15. Prokhorov, A. M., et al., Laser irradiance propagation in turbulent media. *Proceedings of the IEEE*, 1975. **63**(5): 790–810.
16. Tatarski, V. I., The effects of turbulent atmosphere on wave propagation. 1971, Jerusalem: Israel Program for Scientific Translations (available on NTIS Technical Translation 68-50464). p. 472.
17. Hutt, D. L. and D. H. Tofsted, Effect of atmospheric turbulence on propagation of ultraviolet radiation. *Optics & Laser Technology*, 2000. **32**: 39–48.
18. Walters, D. L., D. L. Favier, and J. R. Hines, Vertical path atmospheric MTF measurements. *Journal of the Optical Society of America*, 1979. **69**(6): 828–837.
19. Searles, S. K., et al., Laser beam propagation in turbulent conditions. *Applied Optics*, 1991. **30**(4): 401–406.
20. Laserna, J. J., et al., Study on the effect of beam propagation through atmospheric turbulence on standoff nanosecond laser induced breakdown spectroscopy measurements. *Optics Express*, 2009. **48**(12): 10265–10276.
21. Hanson, F., et al., Laser propagation at 1.56 mm and 3.60 mm in maritime enviroments. *Applied Optics*, 2009. **48**(21): 4149–4157.
22. Mukherjee, A., S. Porten, and C. K. N. Patel, Standoff detection of explosive substances at distances of up to 150 m. *Applied Optics*, 2010. **49**(11): 2072–2078.
23. Hays Parks, W., *Trauvaux preparatoires* and legal analysis of blinding laser weapons protocol. *The Army Lawyer*, 1997. **June**: 33–41.
24. Office of Treaty Compliance & Homeland Defense. Treaty compliance. Office of the Under Secretary of Defense for Acquisition, Technology, and Logistics. www.dod.gov/acq/acic/treaties/ccwapl_laser.htm.

25. Schulmeister, K. The upcoming new editions of IEC 60825-1 and ANSI Z136.1—Examples on impact for classification and exposure limits. *Proceedings of the International Laser Safety Conference*, 2013. Orlando, FL: Laser Institute of America.
26. ICNIRP, ICNIRP guidelines on limits of exposure to laser radiation of wavelengths between 180 nm and 1,000 μm. *Health Physics*, 2013. **105**(3): 271–295.
27. Schulmeister, K., et al., Review of thresholds and recommendations for revised exposure limits for laser and optical radiation for thermally induced retinal injury. *Health Physics*, 2011. **100**(2): 210–220.
28. Oliver, J. W. S., et al., Visible lesion threshold in cynomolgus (*Macaca fascicularis*) retina with 1064-nm, 12-ns pulsed laser. *Proceedings of SPIE*, 2007. **6435**: 64350Q-1–64350Q-12.
29. Mihran, R. T., Interaction of laser radiation with structures of the eye. *IEEE Transactions on Education*, 1991. **34**(3): 250–259.
30. Schulmeister, K., Concepts in dosimetry related to laser safety and optical radiation hazard evaluation. *Proceedings of SPIE*, 2001. **4246**: 104–116.
31. Saunders, L. L., T. E. Johnson, and T. A. Neal, A review of infrared laser energy absorption and subsequent healing of the cornea. *Proceedings of SPIE*, 2004. **5319**: 349–354.
32. McCally, R. L., et al., Corneal endothelial injury thresholds for exposures to 1.54 micro m radiation. *Health Physics*, 2007. **92**(3): 205–211.
33. Schulmeister, K., Review of exposure limits and experimental data for corneal and lenticular damage for short pulsed UV and IR laser radiation. *Journal of Laser Applications*, 2008. **20**(2): 98–105.
34. Boettner, E. A., Spectral transmission of the eye. 1967, Ann Arbor, MI: The University of Michigan. p. 43.
35. Vogel, A. and V. Venugopalan, Mechanisms of pulsed laser ablation of biological tissues. *Chemical Reviews*, 2003. **103**: 577–644.
36. Zuclich, J. A., Ultraviolet-induced photochemical damage in ocular tissues. *Health Physics*, 1989. **56**(5): 671–682.
37. Zuclich, J. A. and J. Taboada, Ocular hazard from UV laser exhibiting self-mode-locking. *Applied Optics*, 1978. **17**(10): 1482–1484.
38. Zuclich, J. A. and J. S. Connolly, Ocular damage induced by near-ultraviolet laser radiation. *Investigative Ophthalmology*, 1976. **15**(9): 760–764.
39. Lund, D. J., P. Edsall, and B. E. Stuck, Wavelength dependence of laser-induced retinal injury. *Proceedings of SPIE*, 2005. **5688**: 383–393.
40. Lund, D. J., B. E. Stuck, and P. Edsall, Retinal injury thresholds for blue wavelength lasers. *Health Physics*, 2006. **90**(5): 477–484.
41. Zuclich, J. A., et al., Variation of laser induced retinal-damage threshold with retinal image size. *Journal of Laser Applications*, 2000. **12**(2): 74–80.
42. Lund, D. J., et al., Variation of laser-induced retinal injury thresholds with retinal irradiated area: 0.1 s duration, 514 nm exposures. *Journal of Biomedical Optics*, 2007. **12**(2): 024023-1–024023-7.
43. Schulmeister, K., et al., Ex vivo and computer model study on retinal thermal laser-induced damage in the visible wavelength range. *Journal of Biomedical Optics*, 2008. **13**(5): 054038-1–054038-13.
44. Zuclich, J. A., et al., New data on the variation of laser induced retinal-damage threshold with retinal image size. *Journal of Laser Applications*, 2008. **20**(2): 83–88.
45. Lund, D. J., P. Edsall, and B. E. Stuck, Ocular hazards of Q-switched blue wavelength lasers. *Proceedings of SPIE*, 2001. **4246**: 44–53.
46. Lund, D. J. and P. Edsall, Action spectrum for retinal thermal injury. *Proceedings of SPIE*, 1999. **3591**: 324–334.
47. Lund, D. J., P. Edsall, and B. E. Stuck, Spectral dependance of retinal thermal injury. *Journal of Laser Applications*, 2008. **20**(2): 76–82.

48. Rockwell, B. A., T. J. Thomas, and A. Vogel, Ultrashort laser pulse retinal damage mechanisms and their impact on thresholds. *Medical Laser Application*, 2010. **25**: 84–92.
49. Lund, D. J. and E. S. Beatrice, Near infrared laser ocular bioeffects. *Health Physics*, 1989. **56**(5): 631–636.
50. Cain, C. P., et al., Ultrashort pulse laser effects in the primate eye, AL/OE-TR-1994-0141. 1994, PN, San Antonio, TX: Analytic Sciences Corp.
51. Cain, C. P., et al., Retinal damage and laser-induced breakdown produced by ultrashort-pulse lasers. *Graefe's Archive for Clinical and Experimental Ophthalmology*, 1996. **234**: S28–S37.
52. Lund, D. J., et al., Bioeffects of near-infrared lasers. *Journal of Laser Applications*, 1998. **10**(3): 140–143.
53. Rockwell, B. A., et al., Ultrashort laser pulse bioeffects and safety. *Journal of Laser Applications*, 1999. **11**(1): 42–44.
54. Cain, C. P., et al., Thresholds for visible lesions in the primate eye produced by ultrashort near-infrared laser pulses. *Investigative Ophthalmology & Visual Science*, 1999. **40**(10): 2343–2349.
55. Cain, C. P., et al., Visible lesion threshold dependence on retinal spot size for femtosecond laser pulses. *Journal of Laser Applications*, 2001. **13**(3): 125–131.
56. Cain, C. P., et al., Thresholds for retinal injury from multiple near-infrared ultrashort laser pulses. *Health Physics*, 2002. **82**(6): 855–862.
57. Lund, D. J., P. Edsall, and B. E. Stuck, Ocular hazards of Q-switched near-infrared lasers. *Proceedings of SPIE*, 2003. **4943**: 44–53.
58. Cain, C. P., et al., *Near-Infrared Ultrashort Pulse Laser Bioeffects Studies*, AFRL-HE-BR-TR-2003-0029. 2003, Brooks, TX: US Air Force Research Laboratory.
59. Cain, C. P., et al., Sub-50-fs laser retinal damage thresholds in primate eyes with group velocity dispersion, self-focusing and low-density plasmas. *Graefe's Archive for Clinical and Experimental Ophthalmology*, 2005. **243**: 101–112.
60. Zuclich, J. A., et al., Ocular effects of penetrating IR laser wavelenths. *Proceedings of SPIE*, 1995. **2391**: 112–125.
61. Zuclich, J. A., et al., High-power lasers in the 1.3–1.4 μm wavelength range: Ocular effects and safety standard implications. *Proceedings of SPIE*, 2001. **4246**: 78–88.
62. Vincelette, R. L., et al., Trends in retinal damage thresholds from 100-millisecond near-infrared laser radiation exposure: A study at 1,110, 1,130, 1,150, and 1,319 nm. *Lasers in Surgery and Medicine*, 2009. **41**: 382–390.
63. McCally, R. L., R. A. Farrell, and C. B. Bargeron, Cornea epithelial damage thresholds in rabbits exposed to Tm:YAG laser radiation at 2.02 microns. *Lasers in Surgery and Medicine*, 1992. **12**: 598–603.
64. Bargeron, C. B., et al., Corneal damage from exposure to IR radiation: Rabbit endothelial damage thresholds. *Health Physics*, 1981a. **40**: 855–862.
65. Lund, D. J., et al., Ocular hazards of Q-switched erbium laser. *Investigative Ophthalmology*, 1970. **9**(6): 463.
66. Brownell, A. S. and B. E. Stuck. Ocular and skin hazards from CO_2 laser radiation. *Proceedings of the 9th Army Science Conference*, 1974. West Point, NY: US Military Academy.
67. Mueller, H. A. and W. J. Ham, The ocular effects of single pulses of 10.6 μm and 2.5–3.0 μm Q-switched laser radiation. A Report to the Los Alamos Scientific Laboratory L-Division, Department of Biophysics, Virginia Commonwealth University, Richmond, VA. 1976.
68. Avdeev, P. S., et al., Experimental determination of maximum permissible exposure to laser radiation of 1.54 μ wavelength. *Kvantovaya Elektronika*, 1978. **5**(1): 220–223.
69. Stuck, B. E., D. J. Lund, and E. S. Beatrice, Ocular effects of holmium (2.06 μm) and erbium (1.54 μm) laser radiation. *Health Physics*, 1981. **40**: 835–846.

70. Zuclich, J. A., et al., Corneal damage induced by pulsed CO_2-laser radiation. *Health Physics*, 1984. **47**(6): 829–835.
71. Bargeron, C. B., et al., Epithelial damage in rabbit corneas exposed to CO_2 laser radiation. *Health Physics*, 1989. **56**(1): 85–95.
72. Farrall, R. A., et al., Structural alterations in the cornea from exposure to infrared radiation. Final Report to the US Army Medical Research and Development Command, Fort Detrick, MD, ADA215340. 1989.
73. Farrell, R. A., et al., Corneal effects produced by IR laser radiation. *Proceedings of SPIE*, 1990. **1207**: 59–70.
74. McCally, R. L. and C. B. Bargeron, Epithelial damage thresholds for sequences of 80 ns pulses 10.6 micron laser radiation. *Journal of Laser Applications*, 1998. **10**: 137–139.
75. McCally, R. L. and C. B. Bargeron, Epithelial damage thresholds for multiple-pulse exposures to 80 ns pulses of CO2 laser radiation. *Health Physics*, 2001. **80**(1): 41–46.
76. Clarke, T. F., et al., Corneal injury threshold in rabbits for the 1540 nm infrared laser. *Aviation Space and Environmental Medicine*, 2002. **73**(8): 787–790.
77. McPherson, N. A., T. E. Eurell, and T. E. Johnson, Comparison of 1540-nm laser-induced injuries in *ex vivo* and *in vitro* rabbit corneal models. *Journal of Biomedical Optics*, 2007. **12**(6): 064033.
78. Courant, D., et al., *In vivo* and *in vitro* evaluation of the corneal damage induced by 1573 nm laser radiation. *Journal of Laser Applications*, 2008. **20**(2): 69–75.
79. McCally, R. L., J. Bonney-Ray, and C. B. Bargeron, Corneal epithelial injury thresholds for exposures to 1.54 mm radiation—Dependence on beam diameter. *Health Physics*, 2004. **87**(6): 615–624.
80. Fiorino, S. T., et al., The HELEEOS atmospheric effects package: A probabilistic method for evaluating uncertainty in low-altitude high energy laser effectiveness. *Journal of Directed Energy*, 2006. **1**: 347–360.
81. Pohl, K. R., A. R. Ford, and R. D. Waterbury, *Optical Hazard Avoidance and Method*. 2012, Largo, FL: Alakai Defense Systems. p. 1–10.
82. Hecht, J., Nonlethal lasers deter attacks and warn away noncombatants. *Laser Focus World*, 2013. **49**(4): 45–48.
83. Laser Institute of America, *American National Standard for Safe Use of Lasers*, ANSI ZJ36.1-2014. 2014, Orlando, FL: Laser Institute of America.

3 Standards for Standoff Optical-Based Explosives Detection

Greg Gillen and Jennifer R. Verkouteren

CONTENTS

3.1 INTRODUCTION

Screening technologies that detect traces of explosives are an important and effective tool against terrorism and are widely used by federal, state, and local agencies tasked with protecting the public. Emerging priorities for noncontact, high-throughput, and noninvasive screening have focused attention on new standoff detection techniques utilizing optical- and laser-based spectroscopies. Significant progress has been made in developing a standards infrastructure around trace explosives detection, and now this is expanding to include optical-based detection methods. In developing standard test materials for standoff optical detection, additional factors must be controlled in order to appropriately evaluate efficacy and stimulate technique development. The challenge is not only to produce chemical standards with a known mass per unit area but also to control physical properties such as particle size, shape, and area distribution. In recent years, several different approaches for production of such standards for optical-based trace explosives detection technologies have been developed, and factors relevant to their preparation and use will be reviewed in this chapter.

The detection of chemical signatures resulting from trace quantities of explosives is one approach to identifying terrorist activity.[1,2] Micrometer-sized particles of explosives can be found in trace residues on people carrying or involved in the manufacture of an explosive device, on objects used to conceal explosive devices, and on surfaces in secondary contact with such devices or people.[3–10] Currently deployed methods for trace explosives detection most often involve the use of ion mobility spectrometry (IMS)-based explosive trace detection instruments (ETDs) to sample and analyze these particle residues.[11,12] In these instruments, the target residue is collected by surface swiping followed by heating of the collection wipe to convert the solid particles into neutral gas-phase vapor, which is introduced into the IMS system. This vapor is subsequently ionized and the atmospheric gas-phase mobility of ions characteristic of a given explosive are used as the basis for identification. Many instruments also have a direct vapor-sampling mode to detect trace-level vapors emanating from either a trace solid residue or, more likely, a bulk source such as a hidden device. However, due to the low vapor pressure of the majority of common explosives,[13,14] direct vapor detection is limited to a small subset of existing explosive threats.

Swipe-based ETD analysis is widely used, robust, and relatively low cost, but the requirement to physically interrogate each sample surface has limitations. These include the invasive nature of the sampling, the time required for sampling, and the fact that it places security screeners and checkpoints in direct contact with possible threats. These limitations have resulted in increased interest in laser- and optical-based detection technologies that would potentially allow for *in situ* chemical analysis of particle residues on surfaces, leading to noninvasive and high-throughput screening.[15–19] In a National Academies Report,[15] standoff optical-based methods were defined as "sensing the presence of explosive devices when vital assets and those individuals monitoring, operating, and responding to the means of detection are physically separated from the explosive device. The physical separation should put the individuals and vital assets outside the zone of severe damage from a potential detonation of the device." The distance is scenario dependent but would typically range from 10 to 100 m.[15]

Standoff explosives detection technologies are typically divided into bulk and trace detection methods. Bulk standoff detection involves identification of specific characteristics of an explosive device (wires, detonators) or the explosive material in the device itself. Trace standoff detection is, as described previously, the identification of the trace chemical signals resulting from contamination of a surface or release of vapor by a device. Relevant analytical techniques include, but are not limited to, a variety of spectroscopies, including laser-induced breakdown spectroscopy (LIBS) (refer to Chapter 7), Raman and resonance Raman spectroscopy (refer to Chapter 5), infrared (IR) spectroscopy (refer to Chapters 8–11), ultraviolet (UV) spectroscopy (refer to Chapter 4), photothermal and photo-acoustic spectroscopy (refer to Chapter 10), and laser-induced fluorescence spectroscopy (refer to Chapter 6). These techniques are discussed in this book as noted, and there are several recent reviews of the detection of explosive materials.[15–19]

One common need for any explosives trace detection technology is the availability of appropriate and realistic test materials. Test materials are required to adequately evaluate the performance of field-deployed systems and to support development and validation of new technologies. Test materials also provide a framework for certification and procurement of equipment. For conventional IMS swipe-based systems, the analytical response of the detection system is evaluated using two complementary approaches: (1) direct measurement of detector response from an introduced known mass of explosives or (2) a more comprehensive testing approach including swipe sampling of the test material. Direct measurements of detector response can be accomplished by solution deposition, either by pipetting, dispensing from calibrated dropper bottles, or ink-jet printing.[20] To provide standardization in testing protocols, a consensus standard from ASTM International (an international standards organization based in West Conshohocken, Pennsylvania) is available that defines the minimum level of performance of an ETD to a known mass of solution-deposited explosive.[20] A second ASTM standard provides robust methods for calculation of the limits of detection (LODs) for a given ETD based on the instrument response to direct insertion of varying masses of solution-deposited explosives.[21] A more inclusive testing protocol includes the requirement to collect the sample from surfaces using swiping. For this purpose, the use of the "dry transfer" test method is commonly used.[22] This approach involves depositing a known amount of explosive in solution onto a flexible, nonstick, polytetrafluoroethylene (PTFE) surface and allowing it to dry. The dry sample is transferred to a secondary surface for testing, and this transferred sample is collected by swiping for analysis in the ETD. One challenge with this technique is reproducibly controlling the transfer efficiency of explosives to the test surfaces and from the test surfaces to the collection swipe. Potentially large uncertainties in the mass of the deposited explosives on a test article can be minimized by direct chemical analysis of explosive remaining on the PTFE surface, as recently demonstrated by Tam et al.[23] For conventional IMS-based ETDs, the primary requirements for test materials are the mass and type of explosive and to a lesser extent the particle size, as it relates to either swipe-based collection or desorption efficiency. For swipe-based collection, the adhesion of particles to surfaces may also be a critical factor, but one that is very difficult to control or measure; this continues to be an active area of research.

The same factors are important to testing of optical-based standoff techniques (with the exception of particle adhesion), but a number of additional factors must be considered. Unlike thermal desorption-based IMS, optical techniques are very sensitive to the type of surface and the physical form of the deposit, particularly the size, shape, and area coverage (fill factor) of the explosives particles. In addition, optical techniques may be sensitive to different polymorphic forms of the explosive compounds, a factor that does not generally affect conventional IMS-based detection. Appropriate test materials for these techniques should not only provide reproducible values of the total mass of explosive but also provide control over these other factors. In this chapter, we review several of the approaches currently used to prepare standard test materials for standoff optical-based detection systems. Because of the challenges involved in creating such complex materials, there is no single approach that will satisfy all needs. We will discuss the advantages and limitations of each approach and examine concepts for further development of realistic test materials.

3.2 CHARACTERISTICS OF TRACE RESIDUES RELEVANT TO STANDOFF DETECTION

As mentioned above, although exceptions exist, the majority of high explosives have vapor pressures that are too low for practical standoff vapor detection using current technologies.[16] At the present time, the majority of standoff trace detection systems under development are targeted at direct analysis of threat particles on a surface. The amount of explosives that may be present as trace residues under different real-world scenarios is an active area of investigation, but the information is typically not available in the open literature due to security constraints. One study by Oxley et al.[4] examined contamination levels resulting from handling a variety of different explosives, including 2,4,6-trinitrotoluene (TNT), triacetone triperoxide (TATP), octahydro-1,3,5,7-tetranitro-1,3,5,7-tetrazocine (HMX), ammonium nitrate (AN), hexahydro-1,3,5-trinitro-1,3,5-triazine (RDX), and pentaerythritol tetranitrate (PETN), in several different forms in a laboratory setting. For most explosives, residual contamination on hands varied from 2.2 to 70 µg, with an average of 18 µg (excluding AN, which had a value of 895 mg). Individual fingerprints made after contact with Composition C-4 plastic explosive were found to contain 1–10 µg of RDX in the first fingerprint, decreasing to low nanogram levels for higher-generation fingerprints.[6,7]

In addition to total mass, critical information that must be known about the target residue includes the size, shape, and area coverage (fill factor) of the particles. The bulk materials that are handled to make devices will, in part, define the characteristics of the residues. Bulk explosives may be pure particulate materials (powders), or they may be more complex formulations containing one or more explosive compounds in a composite polymer binder or mixed with other components such as fuel oil. The particle size of the explosive compounds in bulk materials will vary depending on the size distribution in the original stock material and any changes produced by processing and handling. Even for a given pure explosive, the particle size distribution and particle morphology can be quite variable between manufacturers and will also depend on the intended use of the material. For example, the United States Military Standard (Mil-Spec) for RDX lists eight different classes of RDX, each with different particle size distributions (granulation requirements), depending on the eventual use of the powdered feedstock.[24] Scanning electron and optical micrographs of three plastic-bonded explosives and an ammonium nitrate–fuel oil

(ANFO) prill (i.e., pellet) and their corresponding optical micrographs are shown in Figure 3.1 to highlight this variability in size and morphology. The optical micrographs were taken from samples of the bulk material prepared with minimal grinding, although very large particles (as in the Semtex and C-4) are excluded.

While there are a number of possible scenarios for secondary transfer of a trace residue to a surface, the prototypical model is that of a fingerprint deposited on a secondary surface after the handling of a bulk explosive or powder. Typically, this would be in the form of a latent print, which implies that the deposit would not be visible to the unaided eye. Some limited data are available for particle sizes in fingerprints containing RDX, PETN, and C-4.[3,5–7] The particle sizes range from submicrometer to hundreds of micrometers in diameter, but the target population for mass detection techniques, such as IMS, is typically in the 10–20 μm size range for C-4.[6] This targeted size was based on the prevalence of particles of that size range throughout a series of 50 fingerprints (made with a single contaminated finger) and the mass represented by those particles. Particles smaller than 10 μm have a higher frequency and may provide a richer target for optical-based techniques.

The size and spatial distributions of crystalline explosive particles in fingerprints can be mapped with polarized light microscopy. The area of each particle is summed to calculate the coverage, or fill factor, over the area of the fingerprint. The example in Figure 3.2a shows a fingerprint map with the location of each RDX particle (distance from the origin) coded for particle size in a C-4 fingerprint. The particle heights are more difficult to measure, but were estimated to be approximately one-third of the diameter of each particle in C-4 fingerprints.[6] The calculated fill factor for all particles 4 μm in diameter or larger is 0.1%. The particles will probably be transferred in the ridges of the fingerprint (Figure 3.2b), and the ridge area alone may represent 50% or less of the area occupied by the latent print. In the example shown in Figure 3.2b, the ridges occupy approximately 47% of the area and the particles less than 1%, with the remainder (53%) representing blank substrate. This suggests

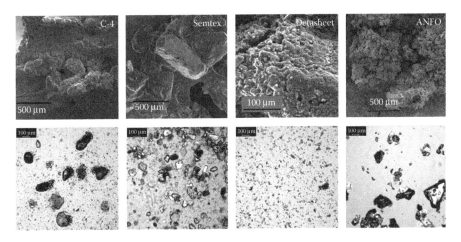

FIGURE 3.1 Montage of particle morphology. Scanning electron micrographs (top) of explosive formulations with polarized light micrographs (bottom). (NIST unpublished data.)

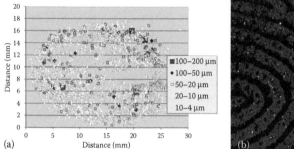 

FIGURE 3.2 (a) RDX particles in C-4 fingerprint represented by their x–y coordinates with symbols based on size. (b) 20 μm fluorescent spheres imaged in latent fingerprint. (NIST unpublished data.)

the critical nature of the fingerprint matrix and substrate in possibly dominating the analytical response.

The discussion above is based on the analysis of residues produced under well-controlled conditions in a laboratory setting. In practice, once a residue is deposited onto a surface, aging effects can modify both the particle size and distribution as well as the chemical composition. Particles of low-vapor-pressure explosives, such as RDX and PETN, are expected to be relatively stable in size and shape with time and exposure to ambient conditions. Higher vapor pressure explosives particles, such as TNT or TATP, can be lost to sublimation or modified by chemical degradation due to UV radiation exposure.[9] The situation becomes even more complex for water-soluble and hygroscopic explosive materials such as AN and urea nitrate (UN). Such materials may dissolve in the sweat (eccrine) component of the fingerprint (fresh fingerprints contain ~99% by weight of water in the eccrine secretions[25]) and precipitate upon drying, giving a larger area of coverage than would be expected from transfer of solid particles. An example of this is shown in Figure 3.3a, where the bright AN particles define the ridge pattern, implying transfer in the sweat component with

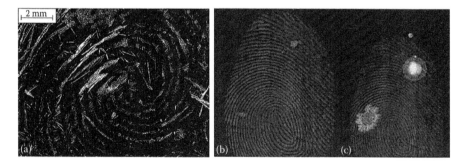

FIGURE 3.3 Optical micrographs of AN fingerprints. (a) Polarized light image with AN as bright particles (scale bar 2 mm). Two AN particles embedded in fingerprint and imaged (b) immediately and (c) after 1 h in 80% RH. (NIST unpublished data.)

subsequent recrystallization. AN is also hygroscopic and can become deliquescent at a relative humidity (RH) of 62% or higher.[26] This can result in enhanced mobility under ambient conditions, and the fill factor may change with time, as seen for the two AN particles in Figure 3.3b and c. The two images were collected 1 h apart after storing the print under conditions of 80% RH.

The fingerprint, as previously alluded to, is a complicated chemical matrix and is expected to contain hundreds of different organic and inorganic materials from the eccrine and sebaceous glands. While there is a high degree of variability between individuals, the typical mass of sebaceous and eccrine components in a fresh fingerprint is ~50 μg,[25] suggesting that a typical trace sample may contain minute quantities of explosives surrounded by similar or even higher levels of the fingerprint materials (as shown for the simulant particles in Figure 3.2b). Table 3.1 shows some of the common components of fingerprints.[27] For other types of residues or secondary transfer, the chemical environment may contain various types of interferents

TABLE 3.1
Components of Fingerprints

Eccrine Inorganic Component	Concentration (mg/L)	Sebaceous Component	Concentration (wt%)
Cl^-	520–7000	Free fatty acids	30–40
HCO_3^-	915–1220	Glycerides	15–25
SO_4^{2-}	7–190	Wax esters	20–25
PO_4^{3-}	10–17	Squalene	10–12
F^-	0.2–1.18	Cholesterol esters	2–3
Br^-	0.2–0.5	Cholesterol	1–3
I^-	0.005–0.012	Aldehydes	Trace
Na^+	750–6100	Ketones	Trace
K^+	190–340	Amines ،	Trace
Ca^{2+}	100–150	Amides	Trace
NH_4^+	10–150	Alkanes	Trace
NII_4^+	9–144	Alkenes	Trace
$Fe^{2+/3+}$	1–70	Alcohols	Trace
Mg^2	Trace	Phospholipids	Trace
Zn^{2+}	Trace	Pyrroles	Trace
$Cu^{+/2+}$	Trace	Pyridine	Trace
Co^{2+}	Trace	Piperidines	Trace
Mn^{2+}	Trace	Pyrazines	Trace
Mo^{2+}	Trace	Furans	Trace
$Sn^{2+/4+}$	Trace	Haloalkanes	Trace
$Hg^{+/2+}$	Trace	Mercaptans	Trace
		Sulfides	Trace

Source: Sisco, E. R. Methods and standards for the analysis and imaging of latent fingerprints and trace contraband using ambient ionization mass spectrometry and secondary ion mass spectrometry, PhD dissertation, University of Maryland, College Park, MD, 2014.

including dirt, dust, and soil. Test materials are currently being developed to address this issue.[28]

3.3 FACTORS TO CONSIDER IN TEST MATERIALS FOR STANDOFF TRACE DETECTION

Ideally, test materials for evaluation of any detection technique would represent all the physical and chemical characteristics relevant to the targeted real-world samples. However, this is generally unrealistic, and typically only the most important characteristics are captured, although the importance may change based on the requirements of a given analytical technique. Historically, the analytical response to a given mass of explosives per unit area, in micrograms per centimeter squared, has been used as the primary metric for evaluation of optical-based trace detection techniques. Understanding and controlling how the material is distributed across this sampled area, however, is also critically important, particularly for optical-based techniques. Consider, for example, Raman spectroscopy, in which the analytical response will depend on such factors as the wavelength of the excitation laser, Raman scattering cross section, the spectrometer depth of focus, sample absorption, and scattering.[29] Depending on the parameters of a given experimental configuration, the sampling depth may vary from millimeters to nanometers. Some configurations would perform best when the analyte is distributed homogeneously over the defined area, as in a film. Other configurations with greater sampling depths may produce better responses for discrete and thicker particles. These factors need to be understood in order to evaluate the efficacy and relevance of a new technique.

Based on these issues, we can define two classes of test materials for standoff detection. The first, appropriate for early technology development, would focus on providing a reproducible distribution of explosive materials, probably in a pure form, that could be deposited in a manner to allow for evaluation of such parameters as sampling depth and quantitative response and assist in development of detection algorithms and spectral libraries. Such standards would be helpful in identifying fundamental limitations in a given technology at an early stage of development. The second type of standard test material would be more appropriate for end-stage evaluation of more mature technologies and would be as close as possible to real-world residue with appropriate particle sizes and distributions and fill factors as well as appropriate background species. The following examples illustrate some of these approaches.

3.4 PREPARATION TECHNIQUES FOR TEST MATERIALS

3.4.1 SOLUTION DEPOSITION METHODS: DROP-CASTING, SAMPLE SMEARING, AND ANTISOLVENT CRYSTALLIZATION

3.4.1.1 Drop-Casting

One of the most direct means of producing test materials containing a known mass of explosive is to dissolve the explosive in solution and pipette a known volume onto a surface. The cast drop is allowed to dry, and a solid deposit is formed. This method of drop-casting is simple to achieve, low cost, and requires no specialized equipment, but it does have limitations. Chief among these is the inability to control the physical characteristics of the solid deposit, including particle size and particle distribution. The deposit that is formed will be highly dependent on the characteristics of the surface and the wetting conditions and evaporation rate of the solvent. In many cases, drop-casting methods often lead to the so-called "coffee-ring" texture, characterized by a nonuniform distribution of crystallized explosive particles with the higher concentrations at the periphery of the dried deposit (an example is shown in Figure 3.4). This phenomenon is thought to result from capillary flow produced by differential evaporation rates at the drop periphery.[31] The spread of the drop will also help define where material is deposited and the thickness of the deposit. Drop spreading will be dependent on the wetting of the selected solvent on the surface, which makes it difficult to prepare similar deposits on a wide variety of surfaces.

The crystallization processes occurring in evaporating drops can be quite complex and unpredictable, resulting in different particle shapes or even different polymorphs. Goldberg and Swift have shown that β-RDX, previously thought to be metastable, can be stabilized by drop-casting, and that the surface characteristics, solvent, and solution dilution factor are all contributing factors.[32] The formation of polymorphs may be relevant for certain types of techniques, particularly Raman spectroscopy.[33] In the example shown in Figure 3.5, the solution-cast deposit was mapped to determine the distribution of the α and β polymorphs of

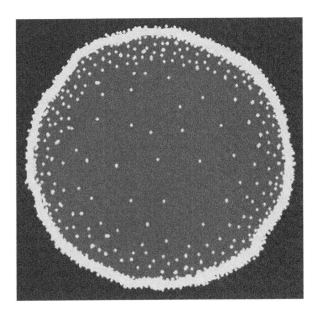

FIGURE 3.4 Deposit illustrating "coffee-ring" texture. (From Holthoff, E. L., Hankus, M. E., Tober, K. Q., and Pellegrino, P. M. Investigating a drop-on-demand microdispenser for standardized sample preparation. ARL-TR-5726. Army Research Laboratory, Aberdeen Proving Ground, MD, Weapons and Materials Research Directorate, 2011.)

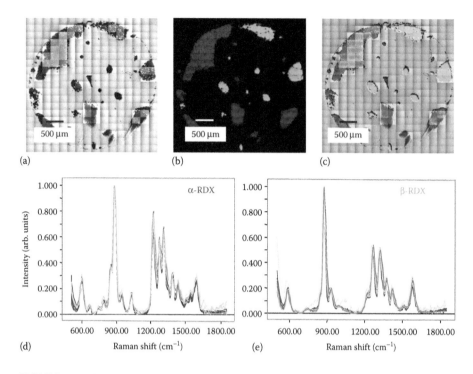

FIGURE 3.5 Polymorph distributions in deposits of RDX. (a) Brightfield image of a 3 μL droplet of 15 mg/mL RDX (military grade) in acetonitrile deposited on a quartz slide; (b) Raman chemical imaging (RCI) of the dried deposit. The areas colored red are α-RDX, and the areas colored green are β-RDX. (c) Overlay of the RCI on the brightfield image; (d) representative single-pixel spectra of regions of the RCI containing α-RDX; and (e) representative single-pixel spectra of regions of the RCI containing β-RDX. (From Emmons, E. D., et al., *Applied Spectroscopy* 66, 628–635, 2012.)

RDX, and the ratio of the two was found to be dependent on solution concentration. There have also been reports that solution-cast deposits do not always dry completely and may contain residual solvent or form metastable glass-like solids.[34]

3.4.1.2   Sample Smearing

The simplest form of drop-casting is to deposit microliter, or larger, volumes by pipetting, allowing the solution to dry without intervention. A modification of this technique, called sample smearing, was developed at the University of Puerto Rico.[35] A droplet of ~20 μL of solution is deposited onto the surface and then smeared using a single pass of an inclined Teflon sheet and allowed to air dry at room temperature for 10–15 min. This modification is meant to address the sample distribution issues and produce a more homogenous deposit. This is the same type of approach used to form films from solution, for example, through spin casting. Challenges with this modification include potential loss of solution, thereby compromising the total mass of explosive in the deposit.

3.4.1.3 Antisolvent Crystallization

Crystallization can be induced in drop-cast solutions by means other than evaporation. One such process is referred to as antisolvent crystallization, where a second solvent, in which the explosive is not soluble, is added. The second solvent, or antisolvent, must be miscible in the solution, and must reduce the solubility of the explosive sufficiently to initiate nucleation. An example of an antisolvent crystallization is shown in Figure 3.6, where water has been added to a drop-cast solution of RDX in ethanol and well-formed, discrete crystals with a limited range of sizes have formed. This approach holds the promise to control some of the physical characteristics of the deposits, but reproducibility remains a challenge. Other methods for crystallization from solutions include cooling or ultrasonic agitation, but these have not been attempted in drop-cast solutions to our knowledge.

3.4.2 FILM DEPOSITION METHODS: ELECTROSPRAY AND SPRAY DRYING

Several research groups have experimented with film deposition methods, using pneumatically assisted nebulization and an electrospray process[36] or aerosol spray brush deposition.[37] In both cases, the liquid spray plume rapidly dries into submicrometer-sized particles, preventing the classic coffee-ring effects observed with solution deposition methods and leading to a more evenly distributed film of controllable mass loading. In one implementation, a commercial air spray gun was used to dispense controlled doses of dissolved analyte with a mass loading reproducibility of $\pm 5\%$.[36] The deposited area was controlled by physically masking the sample substrate to provide a known mass per centimeter squared. To produce films

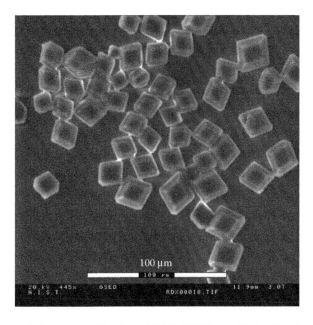

FIGURE 3.6 Antisolvent crystallization of RDX. (NIST unpublished data.)

over larger areas, a stationary airbrush spray can be combined with a two-axis stage to translate the sample under the spray, as shown in Figure 3.7. The spray approaches do not have the spatial control of ink-jet printing (discussed in Section 3.4.4) but offer a robust and rapid method for coating a known amount of explosive over relatively large areas with higher uniformity than can be achieved by drop-casting. These approaches are very complimentary to ink-jet printing and offer the capability of producing samples with relatively high mass loadings (150–2000 µg/cm^2) over larger areas. Quantification of the deposited material is usually achieved by removal and chemical analysis of the film after preparation.

3.4.3 PARTICLE DEPOSITION METHODS: SIEVING, SUSPENSION, PACKING, AND DESIGNER PARTICLES

3.4.3.1 Sieving

In order to avoid the complexities of particle formation from drop-cast solutions, an alternative approach is to use preformed particles and disperse them onto a surface. Nguyen et al. describe a method utilizing stock powders that are ground with a mortar and pestle to achieve a desired range of particle sizes.[34] This material is then passed through a stack of sieves in order to select a fixed particle size range as shown in Figure 3.8. Optical microscopy with customized image processing is used to gauge the efficacy of the sieving process and to adjust for the particle load, as shown in Figure 3.9. The sieving approach has been used successfully for a number of explosives including RDX, TNT, and AN. Quantification of the mass loading is validated by removal of the deposit with subsequent analysis using UV–visible (UV–Vis) spectroscopy. This method offers the advantage of controlled particle sizes and crystalline purity (including selection of polymorph), but its disadvantage is a less direct method of controlling the mass of explosive in the deposit. Additional problems arise from handling bulk powders of explosives to generate the samples, including potential

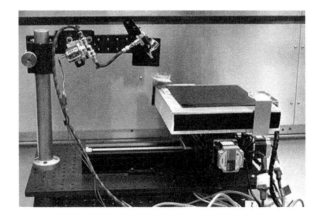

FIGURE 3.7   Spray system for explosives standard production. (Courtesy of US Army, Edgewood Chemical Biological Center, Aberdeen Proving Grounds, MD; from Nordberg, M., Ceco, E., Wallin, S., and Östmark, H., In *Proceedings of SPIE*, 8357, 2012.)

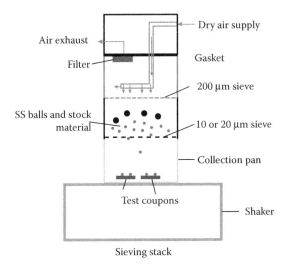

Sieving stack

FIGURE 3.8 Particle sieving apparatus. (From Nguyen, V., et al., In *Proceedings of SPIE Defense, Security, and Sensing*, p. 87100, International Society for Optics and Photonics, 2013.)

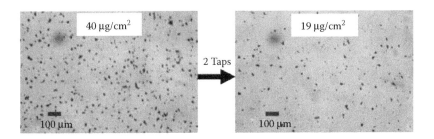

FIGURE 3.9 Optical micrographs of test coupons produced by particle sieving. Test coupon (left) loaded with 40 µg/cm^2 sieved TNT particles (20 µm sieve) and adjusted to 19 µg/cm^2 (right) by tapping twice on the base of the coupon with a pair of metal tweezers. (From Nguyen, V., et al., In *Proceedings of SPIE Defense, Security, and Sensing*, p. 87100, International Society for Optics and Photonics, 2013.)

safety hazards and the possibility of contamination in areas meant for trace sample preparation. This method is generally applicable to a range of explosive materials.

3.4.3.2 Powder Suspensions

Instead of dispensing explosives in solution, particle suspensions can be used, with particle concentrations and size distributions determined prior to deposition, and the mass loading controlled by the volume of suspension dispensed. The area distribution of particles for a water-based suspension can be controlled by modifying the substrate with hydrophobic/hydrophilic regions to constrain the solution. An example of a surface modified in this manner is shown in Figure 3.10. In the example, the particles of RDX are formed through an antisolvent process, as described earlier, rather than deposited as an existing suspension, illustrating an alternative approach.

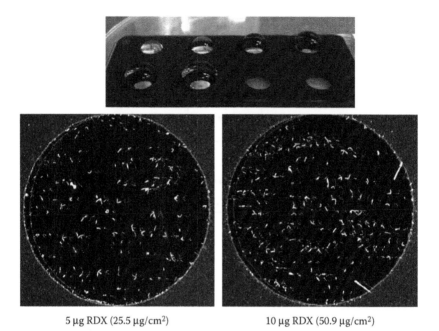

5 µg RDX (25.5 µg/cm²)   10 µg RDX (50.9 µg/cm²)

FIGURE 3.10   RDX particles formed on 5 mm hydrophobic wells on glass slides using a solution of RDX in acetonitrile mixed 50:50 with water. (NIST unpublished data.)

The sample substrates are low cost and easily transferable, and the ability to either keep the particles as a suspension in water or form them *in situ* simplifies transport, handling, and safety issues.

3.4.3.3   Powder Packing

Another approach for using powder samples is to fabricate a series of wells that are then packed with explosive powders.[38] In the example shown in Figure 3.11, wells were drilled in aluminum test plates with varying diameters and depths and packed with powdered TNT. The mass was determined by estimating the volume of the well based on its dimensions and assuming the density was that of the bulk material.[38]

r = 155 µm, m = 15 µg   r = 115 µm, m = 1.6 µg   r = 80 µm, m = 0.5 µg

FIGURE 3.11   Powder-packed wells with measured radii (r) containing an estimated mass (m) of TNT as indicated. (Adapted from Nordberg, M., Ceco, E., Wallin, S., and Ostmark, H., In *Proceedings of SPIE*, 8357, 2012.)

In this type of approach, the particle sizes within the powdered feedstock are only relevant in terms of packing characteristics, as the goal is to provide a homogenous deposit within the confines of the well.

3.4.3.4 Designer Particles

Use of commercially available feedstock powders for any of the particle deposition methods described above requires extra steps, such as sieving, grinding, and measuring, to control particle size. Therefore, there is an advantage to fabricating designer contraband particles with well-controlled properties[39–41] that could then be used in conjunction with one of the methods described above. Designer particles have been prepared using an emulsion process[40,41] with a fixed shape (spherical) and a tunable size, which allows them to be used without sieving to isolate the desired size range (Figure 3.12a). The polymer coating also reduces the safety hazards, stabilizes high-vapor-pressure explosives, and is a simulant for the matrix of plastic explosives such as Semtex and C-4. The particles can be transferred to surfaces using any of the approaches described above for bulk powders. Another approach for creating designer particles is to use a variation of ink-jet printing (discussed in Section 3.4.4) but with the deposition surface at a large distance from the printing device, allowing the droplet to pass through a drying tube before contacting the surface. Particles of RDX (Figure 3.12b) and AN (Figure 3.12c) have been prepared in this manner.

3.4.4 DROP-ON-DEMAND INK-JET MICRODISPENSING

Of all of the methods described thus far, the one that is gaining the most traction among multiple research groups for production of explosives standards is Drop-on-Demand (DOD) ink-jet printing.[30,42–54] Deposits are formed by dispensing solutions, but the droplet volumes are five to six orders of magnitude smaller than the microliter volumes delivered by pipette. These small volumes are key to mitigating some of the problems inherent to drop-casting, specifically the formation of "coffee rings" and the inability to obtain a homogeneous distribution of material over a defined area. Droplets are dispensed on demand at defined locations with spatial accuracies of micrometers, allowing for control over the distribution of material. In addition to these advantages, there is the potential for completely automated production of materials.

FIGURE 3.12 (a) Simulated C-4 particles produced by an emulsion process. (From Staymates, M., Fletcher, R., Staymates, J., Gillen, G., and Berkland, C., *Journal of Microencapsulation*, 27(5), 426–435, 2010.) Particles of (b) RDX and (c) AN produced by an ink-jet printing and drying-tube process. (NIST unpublished data.)

Most industrial printing devices use piezoelectrically or thermally generated pressure pulses to eject droplets from a fluid cavity, with a single droplet ejected for each triggered voltage pulse. For piezoelectric devices, droplet volumes are on the order of 10–100 pL, depending on the diameter of the cavity orifice, the characteristics of the piezoelectric driving waveform, the fluidic pressure, and the rheological properties of the fluid. Droplets are produced in bursts at each stage location by selecting the number of pulses (drops) at a specific frequency (ca. 100 Hz–20kHz). The selected pulse frequency can affect the size of individual droplets within the burst because of the interaction of sequential pressure waves within the cavity, resulting in resonant or antiresonant effects.[52] Between each burst of drops there is sufficient time for the pressure pulses to dissipate, and the ink-jet nozzle returns to a quiescent state prior to the next trigger. An alternative printing approach utilizes *print-on-the-fly*, where a pulse frequency is not selected. Instead, the selection includes the total number of drops and the area over which they are to be deposited. The program then determines the pulse frequency, which may vary throughout printing depending on the speed of the stage movement. "Print-on-the-fly" is much faster, as the stage is continually moving, but the drop size is more difficult to calibrate as it may change depending on pulse frequency.

Driving waveforms in DOD ink-jet printing are typically determined by imaging drops under stroboscopic illumination to find optimal conditions for single droplet formation. The waveform can be a simple trapezoidal form with input parameters of applied voltage and the times required to ramp up and down in voltage (rise and fall times) and hold the pulse (dwell time). The dwell time is generally a function of the length of the fluid cavity, and the rise and fall times are typically limited by the electronics. Applied voltage must be determined for the individual fluid, and drop size is known to be linear with voltage within the range appropriate for droplet formation. For some fluids, particularly low-viscosity fluids (<1 cP), more complicated waveforms are necessary, and bipolar waveforms including echo (negative) voltage pulses may be used. Holthoff et al.[50] have reported success with printing some of the low-viscosity solvents useful for explosives standards development, including acetonitrile and a solution of 2:1 methanol:water, with bipolar waveforms.

3.4.4.1 Dispensing Systems

There are several types of commercial ink-jet dispensing system that have been used for the production of ink-jet standards. Figure 3.13 shows a commercial single-channel JetLab printer from MicroFab, Inc. These "material deposition printers" are single-channel piezoelectric devices. The printhead is kept in a fixed location (adjustable z) and an x, y stage is moved below the printhead to allow location-specific deposition. Cameras are included for visualizing both the ejected drop and the printed surface. Variations on the basic design shown in Figure 3.14 for a GravJet include the addition of a microbalance, a corresponding additional stage movement to access the balance, and larger stage areas.

While precise control and visualization of the droplet formation process are possible in the single-channel systems, one limitation is throughput. For samples requiring higher sample loadings over larger areas, a multiorifice printhead has been used,[49] as shown in Figure 3.15. This system allows large areas to be covered in a single pass, facilitating high-throughput standards production.

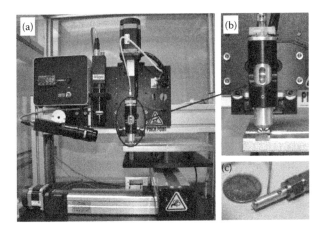

FIGURE 3.13 JetLab drop-on-demand ink-jet printer. Photographs of (a) JetLab 4 drop-on-demand ink-jet printing platform; (b) dispensing device and ink solution encasement; and (c) printhead assembly. (From Holthoff, E. L., Farrell, M. E., and Pellegrino, P. M., *Sensors*, 13(5), 5814–5825, 2013.)

FIGURE 3.14 JetLab GravJet System with integrated balance. (Courtesy of NIST.)

3.4.4.2 Quantification of Dispensed Mass

Quantification of ink-jet-based test materials requires knowledge of both the concentration of explosive in the starting solution and the volume of solution dispensed onto a test article surface. Starting solutions with known concentration are most easily obtained using commercially available solution standards, provided the rheological properties of the ink allow for direct ink-jet deposition. If this is not the case, solutions must be made from neat materials or by reconstituting in an appropriate ink-jet solvent and then analyzed using standard analytical techniques such as gas chromatography/mass spectrometry (GC/MS) or UV–Vis spectroscopy.[46,50] More challenging is the requirement to control and measure the volume dispensed. Classically, droplet size is measured from stroboscopic images by determining droplet diameter,

FIGURE 3.15 Thermal ink-jet printer used for fabrication of standoff explosive test material. (From Moon, R. P., et al., Preparation of chemical samples on relevant surfaces using inkjet technology. ECBC-TR-1056. Edgewood Chemical Biological Center, Army Research Laboratory, Aberdeen Proving Ground, MD, April, 2013.)

but measurements from optical images generally result in uncertainties in droplet volume of 10% or more.[55] A popular method is a gravimetric approach for droplet measurement, which results in uncertainties in droplet volume based on measured droplet mass of less than 1%,[46,51,52] as shown in Figure 3.16. Custom systems from MicroFab have been developed to support this implementation, as shown in Figure 3.14. The measurements utilize repeated bursts of drops, as single droplets are typically below the resolution of a microbalance. One of the advantages of the approach is that it utilizes the same parameters used during printing, specifically the

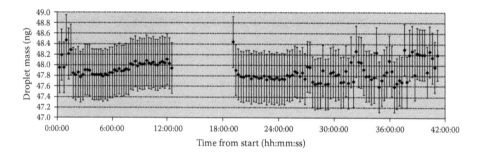

FIGURE 3.16 Repeatability of ink-jet deposition as measured by gravimetry across 42 h. Measurements were repeated every 16 m except during a scheduled break of about 6 h, and the imprecision is less than 0.4%. Error bars are estimated combined standard uncertainties of individual measurements. (From Verkouteren, M., Verkouteren, J., and Zeissler, C., Gravimetric traceability for optical measurements of droplets-in-flight. *Nanotech 2013*, vol. 2, pp. 224–227. CRC Press.)

number of drops in the burst. In this manner, if there are differences in drop size within the burst, they are accounted for in the measurement.

Alternative approaches for quantification include UV–Vis spectroscopy[50] or GC/MS[46] of a collection of droplets. The UV–Vis approaches offer excellent reproducibility and have the advantage of verification of the chemical purity of the dispensed material. Typical results are shown in Figure 3.17 for multiple explosives. In some cases, especially when using multinozzle thermal ink-jet printers, a mass-based approach may not be feasible, and, similar to the spray deposition methods, quantification is achieved by postdeposition mass measurements of the substrate or by chemical analysis after extraction from the printed surface.[49]

3.4.4.3   Control over Characteristics of Printed Deposit

Ink-jet printing allows precise positioning of ink-jet droplets in selected locations and therefore offers a mechanism for delivering a fixed mass of explosive per unit area. This is very useful for obtaining a known area coverage in a uniform and repeatable way. It is also possible, in certain cases, to modify the surface to control the characteristics of the ink-jet deposits, specifically the spreading of the deposited solution over the surface. One such example is shown in Figure 3.18. In this case one or more microdrops of an aqueous AN solution is printed onto a hydrophobic silicon substrate.[54] The hydrophobic substrate constrains the dispensed liquid into a hemispherical droplet such that a solid particle is formed as the solution evaporates. With this method, particle arrays of specific size, shape, and fill factor can be prepared. A variety of substrates, including glass, aluminum, silicon, and Teflon, can be modified with an appropriate surface treatment for applicability to this approach. For water-insoluble explosives such as RDX and PETN, an oleophobic surface can be applied for the same purpose.[54] A particular advantage of this type of approach is the ability to characterize the response of a given optical technique to explosive particles of varying height and diameter. Figure 3.18c shows Raman spectra from a series of particles that range in height from 8 to 96 μm with corresponding masses of 1.23 ng–1.23 μg.

The uniformity of the deposit over the surface test area may be controlled by means other than chemically modifying the substrate. When printing aqueous solvents that do not dry rapidly, adjacent drops can merge, creating unwanted variation in deposit uniformity and particle size. To mitigate this effect, sample substrate heating can be employed, as shown in Figure 3.19 for potassium chlorate. In this example, the deposits within the imaged area were much smaller and more uniformly distributed when printed on the heated surface compared with the unheated surface.[49]

3.4.4.4   Printing High-Viscosity Materials

Most of the ink-jet printers described above operate stably only when printing solutions within a relatively narrow range of viscosities (1–20 cP). However, to increase the complexity of test materials, such as by incorporating matrix components from plastic-bonded explosives or latent fingerprint components, more viscous materials may need to be printed. In such cases, other types of dispensing technologies

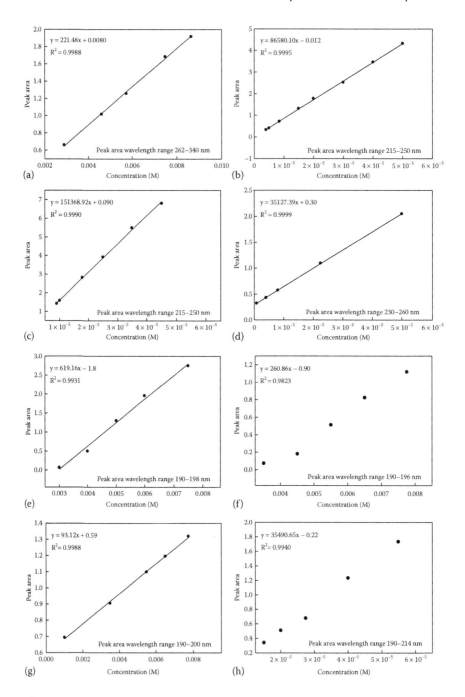

FIGURE 3.17 UV–Vis calibration curves for ink-jet solution quantification. Calibration curve and R^2 value from one UV–Vis data set at various analyte concentrations for (a) AN; (b) TNT; (c) HMX; (d) RDX; (e) urea; (f) potassium chloride; (g) sugar; and (h) PETN. (From Holthoff, E. L., Farrell, M. E., and Pellegrino, P. M., *Sensors* 13(5), 5814–5825, 2013.)

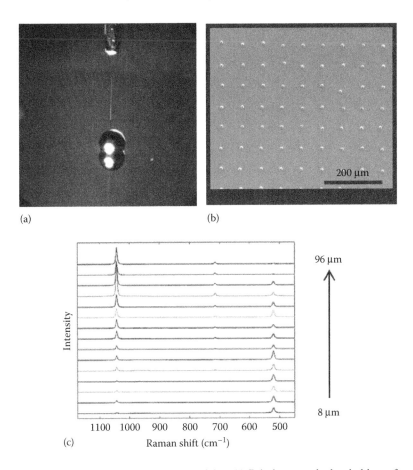

(a)   (b)

(c)   Raman shift (cm^{-1})

FIGURE 3.18 Ink-jet printing to form particles. (a) Printing onto hydrophobic surface to form a hemispherical drop; (b) dried deposits of hemispherical particles; (c) Raman spectra with particle height increasing from bottom to top. (From Gillen, G. et al., In preparation for submission to *Sensors*, 2015.)

could be employed. One of these is the DJ9000 dispense jet system from Nordson ASYMTEK (Carlsbad, CA). This system utilizes a pneumatically controlled piston with a ball-shaped tip on the end that contacts a seat within the nozzle assembly. The movement of the piston results in the ejection of the sample fluid through a narrow orifice at the nozzle tip. The size of the orifice as well as the throw of the piston and the rate at which the piston strikes the seat can be changed to adjust the droplet size (typically 100–300 μm in diameter). Similar to the GravJet system described earlier, this system incorporates a mass balance to determine the mass of the dispensed droplets. Figure 3.20a shows a photograph of this system, and Figure 3.20b shows arrays of artificial fingerprint material containing RDX particles dispensed in uniform arrays on a simulated painted car surface as a prototype standoff detection test material.[56]

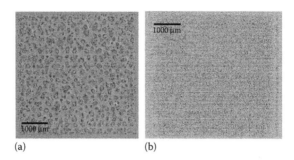

(a)   (b)

FIGURE 3.19 Effects of substrate heating on printed $KClO_3$: (a) no heating, 54 µg/cm²; (b) heated to 80°C, 45 µg/cm². Text under scale bars shows length = 1000 µm. (From Moon, R. P., et al., Preparation of chemical samples on relevant surfaces using inkjet technology. ECBC-TR-1056. Edgewood Chemical Biological Center, Army Research Laboratory, Aberdeen Proving Ground, MD, April, 2013.)

(a)   (b)

FIGURE 3.20 Printing of viscous materials: (a) DJ90000 dispense jet system (Courtesy of NIST.) and (b) printed arrays of artificial fingerprint material. (From Staymates, J. L., Staymates, M. E., and Gillen, G., *Analytical Methods*, 5(1), 180–186, 2013.)

3.5 CONCLUSIONS

This chapter included a brief review of some of the common methods currently in use for preparation of test materials for evaluating standoff optical detection systems. Clearly, the need for standards for trace detection technologies is critical, both for evaluation of already deployed technologies and to support new technologies in an early stage of development. Unlike the commonly used swipe-based trace detection systems based on total mass collected, optical-based standoff typically directly interrogates the analytical surface, requiring a higher degree of complexity for the standards and more strict control over particle size, distribution, and fill factor. No one approach will be relevant for all technologies, and different methods may be appropriate for different stages in the technology development process. Progress to date in developing these standards has been significant, and there is every reason to believe that it will continue at the same pace in the future.

DISCLAIMER

Certain commercial equipment, instruments, or materials are identified in this paper to specify adequately the experimental procedure. Such identification does not imply recommendation or endorsement by the National Institute of Standards and Technology, nor does it imply that the materials or equipment identified are necessarily the best available for the purpose.

ACKNOWLEDGMENTS

The Science and Technology Directorate of the US Department of Homeland Security sponsored portions of this material under Interagency Agreements HSHQPM-12-X-00070 and HSHQDC-12-X-00024 with the National Institute of Standards and Technology (NIST).

We would like to thank members of the NIST Surface and Trace Chemical Analysis Group, specifically Dr. Tom Forbes and Mr. Matthew Staymates, for providing some of the figures used in this document, and Dr. Edward Sisco for providing Table 3.1.

REFERENCES

1. Hallowell, S. F. Screening people for illicit substances: A survey of current portal technology. *Talanta* 54(3): 447–458 (2001).
2. Moore, D. S. Instrumentation for trace detection of high explosives. *Review of Scientific Instruments* 75(8): 2499–2512 (2004).
3. Carmack, W. J. and Hembree, P. B. Particle size analysis of prepared solutions and fingerprint deposits of high explosive materials. INEEL/EXT–97-01202. Lockheed Martin Idaho Technologies Company, Idaho National Engineering and Environmental Lab., Idaho Falls, ID (1998).
4. Oxley, J. C., Smith, J. L., Resende, E., Pearce, E., and Chamberlain, T. Trends in explosive contamination. *Journal of Forensic Sciences* 48(2): 334–342 (2003).
5. Verkouteren, J. R. Particle characteristics of trace high explosives: RDX and PETN. *Journal of Forensic Sciences* 52(2): 335–340 (2007).
6. Verkouteren, J. R., Coleman, J. L., and Cho, I. Automated mapping of explosives particles in composition C-4 fingerprints. *Journal of Forensic Sciences* 55(2): 334–340 (2010).
7. Gresham, G. L., et al. Development of particle standards for testing detection systems: Mass of RDX and particle size distribution of composition 4 residues. In *Proceedings of SPIE's 1994 International Symposium on Optics, Imaging, and Instrumentation*, pp. 34–44. International Society for Optics and Photonics (1994).
8. Miller, C. J. and Yoder, T. S. Explosive contamination from substrate surfaces: Differences and similarities in contamination techniques using RDX and C-4. *Sensing and Imaging: An International Journal* 11(2): 77–87 (2010).
9. Kunz, R. R., Gregory, K. E., Aernecke, M. J., Clark, M. L., Ostrinskaya, A., and Fountain, A. W., III. Fate dynamics of environmentally exposed explosive traces. *The Journal of Physical Chemistry A* 116(14): 3611–3624 (2012).
10. Kunz, R. R., Gregory, K. C., Hardy, D., Oyler, J., Ostazeski, S. A., and Fountain, A. W., III. Measurement of trace explosive residues in a surrogate operational environment: Implications for tactical use of chemical sensing in C-IED operations. *Analytical and Bioanalytical Chemistry* 395(2): 357–369 (2009).

11. Ewing, R. G., Atkinson, D. A., Eiceman, G. A., and Ewing, G. J. A critical review of ion mobility spectrometry for the detection of explosives and explosive related compounds. *Talanta* 54(3): 515–529 (2001).

12. Mäkinen, M., Nousiainen, M., and Sillanpää, M. Ion spectrometric detection technologies for ultra-traces of explosives: A review. *Mass Spectrometry Reviews* 30(5): 940–973 (2011).

13. Östmark, H., Wallin, S., and Ang, H. G. Vapor pressure of explosives: A critical review. *Propellants, Explosives, Pyrotechnics* 37(1): 12–23 (2012).

14. Ewing, R. G., Waltman, M. J., Atkinson, D. A., Grate, J. W., and Hotchkiss, P. J. The vapor pressures of explosives. *TrAC Trends in Analytical Chemistry* 42: 35–48 (2013).

15. Anderson, J. L., Cantu, A. A., Chow, A. W., Fussell, P. S., Nuzzo, R. G., Parmeter, J. E., Sayler, G. S. et al. *Existing and Potential Standoff Explosives Detection Techniques*. The National Academies Press, Washington, DC (2004).

16. Wallin, S., Pettersson, A., Östmark, H., and Hobro, A. Laser-based standoff detection of explosives: A critical review. *Analytical and Bioanalytical Chemistry* 395(2): 259–274 (2009).

17. Hummel, R. and Dubroca, T. Laser-and optical-based techniques for the detection of explosives. In *Encyclopedia of Analytical Chemistry*. R. A. Meyers (ed.), pp. 1–32, John Wiley and Sons, Hoboken, NJ (2013).

18. Fountain, A. W., III, Guicheteau, J. A., Pearman, W. F., Chyba, T. H., and Christesen, S. D. Long-range standoff detection of chemical, biological, and explosive hazards on surfaces. In *Proceedings of SPIE Defense, Security, and Sensing*, pp. 76790 H–76790 H. International Society for Optics and Photonics (2010).

19. López-López, M. and García-Ruiz, C. Infrared and Raman spectroscopy techniques applied to identification of explosives. *TrAC Trends in Analytical Chemistry* 54: 36–44 (2014).

20. ASTM Standard 2520-07, Standard practice for verifying the minimum acceptable performance of trace explosive detectors, ASTM International, West Conshohocken, PA, (2007), DOI: 10.1520/E2520-07, www.astm.org.

21. ASTM Standard E2677-14, Standard test method for determining limits of detection in explosive trace detectors, ASTM International, West Conshohocken, PA, (2007), DOI: 10.1520/E2677, www.astm.org.

22. Chamberlain, R. T. Dry transfer method for the preparation of explosives test samples. US Patent 6,470,730, issued October 29 (2002).

23. Tam, M., Pilon, P., and Zaknoun, H. Quantified explosives transfer on surfaces for the evaluation of trace detection equipment. *Journal of Forensic Sciences* 58(5): 1336–1340 (2013).

24. Department of the Army Technical Manual, TM9-1300-214, Military Explosives, 20 September (1984).

25. Yamashita, B. and French, M. Latent print development. In *Fingerprint Sourcebook*. US Department of Justice and National Institute of Justice (special publication 22530), CreateSpace (2011).

26. Lightstone, J. M., Onasch, T. B., Imre, D., and Oatis, S. Deliquescence, efflorescence, and water activity in ammonium nitrate and mixed ammonium nitrate/succinic acid microparticles. *Journal of Physical Chemistry A* 104: 9337–9346 (2000).

27. Sisco, E. R. Methods and standards for the analysis and imaging of latent fingerprints and trace contraband using ambient ionization mass spectrometry and secondary ion mass spectrometry, PhD dissertation, University of Maryland, College Park, MD (2014).

28. Verkouteren, J. R. Trace explosives: Reference materials and process optimization through training. *Defense Standardization Program Journal*, Triagency Standardization 37–42, April/September (2012).

29. McCreery, R. L. Photometric standards for raman spectroscopy. In *Handbook of Vibrational Spectroscopy*. J. M. Chalmers and P. R. Griffiths (eds), John Wiley and Sons, Chichester, UK (2002).

30. Holthoff, E. L., Hankus, M. E., Tober, K. Q., and Pellegrino, P. M. Investigating a drop-on-demand microdispenser for standardized sample preparation. ARL-TR-5726. Army Research Laboratory, Aberdeen Proving Ground, MD, Weapons and Materials Research Directorate, 2011.
31. Deegan, R. D., Bakajin, O., Dupont, T. F., Huber, G., Nagel, S. R., and Witten, T. A. Capillary flow as the cause of ring stains from dried liquid drops. *Nature* 389(6653): 827–829 (1997).
32. Goldberg, I. G. and Swift, J. S. New insights into the metastable β form of RDX. *Crystal Growth and Design* 12: 1040–1045 (2012).
33. Emmons, E. D., Farrell, M. E., Holthoff, E. L., Tripathi, A., Green, N., Moon, R. P., Guicheteau, J. A., Christesen, S. D., and Pellegrino, P. M. Characterization of polymorphic states in energetic samples of 1,3,5-trinitro-1,3,5-triazine (RDX) fabricated using drop-on-demand inkjet technology. *Applied Spectroscopy* 66(6): 628–635 (2012).
34. Nguyen, V., Papantonakis, M., Furstenberg, R., Kendziora, C., and McGill, R. A. Real-world particulate explosives test coupons for optical detection applications. In *Proceedings of SPIE Defense, Security, and Sensing*, p. 87100T. International Society for Optics and Photonics (2013).
35. Primera-Pedrozo, O. M., Pacheco-Londono, L. C., De la Torre-Quintana, L. F., Hernandez-Rivera, S. P., Chamberlain, R. T., and Lareau, R. T. Use of fiber optic coupled FT-IR in detection of explosives on surfaces. *Proceedings of SPIE* 5403: 237–245 (2004).
36. Yasuda, K., Woodka, M., Polcha, M., and Pinkham, D. Reproducible deposition of trace explosives onto surfaces for test standards generation. No. RDER-NV-TR-265. Army Communications-Electronics Command, Fort Belvoir, VA. Night Vision and Electronics Sensors Directorate (2010).
37. Barreto-Cabán, M. A., Pacheco-Londoño, L., Ramírez, M. L., and Hernández-Rivera, S. P. Novel method for the preparation of explosives nanoparticles. In *Defense and Security Symposium*, pp. 620129–620129. International Society for Optics and Photonics (2006).
38. Nordberg, M., Ceco, E., Wallin, S., and Östmark, H. Detection limit of imaging Raman spectroscopy. *Proceedings of SPIE* 8357 (2012).
39. Davies, J. P., Hallowell, S. F., and Hoglund, D. E. Particle generators for the calibration and testing of narcotic and explosive vapor/particle detection systems. *Proceedings of SPIE* 2092: 137–144 (1994).
40. Fletcher, R. A., Brazin, J. A., Staymates, M. E., Benner, B. A., Jr, and Gillen, J. G. Fabrication of polymer microsphere particle standards containing trace explosives using an oil/water emulsion solvent extraction piezoelectric printing process. *Talanta* 76(4): 949–955 (2008).
41. Staymates, M., Fletcher, R., Staymates, J., Gillen, G., and Berkland, C. Production and characterization of polymer microspheres containing trace explosives using precision particle fabrication technology. *Journal of Microencapsulation* 27(5): 426–435 (2010).
42. Verkouteren, R. M., Gillen, G., and Taylor, D. W. Piezoelectric trace vapor calibrator. *Review of Scientific Instruments* 77(8): 085104 (2006).
43. Verkouteren, M., Windsor, E., Fletcher, R., Maditz, R., Smith, W., and Gillen, G. Inkjet metrology and standards for ion mobility spectrometry. *International Journal for Ion Mobility Spectrometry* 9(1): 19–23 (2006).
44. Windsor, E., Najarro, M., Bloom, A., Benner, B., Jr, Fletcher, R., Lareau, R., and Gillen, G. Application of inkjet printing technology to produce test materials of 1,3,5-trinitro-1,3,5 triazcyclohexane for trace explosive analysis. *Analytical Chemistry* 82(20): 8519–8524 (2010).
45. Wrable-Rose, M., Primera-Pedrozo, O. M., Pacheco-Londoño, L. C., and Hernandez-Rivera S. P. Preparation of TNT, RDX and ammonium nitrate standards on gold-on-silicon surfaces by thermal inkjet technology. *Sensing and Imaging: An International Journal* 11(4): 147–169 (2010).

46. Verkouteren, M., Gillen, G., Staymates, M., Verkouteren, J., Windsor, E., Walker, M., Zeissler, C., et al. Ink jet metrology: New developments at NIST to produce test materials for security applications. In *NIP & Digital Fabrication Conference*, pp. 705–708. Society for Imaging Science and Technology (2011).
47. Reichardt, T. A., Bisson, S. E., and Kulp, T. J. Standoff ultraviolet Raman scattering detection of trace levels of explosives. SAND2011-7955. Sandia National Laboratories, Livermore, CA (2011).
48. De Lucia, F. C., Jr. Producing known quantities of RDX for LIBS limit of detection study. ARL-TR-6002. Army Research Laboratory, Aberdeen Proving Ground, MD, Weapons and Materials Research Directorate (2012).
49. Moon, R. P., et al. Preparation of chemical samples on relevant surfaces using inkjet technology. ECBC-TR-1056. Edgewood Chemical Biological Center, Army Research Laboratory, Aberdeen Proving Ground, MD, April (2013).
50. Holthoff, E. L., Farrell, M. E., and Pellegrino, P. M. Standardized sample preparation using a drop-on-demand printing platform. *Sensors* 13(5): 5814–5825 (2013).
51. Verkouteren, R. M. and Verkouteren, J. R. Inkjet metrology: High-accuracy mass measurements of microdroplets produced by a drop-on-demand dispenser. *Analytical Chemistry* 81(20): 8577–8584 (2009).
52. Verkouteren, R. M. and Verkouteren, J. R. Inkjet metrology II: Resolved effects of ejection frequency, fluidic pressure, and droplet number on reproducible drop-on-demand dispensing. *Langmuir* 27(15): 9644–9653 (2011).
53. Gura, S., Joshi, M., and Almirall, J. R. Solid-phase microextraction (SPME) calibration using inkjet microdrop printing for direct loading of known analyte mass on to SPME fibers. *Analytical and Bioanalytical Chemistry* 398(2): 1049–1060 (2010).
54. Gillen, G. et al. Particle fabrication using inkjet printing. In preparation for submission to *Sensors* (2015).
55. Verkouteren, M., Verkouteren, J., and Zeissler, C. Gravimetric traceability for optical measurements of droplets-in-flight. *Nanotech 2013: Technical Proceedings of the 2013 NSTI Nanotechnology Conference & Expo*, vol. 2, Chapter 4: Inkjet Design, Materials & Fabrication, pp. 224–227. CRC Press.
56. Staymates, J. L., Staymates, M. E., and Gillen, G. Evaluation of a drop-on-demand micro-dispensing system for development of artificial fingerprints. *Analytical Methods* 5(1): 180–186 (2013).

4 Explosives Detection and Analysis by Fusing Deep Ultraviolet Native Fluorescence and Resonance Raman Spectroscopy

Rohit Bhartia, William F. Hug,
Ray D. Reid, and Luther W. Beegle

CONTENTS

4.1 INTRODUCTION

Raman spectroscopy of chemical, biological, and explosive (CBE) materials provides a high-specificity means to determine whether an unknown material is a CBE compound, is related to or a precursor for a CBE, or is simply nonhazardous/nonthreatening. During the last two decades, advancements in lasers, electronics, optics, and miniaturized computing systems have enabled the development of compact Raman instruments that mobilize this capability from the lab to theater. However, even with these advancements, the traditional visible and near-infrared (NIR) Raman instruments are still plagued with two major issues: (1) naturally occurring and material-related fluorescence emissions that interfere with and obscure the Raman scattering, and (2) an overall low sensitivity of Raman measurement. These two problems reduce the probability of detection and increase false negatives.

A solution to obscuration by the fluorescence background and the increased probability of false negatives is the use of deep ultraviolet (UV) (<250 nm) excitation sources that enable a fluorescence-free Raman region and increase sensitivity to materials from Rayleigh and resonance Raman effects. However, one of the more interesting results of using deep UV is that the traditionally obscuring native fluorescence can be used as an orthogonal means of detection. Since the fluorescence is many orders of magnitude more sensitive than Raman and resonance Raman, it acts as a means to increase the sensitivity and probability of detection. While the probability of false positives may be higher, coupling the native fluorescence with deep-UV Raman allows for both rapid searching over large areas, a means to down-select areas of potential concern, and the use of the Raman effect to provide the high specify—a search analysis not possible with traditional visible and NIR Raman spectrometers.

This chapter focuses on explosives detection using the combination of deep-UV native fluorescence and Raman spectroscopy using deep-UV sources that enable compact, low-power-consuming devices. This includes a discussion of the deep-UV lasers that are the core of these fused instruments, the deep-UV Raman/fluorescence spectra using these compact deep-UV lasers, and how the fusion of fluorescence and Raman enables a rapid, high-sensitivity, high-specificity analysis of explosives on surfaces.

4.2 REVIEW OF DEEP-UV SPECTROSCOPY FOR EXPLOSIVES DETECTION

4.2.1 Explosives Detection Overview

What are you looking for, what level of information is required, and how do you intend to look for it?

These are questions commonly asked in the NASA planetary science community when determining what payload a mission to another planet will carry and providing the *target, required information content, and concept of operations (ConOps)*. For example, given the prevalence of perchlorates in the Mars soil, any instrumentation for organic detection and characterization that ingests and processes samples via heating or aqueous methods will not be ideal. In a similar manner, detection of explosives using instrumentation that requires direct contact with the material and/or suffers from ambient light rejection or background interference from the explosive material is not appropriate for explosive detection.

The intent of an active spectroscopic method is to illuminate or excite a target to perturb the chemical structure in a manner that leads a detectable and uniquely identifiable response. As such, excitation of a target with a given wavelength of light leads to a variety of spectral responses, including reflectance/absorption, Raman scatter, fluorescence, and, in some cases, phosphorescence; each provides information specific to the chemical structure. The wavelength of light can affect both the presence and intensity of each type of response. For instance, observing the entirety of the fluorescence spectrum of single aromatic ring compounds requires an excitation wavelength of <260 nm, and, while Raman scattering is technically independent of wavelength, peak intensities can vary as a function of wavelength, especially when considering resonance effects. With all of the possible spectral options and the benefits and challenges of each, there is no one detection method that fits all detection scenarios. It is imperative that the instrument design, capability, *and implementation* fulfill the needs of the explosive detection community. The goals stated below are high-level perspectives for targets, required information, and ConOps.

4.2.1.1 Target

The technique needs to detect a variety of explosive materials, including military-grade explosives, as well as materials used for improvised explosive devices (IEDs) made from either military-grade materials or homemade materials (homemade explosives [HMEs]). This also requires that the sensitivity to the targets is aligned with the ConOps.

4.2.1.2 Required Information Content

While the specific level of information is dependent on the exact nature of the ConOps, a thread common to this is a determination of whether a material is an explosive, a precursor to, or a component of an explosive device. In all cases, the detection method needs to fit the false-positive and false-negative requirements of the ConOps.

4.2.1.3 Concept of Operations

This requirement is perhaps the most elusive as it is in some manner dependent on capabilities of available technology. However, some level of common sense can be applied to this. An ideal *warfighter-level system* would include a compact, low system-mass, rugged, noncontact means of detection from some standoff distance, preferably one that extends to a range outside the hazard zone of a potential device, can operate under ambient light conditions, can handle a wide array of surface types, is simple to operate, and enables a rapid method of detection from bulk to trace concentration of materials. Systems that operate at the *checkpoint level* may allow for larger, less compact instrumentation, but they would still require rapid detection of trace concentrations in ambient light conditions and detection on a variety of surface types. All of these, however, also require that the detection method does not initiate detonation of a potential device and will not be hazardous to the user or civilians.

To the knowledge of the authors, no current instrument fits all the requirements. Some systems, based on 1064, 785, and 532 nm laser-based Raman methods, provide part of the solution and have been very effective in introducing the potential of Raman spectroscopy for explosives detection [1–3]. However, while compact and potentially useful for a warfighter level, there are limitations from an implementation/use perspective that include sample burning, obscuration of Raman signals from background fluorescence, and challenges with detection in ambient light conditions. It is not an issue with the specifics of the instrumentation design, but simply the challenges of operating in the NIR on the target materials of interest and in environments and with operation parameters associated with the ConOps. This is also an issue with UV to visible Raman spectroscopy that includes excitation wavelengths 263, 266, 325, 405, 448, 532, and 633 nm. While ambient light does not obscure Raman scatter collected at 263 and 266 nm, the background fluorescence does still pose a significant problem.

Deep-UV spectroscopy, defined here as excitation at wavelengths <250 nm, provides a potential compact solution for explosives detection as it avoids the potential for sample burning, fluorescence obscuration, and ambient light challenges, while providing increased sensitivity and specificity. However, enabling this technology for field use has required both scientific and technological research and development [4–13]. Deep-UV-based fluorescence and Raman instrumentation has been demonstrated as compact, accessible at low cost, and moving from specialized laboratories to compact handheld sensors with wide-scale production. This happened only after the advent of compact deep-UV laser sources <250 nm, technological developments in optics, and development of understanding on how best to implement these technologies. The range of potential applications includes, but is not limited to, explosives detection, chemical and biological detection, forensic analysis, space exploration, pharmaceuticals, and water quality.

4.2.2 DEEP-UV LASERS

Deep-UV lasers with emission wavelengths below 250 nm have typically been complex in design, expensive, and volumetrically large, with high power requirements. However, pioneering research from the Asher group demonstrated unique benefits, such as resonance enhancements for organics and a fluorescence-free Raman region, both features that capitalize on deep-UV sources <250 nm [14–25,53]. These benefits were of particular interest to the resurrected NASA Astrobiology program in 1998, soon after claims of life in the Martian meteorite (Alan Hills 83001), where there was an obvious need for new life detection methods. Fortuitously, in 1996, Photon Systems Inc. was developing a new, simple, low-cost, compact, low-power requiring, deep-UV laser <250 nm (Figure 4.1) [26–29]. It was the advent of these lasers coupled to high-quality deep-UV optics that enabled a new generation of instruments [30–32]. More recently, it has also been realized that the Raman spectral enhancements, as well as the fluorescence features enabled by the deep UV, are highly applicable to chemical, biological, and explosive counterterrorism efforts as well as microbial detection for contamination assessment [33–39].

The choice of excitation wavelength depends on the availability of lasers with appropriate deep-UV wavelength, emission linewidth, and other performance features such as size, weight, power consumption, and cost appropriate to a miniature, handheld, or small robot-mounted sensor. Laser emission wavelength, output power

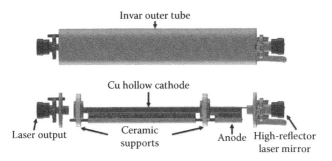

FIGURE 4.1 Schematic of transverse excited hollow cathode (TEHC) neon–copper (NeCu) laser (248.6 nm). The 224.3 nm helium–silver (HeAg) has a similar construction where the Cu cathode is replaced by an Ag one. The robustness of the laser comes from its simplicity in design, tolerance to alignment, and no requirement of temperature sensitive components/materials.

or energy, and linewidth are the dominant criteria to enable the ability to measure Raman emissions from a target, and they are the basis of an orthogonal Raman and fluorescence detection method. As illustrated in Figure 4.1, lasers below 250 nm separate Raman and fluorescence spectral regions for most materials of interest as targets or background materials and for the widest Raman shifts of possible interest.

Lasers that emit in the deep UV below 250 nm employ one of several basic technologies, including gas, solid state (including optical fiber), and semiconductors. Many government and internally funded programs are seeking to develop an "ideal" narrow-linewidth deep-UV laser source compatible with both Raman and native fluorescence methods. Only two laser technologies provide *fundamental emission* wavelengths below 250 nm. These include excimer lasers such as krypton fluoride (KrF) at 248 nm, krypton chloride (KrCl) at 222 nm, argon fluoride (ArF) at 193 nm, and fluorine (F_2) at 157 nm. In addition, there are transverse excited hollow cathode (TEHC) glow discharge lasers such as neon–copper (NeCu) lasers at 248 nm and helium–silver (HeAg) lasers at 224 nm. Wide bandgap semiconductor lasers such as aluminum gallium nitride (AlGaN) have sufficient bandgap to theoretically produce laser emission at wavelengths as low as 200 nm but have not yet demonstrated emission below 342 nm. All other lasers produce deep-UV *emission using harmonic conversion* of their fundamental wavelength using nonlinear optical crystals. Fundamental to the harmonic conversion process is the need for very high spectral radiance, measured in watts per square centimeter per steradian per nanometer. Second harmonic conversion efficiency is proportional to radiance in the nonlinear crystal, with second harmonic power being quadratically proportional to the fundamental source radiance. The geometry of crystals limits the f/# and thus the maximum achievable power density. Power densities of hundreds of megawatts per centimeter squared are commonly needed to achieve reasonable conversion efficiency, over 30%–50%. To achieve this in relatively miniature diode-pumped-solid-state (DPSS) lasers, peak power levels and pulse widths typically less than a few nanoseconds are required with consequences of sample damage. Continuous wave (CW) versions of harmonic generated deep-UV lasers are typically very large and heavy, with high power consumption.

Of the lasers described above, only the TEHC lasers have the size, weight, and power consumption to be compatible with handheld instruments for proximity detection of explosives with the advantages of Raman and fluorescence using deep-UV excitation. These lasers have excellent emission wavelengths at 224.3 and 248.6 nm, have narrow linewidth less than 0.1 wavenumbers, weigh less than 2 lb., including drive and power conditioning electronics and consume less than about 10 W. Excimer lasers also have excellent wavelengths and can be made with narrow linewidths, but they typically weigh about 50 lb. and consume over 50 W of electrical power. As a result, they are not compatible with handheld sensor applications. CW lasers such as argon ion lasers also have excellent wavelengths and narrow linewidths but typically weigh over 100 lb. and consume over 10 kW of electrical power. The only lasers with the potential size, weight, and power consumption to be considered for handheld sensors are 4th and 5th harmonic DPSS lasers. Unfortunately, the emission wavelengths of these lasers, at about 263 nm or 266 nm, are not compatible with fluorescence-free Raman spectroscopy. Versions of DPSS lasers have been proposed at 236 and 228 nm, but these have not been demonstrated in small, low-power consumption package sizes compatible with handheld application.

The lasers employed for all of the Raman and native fluorescence results described here are TEHC lasers. TEHC lasers have been under development at Photon Systems since the beginning of full-time operations in 1996. They are a unique type of laser, similar in construction to a miniature traveling wave tube or klystron, which provide direct CW transitions in the deep UV at 224.3 and 248.6 nm, with peak output power over 500 mW. TEHC lasers have an emission linewidth less than 3 GHz, corresponding to less than 0.1 wavenumbers and emission wavelength stability better than 100 ppb, independent of ambient temperature. TEHC lasers have a high threshold for lasing and high slope efficiency. As a result, in order to reduce average power during operation, the input energy to the laser is commutated with a duty cycle typically less than about 1%. However, since the transition is CW, a 30–100 μs "long" pulse can be generated with an output energy of 10–30 μJ per pulse. Comparatively, traditional pulsed lasers operate with pulse lengths of nanoseconds or less, creating very high peak-power outputs that, consequently, can alter samples through heating.

TEHC lasers (Figure 4.1) are about 30 cm long and 3.8 cm in diameter and weigh less than 1 lb. Average electrical power consumption is between about 1 and 10 W, depending on pulse repetition rate, which is up to about 40 Hz in present configurations, but they have been demonstrated to operate at >200 Hz. Because of the transverse excitation nature of TEHC lasers, hundreds of milliwatts of deep-UV output can be achieved in less than 10 μs after application of electrical power at any ambient temperature from about −130°C to +70°C, without warm-up, heating, cooling, or temperature regulation.

Laser lifetime is presently between 10 and 50 million pulses, corresponding to 10–50 million spectra for efficient emitting target materials. Since these lasers have a virtually instantaneous warm-up, the lasers do not need to be consuming lifetime except when taking Raman or fluorescence data. This dramatically increases the effective field lifetime of these lasers compared to all other types of lasers. TEHC lasers are also unique in that they are, by far, the least expensive lasers of all lasers emitting in the deep UV below 300 nm.

In 2005, the US Army independently rated Photon Systems' TEHC lasers at TRL 5+ during a technical readiness evaluation at Dugway, Utah, and currently they are under NASA being transitioned to TRL 6 for Mars applications. These lasers and related instruments and sensors have been on over a dozen major field expeditions in Antarctica, the Arctic, the deep ocean, Death Valley, and other hostile environments, as well as many other field trials and tests in real environmental circumstances. Recently, these TEHC lasers have been proposed as part of a deep-UV fluorescence/ Raman instrument for a multiyear surface rover mission to the planet Mars in 2020. In preparation for this mission and development to TRL 6 for Mars, the lasers were repeatedly cycled at temperatures between −130°C and +70°C without failure. In addition, the lasers have been submerged in liquid nitrogen, without failure. They have been tested to three times the shock and vibration specification for launch, cruise, and landing on Mars, without failure.

TEHC lasers achieve these performance characteristics by employing sputtering to generate the gain medium. In the case of the 224.3 nm HeAg laser, the gain medium is silver with a pump gas of a mixture of helium and heavier noble gas elements. In the case of the 248.6 nm NeCu laser, the gain medium is copper with a pump gas of neon. In both versions of these TEHC lasers, a glow discharge is generated within the hollow cathode via flutes in the cathode to a lateral brush anode. Radially inward 200 eV electrons generated at the inside surface of the cathode ionize and excite both noble gas atoms and ions, which in turn impact the cathode wall, generating metal atoms and ions to produce via charge transfer the upper state population inversion needed for lasing. This process is fast, taking less than about 10 µs, even in cold ambient conditions. This process is not dependent on ambient temperature or the heating of copper or silver to produce metal vapor. That is the reason these lasers can generate energetic states in the gain medium at extreme ambient temperatures without external heating and associated power consumption.

4.2.3 DEEP-UV NATIVE FLUORESCENCE SPECTROSCOPY

4.2.3.1 Overview of Native Fluorescence

Native fluorescence was first observed in 1565 from the extract of a medicinal wood, matlaline [40]. The variety of compounds that have since been found to have fluorescence is extensive. Most fluorescent compounds incorporate aromatic ring structures, but there are some, such as acetone, that fluoresce without an aromatic ring [41]. In simple terms, fluorescence is the emission of a photon that *may* occur as an electron transitions from an excited stated (S1–2) to a ground state (S0). In most cases, this requires that a photon of a higher energy be absorbed by the molecule such that an electron can jump to the S1 or S2 state. Once excited into the S1 or S2 state, vibrational energy levels consisting of vibrational, rotational, and molecular collisions cause the excited electron to nonradiatively lose energy until it reaches the lowest vibrational level of the excited energy state. This also includes an internal conversion from electrons in the S2 state to fall to the S1 excited state. The gap between the lowest energy level of S1 and the ground state will dictate emission wavelength produced when the electron returns to S0. This emission can be heat or a photon that is less energetic than the initial absorbed photon. The wavelength of this emitted photon provides some information about the electronic levels that were available to the electron. Typically, as the physical size of the compound increases, there are

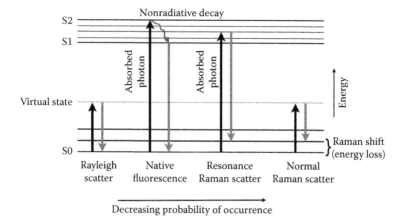

FIGURE 4.2 Electronic transitions that lead to Rayleigh, fluorescence, and Raman. A comparison of the electronic transitions that occur for Rayleigh scatter, native fluorescence, and resonance and normal Raman scattering. These have been ordered in terms of decreasing efficiency from left to right.

more available vibrational states and the emitted photon is of a lower energy (longer wavelength); this is known as a Stokes shift. For example, large sheets of graphene oxide absorb the UV and emit at 600–700 nm, while benzene, the simplest aromatic-ring compound, fluoresces at 260 nm [34,42,43]. These electronic transitions are best described using a Jablonski diagram (Figure 4.2). Phosphorescence is another light-emitting event that has been described. This differs from fluorescence, since the electrons undergo a spin transition to a T1 level in a process called intersystem crossing. Conversion from the T1 to S0 also can result in an emission. This event is much slower than fluorescence and can continue even after the excitation source is removed.

4.2.3.2 Advantages of Deep-UV Native Fluorescence

In 2008, it was demonstrated that deep-UV native fluorescence could detect and differentiate organics from a number of real-world backgrounds [34]. While the data set at the time did not include explosives, it demonstrated that the excitation wavelength affected the ability to differentiate organics and that excitation <250 nm was optimal. The rationale behind this is a result of two main features (advantages) of operating in the deep UV (<250 nm):

4.2.3.2.1 Advantage 1: Observation of the Full Fluorescence Envelope

The fluorescence envelope can be considered as a convolution of a number of fluorescence compounds and includes organics structures such as aromatic compounds (benzene, naphthalene, anthracene, benzanthracene, etc.), nonaromatic organics such as acetone, and materials that are a composite of these compounds (microbes, mammalian cells, diesel soot, gasoline, plastics, etc.), as well as minerals that may fluoresce intrinsically or as a result of trace metals. An example of this fluorescence envelope is shown in Figure 4.3. To observe the entirety of a fluorescence spectrum,

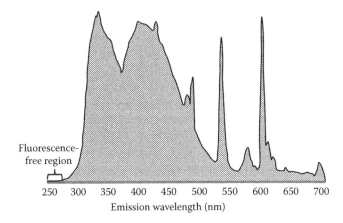

FIGURE 4.3 Native fluorescence "envelope." The data presented are an average of about 1000 naturally occurring materials (organics, microbes, minerals, etc). The excitation wavelength for all the spectra was 248.6 nm. This visually shows the fluorescence-free region <270 nm.

the excitation wavelength needs to be less than the minimum emission wavelength.* As described in a number of papers, benzene, a single-ring aromatic molecule, has some of the lowest fluorescence emission characteristics, with a lower emission wavelength at 270 nm [16,34,44–46,57]. While Advantage 1 suggests that the use of a 270 nm excitation wavelength is sufficient, Advantage 2 explains why the optimal wavelength is actually <250 nm.

4.2.3.2.2 Advantage 2: Separation of Fluorescence and Raman Regions

When illuminating a sample with a laser, a number of spectral phenomena occur nearly simultaneously: Rayleigh scatter, fluorescence, phosphorescence, Raman scatter, and of course heat. The Raman scatter, while rather weak compared to fluorescence, has some strong spectral features that can confound or interfere with the fluorescence spectra. While this sounds counterintuitive, it becomes an issue when considering trace detection. For instance, many organic hydrocarbons will have an intense C–H Raman stretching mode around 2900 cm⁻¹. Similarly, an O–H stretching mode from materials in solution or compounds with hydroxyls will appear around 3400 cm⁻¹. If the background material such as plastic or water is present, these spectral features will confound low-intensity fluorescence features from trace target materials. With an excitation line of 270 nm, these two spectral features appear at 292 and 297 nm respectively, well within the spectral range of a number of single-ring aromatic compounds that constitute potential targets. While the Raman shift (energy loss) is identical for any excitation wavelength, using an excitation at 248.6 nm will shift these vibrational modes to 267 and 271 nm, allowing for a clear separation of Raman and fluorescence features, and avoiding confounding either spectral phenomena.

---

* For the purposes of this discussion, the authors are considering traditional fluorescence spectroscopy. However, there are some cases where this "rule" can be broken (two-photon spectroscopy, anti-Stokes emission from photodissociation followed by laser-induced fluorescence (PD-LIF), etc.).

4.2.3.3   Explosives with Deep-UV Native Fluorescence

4.2.3.3.1   Targets

When considering *target* materials of explosives, there are two groups: *military-grade explosives* and *HMEs*. These both can operate as IEDs; however, HMEs use common materials in unknown concentrations and are more challenging to detect [46]. Few studies have analyzed the source of fluorescence features of military-grade active components, such as trinitrotoluene (TNT), cyclotrimethylene-trinitramine (RDX), pentaerythritol tetranitrate (PETN), and octahydro-1,3,5,7-tetranitro-1,3,5,7-tetrazocine (HMX). Some of the early work on luminescence of RDX and HMX suggests a fluorescence emission associated to charge transfer occurring in the solid state [47]. In the Marinkas report, it was suggested that the fluorescence emission is centered at 465 nm at room temperature and is a result of a condensed state where charge-transfer effects are possible. In a more recent report, from Sandia National Laboratories, the position-to-position fluorescence variability for a variety of explosives led to the conclusion that the fluorescence was a result of photodegradation products such as toluene (for TNT) or a result of contamination [48]. Other reports suggest that 248 nm excitation of RDX will lead to fluorescence emissions from nitrous oxide (NO_2) and hydroxyls (OH) [49]. However, these experiments either used high peak-power nanosecond pulsed sources, used sources in the *near* UV, and/or had limited spectral emission ranges. More recently, fluorescence data of RDX, TNT, and PETN were acquired using a low-energy 248.6 nm laser (NeCu), with no more than 50 mJ/cm^2 at the sample, and yield similar, but highly consistent, fluorescence emissions around 400 nm. The maximum emission of RDX appears to be closer to 433 nm, and observations at various locations show identical spectral features, suggesting that the fluorescence may not be associated with contamination but more likely is associated with the charge-transfer processes that Marinkas indicates. The primary band in TNT is 437 nm and again suggests that minor spectral variations between these materials are consistent with charge-transfer features [47,50]. However, it should be noted that these emission features are not intense, but the fluorescence is still ~100× more intense than the deep-UV Raman response. Furthermore, explosives that use the active components (RDX, PETN, HMX, etc.) in a composite matrix (C4, Plastic Explosive 4 [PE4], or Semtex) contain plasticizers, binders, dyes, and antioxidants that do have strong fluorescence features. The antioxidants in particular are an integral part of the stability of these explosives and contain base aromatic structures [51]. Semtex, for instance, uses the same antioxidant (*N*-phenyl-2-naphthylamine), for all three versions. While this is a common antioxidant used in rubber production, the interaction between it and active energetic components such as PETN and RDX leads to a unique and differentiable fluorescence signature. This composite effect is similar to the fluorescence of microbes using deep-UV native fluorescence [34] where the emission spectrum is a unique combination of fluorophores (aromatic amino acids) and absorbers.

Compared to military-grade explosives, materials used for HMEs are quite varied and many of the components that are integral to the development of the explosive or the explosive end product can provide significant fluorescence emissions. As with all explosives, there is an oxidizer and a reducing component. For example, in ammonium nitrate–fuel oil (ANFO) explosives, the ammonium nitrate is the oxidizer and the fuel oil is the reducing component. While the ammonium nitrate is not natively fluorescent,* the fuel oil, comprised of a variety of hydrocarbons including aromatic compounds, provides a highly structured fluorescence feature. As with the antioxidant fluorescence in military-grade explosives, when fuel oil is mixed with ammonium nitrate, the fluorescence emission is altered from what it would be with the fuel oil alone. This also applies to perchlorate bombs, where fuel oils are also used.

In addition to detecting the final explosive product, native fluorescence can be used to detect components used in forming HMEs. For example, triacetone triperoxide (TATP), a highly unstable explosive commonly used in HMEs, begins with acetone and hydrogen peroxide, where the former is fluorescent when excited at 248 nm. While it has no aromatic ring structure, acetone's electronic states are structured such that they provide a unique fluorescence feature [41,52,53].

4.2.3.3.2 Fluorescence Information Content

In addition to detection of explosive materials either through primary phenomena or a secondary one (i.e., a photodissociation effect), it is equally important to assess whether the response can be used to differentiate explosives from materials in the environment. In the case of fluorescence, the goal is to "classify" the material as hazardous or likely hazardous. Given the broad spectral features associated with fluorescence, this is a more attainable goal compared to identification. To understand the classification potential, the fluorescence spectra of explosives and other naturally and man-made materials can be analyzed by multivariate methods such as principal component analysis (PCA). This method offers a rapid means to determine whether any spectral features can be used to isolate or uniquely identify target materials; that is, how unique are the explosive-related spectral features detected by fluorescence. The traditional PCA model for fluorescence observed from 270 to 400 nm will separate materials based on their aromaticity (number and arrangement). A second-order effect separates small aromatic compounds based on how they may be functionalized (hydroxylated (–OH), methylated (–CH$_3$), chlorinated (–Cl), aminated (–NH$_3$), carboxylated (–COOH), etc.) [34]. The PCA plot should place materials in "groups" with similar spectral features near one another. It should be noted that PCA is not a cluster analysis method, however samples of that group have spectral commonality that can assist in explaining what chemical feature (or component for composite materials) causes it to occupy a particular PCA position. For instance, an organic such as benzene compared to a bacterial spore (containing dityroine) should be closer than chemicals like benzene and anthracene (single-ring versus three-ring aromatics).

Figure 4.4 shows a PCA plot of 27 samples using native fluorescence spectra excited at 248.6 nm using Photon Systems' NeCu laser [8]. The gray arrow describes the overall trend line of aromaticity, beginning with benzene (one-ring aromatic) and anthracene (three-ring aromatic). Other samples in the database follow this line and in the majority of cases are a result of the aromaticity of the material. Composite materials such as bacterial cells and spores are not only separable but appear between the components that drive their fluorescence. Similarly, explosive materials

---

* Similar to RDX and HMX, the condensed state appears to have charge transfer "luminescence" features. This has not been verified and is an area that requires further assessment.

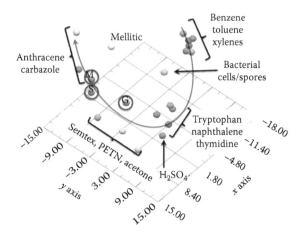

FIGURE 4.4 Fluorescence information content. Fluorescence analysis of the 27 samples that include explosives and components used to develop explosives compared to the organics and materials found in an operational environment. The excitation wavelength was 248.6 nm (NeCu laser). (Adapted from R. Bhartia, W. F. Hug, and R. D. Reid, Improved sensing using simultaneous deep UV Raman and fluorescence detection, *Proceedings of SPIE*, vol. 8358, pp. 83581A, Baltimore, MD, May 2012.)

such as Semtex appear between two-ring compounds (naphthalene) and three-ring compounds (anthracene). This is likely a result of the two overlapping fluorescence sources—the fluorescence of the antioxidant and PETN and/or RDX charge-transfer effects. PETN also appears near Semtex; as it does not include the antioxidant features, it is not grouped with Semtex, but since it is close spatially, it shows relatedness. Figure 4.4 also shows where that acetone appears in the region near the PETN and Semtex. This begins to suggest that there may a spectral "region" that is common to explosives (see Section 4.2.5 for a more complete analysis using a wider array of explosives and interferrants). As the data set is by no means comprehensive, it is possible that nonexplosive materials may appear in this "explosives region." However, it does demonstrate that native fluorescence spectral features of explosives, whether from antioxidants in the sample or the result of condensed state charge-transfer, can be used to differentiate a number of explosives from a wide array of organics and materials found in an operational environment. For example, preliminary analysis of explosives on a variety of surfaces including car panels demonstrated that native fluorescence was capable of detecting and differentiating explosives from potential interferents such as Arizona road dust [7].

4.2.3.3.3   Fluorescence ConOps

The spectral features of explosive materials and the information content that deep-UV native fluorescence provides enable a "search and verify" concept of operations. Another way of describing this is that native fluorescence can be used to reduce the search area for slower and/or less sensitive but more specific detection methods like deep-UV Raman. This ConOps leverages the benefits (sensitivity, speed, standoff distance) of the native fluorescence methods with some of the challenges (specificity).

From the perspective of a warfighter system, a compact, standoff, handheld fluorescence system can be easily envisioned. Unlike high-resolution requirements for Raman, the broader features of the fluorescence require a spectral resolution of no greater than 1–3 nm, and in many cases 10–20 nm bands are sufficient [33]. With this type of instrument, the warfighter would be able to enter an area, either indoors or outdoors, in any normal ambient light environment,* rapidly scan areas of interest, compare spectra with an internal database, and determine whether any features correlate to spectral regions that contain explosive materials or components used to develop explosives. However, an instrument that used native fluorescence alone would require a secondary system to verify whether the detected feature is a potential hazard. Fortunately, one of the benefits of deep-UV spectroscopy is that fluorescence and Raman spectral features can be collected using a single instrument without significantly increasing the size, power requirements, or complexity of the instrument.

With a checkpoint system, native fluorescence enables reduction in the number of areas to target and also extends the detection to a further standoff distance, without the hazards typically associated to higher-powered lasers. This enables a means for early detection and preparation for a potential threat. Again, integrating native fluorescence with deep-UV Raman would allow for verification and identification of the explosive materials as a vehicle, object, or person approaches.

In both scenarios, it is imperative that native fluorescence spectral databases continue to develop and maintain traceability such that they can be integrated with onboard spectral libraries. These databases need to include not only the military and HME grades of explosive materials and precursor components but need to include materials from the operational environments. As these grow, newer more advanced algorithms can be developed to increase the certainty of detecting explosives while reducing false positives.

4.2.4   DEEP-UV RAMAN SPECTROSCOPY OF EXPLOSIVES

4.2.4.1   Overview and Advantages of Raman and Resonance Raman

The Raman scatter effect is an inelastic scattering event first noticed by Sir C.V. Raman in 1928 [54]. For his discovery, he was awarded the Nobel Prize in Physics in 1930. Unlike Rayleigh scatter, this effect occurs when the illuminating photon interacts with bonds in the molecule and loses a small amount of energy defined by the bond. The stretching, bending, or breathing motion and the atomic nature of the bond cause shifts in the returned scatter light. This shift in the excitation energy is independent of excitation wavelength [54] and is measured in terms of distance from the excitation energy in terms of wavenumbers (cm^{-1}). The excitation energy independent of wavelength is zero, and the effect of this light interacting with a polarizable bond is described as energy lost from the excitation wavelength. Bonds such as

---

* This assumes that the spectral range observed by an instrument is <400 nm where direct sunlight during a 60–100 µs laser pulse over the illuminated area would lead to a negligible amount of ambient background. This range is normal for the deep-UV field instruments that have been developed by Jet Propulsion Laboratory and Photon Systems and sufficient to observe CBE materials.

the C–H stretching mode have an energy loss of 2990 cm^{-1}; this will be the amount of energy lost from the incident source, irrespective of the excitation wavelength.

Figure 4.5 shows the Raman regions for various excitation wavelengths in wavelength space. Although the energy lost from the excitation wavelength is constant for a particular vibrational bond mode, since the energy per wavelength is lower with increasing wavelength, that is, as you move from the UV to IR the energy of the photon decreases, the wavelength shift for the same bond increases with excitation wavelength. As a consequence, the Raman region, nominally from 0 to 4000 cm^{-1}, increases with increasing excitation wavelength.

Raman scattering is a fundamentally low-efficiency phenomenon. Compared to fluorescence, Raman bands are many of orders of magnitude less efficient. While Sir C.V. Raman used narrow line emissions from a mercury arc lamp at 435.8 nm and was able to detect faint lines from carbon tetrachloride using film [55], the advent of lasers has allowed researchers to incorporate high-radiance narrow-linewidth lasers that typically range from the visible (532 nm) to the NIR (785 nm). While the wavelength of the laser used does not affect the Raman shift of a particular vibrational band, the cross section (intensity) is related to the excitation wavelength. Thus, the band intensities change as a function of excitation wavelength. This is independent of the Rayleigh law effects and dependent on matching the excitation energy with the bond energy.

However, as the excitation wavelength decreases, the Raman scattering efficiency increases by the Rayleigh law, which is dependent on $1/\lambda^4$. This states that the Raman cross section (efficiency of the Raman scatter) of any Raman band is 20× larger at 248 nm than at 532 nm and 100× larger than at 785 nm. The intensity of the typical Raman bands is defined by the simplified Equation (4.1) [56]:

$$I_R = (I_L \sigma KS) PC \qquad\qquad (4.1)$$

where:
- I_R = detected Raman intensity in a specific band (in photons)
- I_L = intensity of the laser (photons)
- σ = absolute differential Raman cross sections in terms of cm^2/molecule/ steradian
- K = optical component efficiencies
- S = solid angle of collection (steradians)
- P = the depth of focus
- C = concentration in molecules/cm^3

For this equation, the polarizability has been included as a part of the σ-term since these parameters are difficult to determine from first principles. This equation also assumes the Raman cross section is provided for a given wavelength whose energy (in cm^{-1}) is significantly different from the bond energy. However, when the excitation wavelength and the excited vibrational state become similar, the Raman cross section is said to be in preresonance or resonance. The cross section is defined in Equation (4.2) [57]:

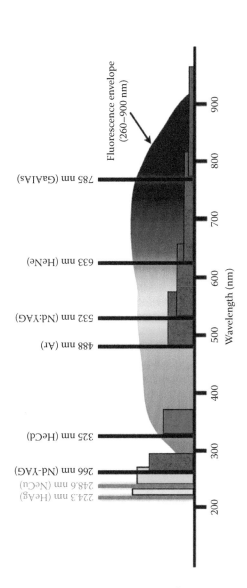

FIGURE 4.5 Raman regions and fluorescence envelope and available laser source. Each vertical line represents the location of a variety of laser wavelengths from the deep UV (<250 nm) to the NIR (785 nm). The green or red boxes next to each line represent the Raman region for the laser wavelength line to the left of it—with increasing excitation wavelengths, the Raman region encompasses a larger wavelength span. The colored region extending from 260 to 900 nm is the fluorescence envelope that includes organic and mineral fluorescence. Only the deep-UV Raman regions (excitation wavelengths <250 nm) have green boxes, indicating that the Raman and fluorescence spectral regions are separate.

$$\sigma = A \cdot v_o \left(v_o - v_R \right)^3 \left[\left(\left(v_e^2 + v_o^2 \right) \middle/ \left(v_e^2 - v_o^2 \right)^2 \right) + B \right] \tag{4.2}$$

where:

v_R	= the Raman frequency (cm^{-1})
v_o	= the laser frequency (cm^{-1})
v_e	= the frequency of the transition to the excited state (cm^{-1})
A and B	= constants

The parameters A, B, and v_e are adjusted to fit the curve to experimental σ versus v_o data. When the difference between the laser excitation frequency, v_o, and the frequency of the transition to the excited electronic state, v_e, goes to zero, the Raman scatter cross section in Equation 4.2 $[(v_e^2+v_o^2)/(v_e^2 - v_o^2)^2]$ tends toward infinity. In practice, the enhancements provide 100–1000× and in special cases, 1×10^6 times gain in the expected Raman cross section [16]. Just between excitation at 532 nm and 248 nm, O–H stretching mode increases by 120× due to Rayleigh and preresonance effects. The Raman cross section of water, including both Rayleigh and preresonance effects, is 570× between 785 nm and 248 nm. This in turn states that the sensitivity achievable with a 248 nm laser for an O–H stretching mode is only matched with a 785 nm laser with 570× more power.

Traditionally, Raman spectroscopy attempts to excite a sample using an excitation wavelength that is not in the absorption region of the molecule(s) of interest [55]. Avoiding an absorption region of the molecule decreases the potential for a fluorescence background that, considering the significant variability in cross section, would obscure the Raman scattering signal. For many organic and biological materials, visible-to-NIR excitation wavelengths result in a fluorescence background and are a problem from both detection and interpretation perspectives [16]. Typically, Raman scattering is 10^4–10^8 times less efficient than fluorescence. Therefore, if any fluorescence occurs within the target molecules or surrounding materials, it will overwhelm the weak Raman emissions. To minimize the fluorescence effects, researchers employ solutions such as time gating to take advantage of the response difference between Raman and fluorescence, or they use an instrument design with a highly confocal optical design such that the collected light is limited to the illuminated volume contributing to the majority of the Raman signal. While time gating has been shown to be effective, the current complexity of the instrumentation (detectors and lasers) prohibits its use as a compact solution for either the warfighter or checkpoint systems. Confocal optics, while highly effective in the lab, are not designed to be ideal for field use where focus-tolerant designs are imperative. Alternative methods to avoid fluorescence are to capitalize on excitation of NIR to IR regions that access a low fluorescence region using 976 and 1064 nm lasers. These require the use of Fourier transform spectroscopy to overcome the thermal noise found in IR detectors. However, this comes at a cost to sensitivity; since Raman scattering follows the Rayleigh law, the Raman cross section in the IR is more than two orders of magnitude lower. Unfortunately, increasing the laser power is not a solution as the potential of thermally altering/burning increases with higher laser power.

The more effective solution to avoid fluorescence is deep UV (<250 nm). As stated in Section 4.2.3.2, excitation below 250 nm can produce Raman scattering in a fluorescence-free or fluorescence-limited zone. For the majority of organics, mineral defects, conduction bands, and so on that contribute to typical fluorescence/luminescence backgrounds, the emissions are greater than ~260 nm. This was further proven in many subsequent publications on deep-UV excitation [32,43,58–60]. The access to new deep-UV laser sources and optics capitalizes on both resonance Raman effects and a fluorescence-free Raman region.

4.2.4.2   Deep-UV Raman of Explosives

4.2.4.2.1   *Targets/Information Content*

Driven partly by a number of new deep-UV laser sources, including NeCu and HeAg TEHC lasers, in the past 8 years, there has been an increase in the number of deep-UV Raman-related papers exploring explosives detection [4–8,13,61–63]. The realization was that deep-UV Raman (excitation <250 nm) provides an increased sensitivity to explosives through resonance effects, without the concern of obscuration from ambient light or fluorescence features stemming from dyes, antioxidants, binders, and aromatic hydrocarbons found in many military-grade and homemade explosives, or surfaces. The research world quickly demonstrated that this enhancement was prevalent for most major explosive materials and many of the primary HMEs [4,5,61,62]. Unlike native fluorescence, deep-UV Raman detects the active energetic components such as PETN, RDX, and TNT, as well as HME components such as NO_3 (nitrates), ClO_4 (perchlorates), ClO_3 (chorates), and so on (see Table 4.1 for deep-UV Raman bands for these materials).

Most of the laboratory experiments to date have demonstrated these results using high-powered deep-UV lasers. However, these are highly unlikely to be used for a warfighter system. In addition, most results use long integration time coupled with high numerical apertures (NA) (low f-numbers). While these establish factors such as the Raman cross section, the obvious question that is posed is whether a more portable, compact, and ruggedized laser, such as the TEHC deep-UV lasers, can be used for resonance Raman explosives detection and can enable a compact implementation that considers the warfighter. For reference, these lasers deposit 3–6 µJ/60 µs (50–100 mW peak power) on the sample per pulse with a repetition rate of up to 40 Hz. However, beyond the compact size of the laser, the implementation for a warfighter requires that the system's f-number is high enough that the design is relatively focus tolerant; that is, it will enable operation similar to a digital single lens reflex (DSLR) camera (simple and rapid autofocusing capabilities). This translates to an increased observational area with a decreased solid angle of collection—exactly the opposite of the traditional method used for sensitive detection.

However, even with these limitations, high-sensitivity detection is still feasible with these lower-powered fieldable lasers—mainly a result of the increased Raman cross sections in the deep UV. To understand this, a photon budget model considers the explosives target's Raman cross section, optical collection and throughput,

TABLE 4.1
Deep-UV Raman Photon Budget Model for PETN Using the 1295 cm^{-1} Resonantly Enhanced Band

Parameter	Units	Lab Instrument	NextGen System	Notes/ References
Parameter 1: Material				
Explosive material		PETN		
Fill factor		100%		
Material Raman cross section	cm²/band/ molecule/sr	3×10^{-27} (1294 cm^{-1})		Ghosh et al. 2012 [61]
Parameter 2: Optics (Illumination/Collection)				
Collection aperture	f-number	4.0	7.0	
Laser photon energy deposited	J/cm²	53	46	
Collection/spectrometer optics throughput	%T	0.08	0.21	
Parameter 3: CCD Parameters				
Collected photons incident on detector	photons/pix	227	542	
Spectrometer CCD		e2v - 42-10		
Detector temperature	QC	−126	−35	
Integration time	s	30	10	
Total noise (sqrt[Dc+RN^2])	e-	2	3	1
Signal detected on CCD	counts	89	232	
Results				
SNR		45	116	

Note: The model incorporates instrument performance from the laser flux at the sample to the CCD and calculates an SNR.

detector sensitivity, and noise profiles, as well as the laser energy deposited on the sample (Table 4.2). The collection volume is stated as fill factor, a term used to describe how much of the illumination beam diameter is occupied by the target material [64]. The NextGen system column in Table 4.2 describes the signal-to-noise ratio (SNR) with a thermoelectrically (TE) cooled charge-coupled device (CCD) rather than the lab standard liquid nitrogen–cooled CCD and a larger f-number (f/7) and a shorter integration time. Both instruments employ shot-noise limited detectors where the dark noise is negligible at the required integration times. However, the reason for the observed difference in SNR between the two instruments is an optimized coupling of the spectrometer and decreased magnification in the NextGen system.

Beyond the theoretical (Table 4.2), this has been demonstrated empirically with PETN, TNT, RDX, RDX, C4, and Semtex using the 248.6 nm NeCu lasers (Figure 4.6). In addition, HME-relevant materials such as NH_4NO_3, $-ClO_4$, and ClO_3 have also been demonstrated (Figure 4.7). Both sets of data were collected on the lab

TABLE 4.2

Deep-UV Raman Bands for a Number of Military and HME Materials

	Primary Raman Bands (wavenumbers) with 248.6 nm Excitation								
Military explosives									
Trinitrotoluene (TNT)					1361			1624	
Pentaerythritol tetranitrate (PETN)	872			1295				1511	1658
Cyclotrimethylene-trinitramine (RDX)		1025	1280			1381	1464		
C4 (RDX)		1025	1280			1381	1464		
Semtex (PETN+RDX)	872			1295		1381	1464		
Homemade explosives (HME)s									
Nitrate			1044	1365					
Perchlorate	932								
Chlorate	935								

instrument with a NeCu 248.6 nm laser (Photon Systems Inc.), with 53 J/cm^2 at the sample, using an f/4 objective lens, with a 100% fill factor and detector integration time of 30 s. It should be noted that the PETN data verify the model and show a 98% correlation between the estimation and actual data.

Visual comparison of the Raman spectra in Figures 4.6 and 4.7 of military-grade and HME-related explosive materials shows that while the Raman cross sections for TNT, PETN, C4, and Semtex are higher than those for the nitrate, perchlorate, or chlorate, the SNRs are inversely related. Since the data were collected with identical acquisition parameters, the relative signal strength can be compared using the intensity of the N_2 (air) line. In the military-grade samples, this band is apparent, while in the HME spectra it cannot be seen when plotted on the same scale as the primary Raman peaks. This appears to contradict the expectation that materials with enhanced Raman cross section lead to decreased SNR. What needs to be taken into consideration is the actual interaction volume and the mass or number of molecules within the interaction volume. The molecular absorptivities at 248 nm of nitrate, PETN, and TNT are negligible, 0.05, and 1.5 ($\times 10^4$ L/mol/cm), respectively [4]. The resulting SNRs from the collected spectra are inversely related with SNRs of 1400 (nitrate), 44 (PETN), and 12 (TNT). This demonstrates that as the excitation wavelength decreases, the Raman cross section may increase, but at some "cost" of a decreased interaction volume (i.e., a reduction in the volume of material from which the signal is detected). This effect is most obvious when looking at bulk samples where sensitivity appears low (low SNR). However, because of the material absorptivity, it limits the number of molecules being observed, even in bulk. Therefore, linearly decreasing the concentration will lead to a nonlinear decrease in SNR. With highly absorbing materials, bulk analyses do not lead to a determination of theoretical detection limits. Furthermore, highly absorbing materials are more amenable to "trace detection" where materials are dispersed in a background matrix or are a thin layer on the background.

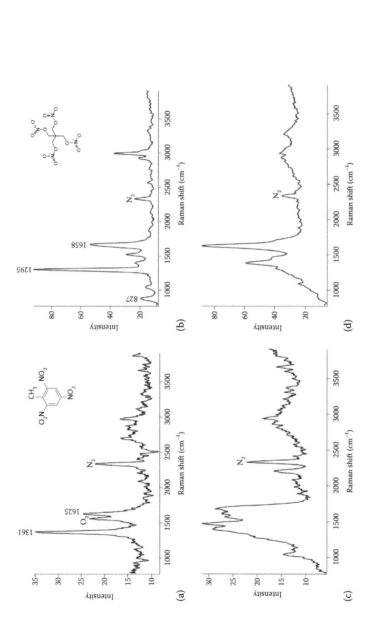

FIGURE 4.6 Deep-UV Raman spectra of military-grade explosives using a compact, ruggedized hollow cathode NeCu laser. (a) TNT, (b) PETN, (c) C4, and (d) Semtex.

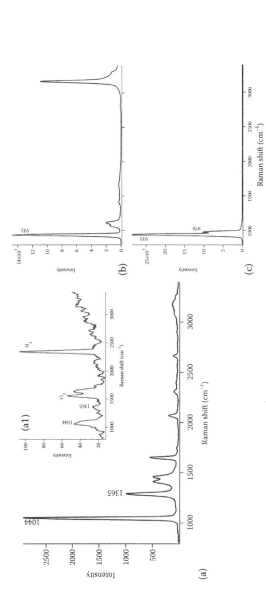

FIGURE 4.7 Deep-UV Raman spectra of HME-related materials. (a, a1) NO_3, (b) ClO_4, and (c) ClO_3 with a NeCu laser. Inset (a1) shows the capability to detect nitrates at 1% in basalt (natural environment).

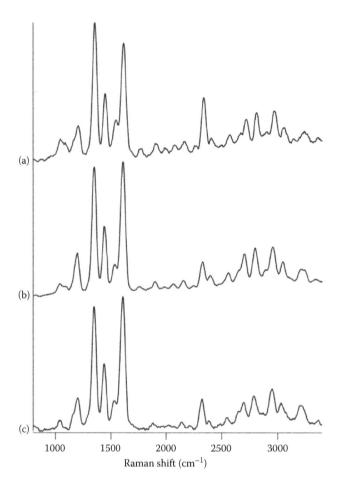

FIGURE 4.8 Bulk vs. trace detection with materials high-molecular absorptivity/advantage of resonance Raman. 248.6 nm deep-UV Raman spectra of phenanthrene in pellets as pure (100%) and mixtures in naturally occurring basalt at 1% and 0.1%. (a) 0.1% Phenanthrene + basalt, (b) 1% phenanthrene in basalt, (c) 100% phenanthrene pellet. The SNR for the pure material is 155, a 1% solution is 199, and the SNR for the 0.1% solution is only 92, i.e., a ~25% decrease in SNRs with an order of magnitude decrease in concentration.

To demonstrate this, an aromatic compound, phenanthrene, was formed into three pressed pellets at concentrations ranging from 100% to 0.1% (weight percent). One percent and 0.1% solid dilutions for the pellets were made in a naturally occurring basalt whose chemistry has been well established [65]. With a molecular absorption value of ~1.3×10^4 L/mol/cm at 250 nm, phenanthrene's depth of penetration will be equal to that of TNT [4,66]. The data collected (Figure 4.8) show deep-UV Raman spectra where the SNRs are 155, 119, and 92 for 100%, 1%, and 0.1% dilutions, respectively. For weakly or nonabsorbing species, the SNRs should follow a linear plot. As shown in Figure 4.7, the nitrate shows this linearity. Potassium nitrate was formed into a pure pellet and a 1% mixture with basalt. In the pure pellet, a 100% fill factor, the SNR was 1400, while the 1% fill factor has an SNR of 22. The phenanthrene data, however, follow a natural log plot showing the effect of transmission of laser into the material. The depth of penetration with a 248 nm laser into pure phenanthrene can be calculated as 400 nm, resulting in a view volume of ~2 pL and an observable mass of phenanthrene of ~2.3 ng. Previous efforts have demonstrated that the depth of penetration in the deep UV in basalt substrate is >30 μm. Therefore, given the 23% decrease in the SNR with a 1% solution of basalt and phenanthrene, the mass of phenanthrene observed should be 1.7 ng and equates to a depth of penetration of 15 μm. With a 0.1% solution, the depth of penetration increases to ~20 μm with an observable mass of 240 pg.

The phenanthrene dilution analysis demonstrates the benefits of resonance effects for trace detection, but it also highlights the nonlinearity as a function of concentration. As a result of high molar absorptivities, the mass of material being detected, even in bulk samples, is limited. The mass of observed material can be extended by increasing the illumination beam diameter, but traditionally there are negative ramifications to the instrument in terms of increased size or reduction in spectral resolution limits increasing the beam diameter. Alternative spectrometer designs such as spatial heterodyne spectrometers (SHS) may offer a means to increase the illumination area without these particular side effects and this is an area of research and development [67]. In the case of the instruments described here, the maximum observable mass is 40 μg/cm². With a 6:1 SNR, the detection limits are <1 ng/cm². Comparatively, excitation wavelengths in the visible or NIR that are poorly absorbed by the material require a large quantity of material in their interaction volume and are not amenable to trace detection. Solutions to increase sensitivities with vis-NIR methods involve increasing laser power that leads to thermal damage and potential fire hazards. Given that lower concentrations of particles are more likely a realistic scenario, the deep-UV excitation offers unique advantages for performing both bulk and trace detection.

4.2.4.2.2 Deep-UV Raman ConOps

As shown in a number of publications, deep-UV Raman spectroscopy offers a high-sensitivity method for explosives detection. However, until recently, this has only been demonstrated with laboratory instruments that would be difficult to transition into field-deployable sensors for the warfighter. For checkpoint ConOps where power and size are less of an issue, existing larger lasers have been implemented and used to demonstrate the utility of the technology. However, as shown with the data above, warfighter ConOps can be enabled and the checkpoint-level ConOps can be enhanced by compact deep-UV fluorescence/Raman devices enabled by these lower-powered deep-UV lasers. While the SNRs may not be equivalent to the laboratory systems, they are more than sufficient for trace detection and could even enable detection with f-numbers that are amenable to hand-held deployments. However, the deep-UV Raman detection methods, as with all Raman-based devices, observe small areas and require a secondary method, such as deep-UV fluorescence, to reduce the search space/target deep-UV Raman analyses.

4.2.5   SPECTRAL FUSION: BENEFITS OF COMBINING DEEP-UV FLUORESCENCE AND RAMAN SPECTROSCOPY

In addition to operating as a means to search for potential explosives, the native fluorescence methods can be fused with the deep-UV resonance Raman information to provide an orthogonal information set. Preliminary tests of the combined method have been successfully demonstrated on explosives and chemical agents, as well as planetary science. In effect, the fluorescence method provides a separation based on electronic characteristics of the material (available energy levels), while the deep-UV Raman provides the vibrational description of the material. As stated in Section 4.2.3.3, the majority of military-grade explosive fluorescence features are a result of charge-transfer-based fluorescence. While explosive materials are not the only class of materials that exhibit this feature, detection of this does reduce the number of potential targets that would require analysis by deep-UV Raman. Additionally, while the fluorescence spectra are not diagnostic for explosives, they can also be used to aid in reducing the number of samples in a Raman database classification step. This helps resolve one of the challenges that Raman algorithms with large databases face. Although the details of these algorithms are proprietary, the basic concept typically compares the spectral features from an unknown sample to a database of samples previously acquired on the instrument or on a master instrument. As the types of materials in the database increase, the features that enable differentiation begin to become increasingly difficult where resolution, stability (e.g., laser position), background suppression, and SNR become increasingly important to enable classification/identification. Any means to reduce the possible range of materials provides an advantage and can reduce the requirements of spectral resolution and enable detection and identification with lower SNRs.

Figure 4.9 graphically represents how the stepwise use of fluorescence followed by deep-UV Raman compares to a Raman-only approach. This analysis assumes that the material in question is Semtex. The database includes toxic industrial chemical (TICs) (xylenes, benzene, sulfuric acid, etc.), biological material (bacteria, amino acids, plant material), minerals, polyaromatic hydrocarbons (PAHs), chemical agent analogs, and military-grade and HME materials. Using fluorescence, Semtex would appear in a cluster near two-ring compounds, close to other explosive materials. The rationale as to why Semtex appears here is its antioxidant chemistry that contains a naphthalene base. However, natural materials that have no explosive-relevant chemistry are also interspersed. It should be noted that this region appears to include spectral features associated to two-ring aromatic compounds (such as naphthalene) and nonaromatic compounds that exhibit charge transfer-like fluorescence (or luminescence) properties in the condensed state. While these other compounds can lead to increased potential of false positives, in a potential ConOps, a warning can appear stating that further investigation is required, at which point the area in question is targeted and deep-UV Raman data area collected. However, rather than comparing the acquired results to the entire database, the data are only compared to the subset of samples close to Semtex (in PCA-fluorescence space). The results of the subset analysis show an "explosives"

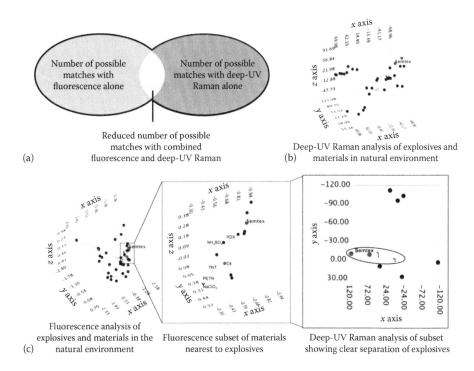

(a)

Reduced number of possible matches with combined fluorescence and deep-UV Raman

(b) Deep-UV Raman analysis of explosives and materials in natural environment

(c) Fluorescence analysis of explosives and materials in the natural environment

Fluorescence subset of materials nearest to explosives

Deep-UV Raman analysis of subset showing clear separation of explosives

FIGURE 4.9 Improved detection with deep-UV Raman analysis preceded by deep-UV fluorescence analysis. The Venn diagram pictorially describes the benefit of combining fluorescence and deep-UV Raman. Using either of the spectral phenomena alone requires differentiating an unknown material from a large number of possible materials. (b) Example Raman of explosives and common materials in the natural environment. By combining the two spectral features, the number of possible materials decreases (c). In this case, combining is an iterative process that begins with fluorescence to reduce the possible materials to a small subset and then Raman analysis of the subsets. PCA was used for this analysis for simplicity and traceability to previous work. In (c), the gray dotted line in the fluorescence analysis defines increasing aromaticity (single-ring aromatic on the left). In all panels, C4 and Semtex (labeled) are the red spheres and RDX, PETN, and TNT are the yellow spheres. For the fluorescence analysis, explosive compounds appear in a cluster near two-ring aromatic compounds. This region is extracted as a subset of the database and used for deep-UV Raman analysis. Visual comparison between the methods shows grouping of the explosive materials is better in the combined method (c).

group forming where it easily separates from the other nonexplosive materials. Comparatively, if deep-UV Raman alone were used, Figure 4.9 shows how Semtex and other explosives (red and yellow spheres) are interspersed between biological compounds and nonhazardous components. Theoretically, by increasing this, the PCA space for the Raman-only analysis would lead to increases in separation of the materials; however, it is likely that they will still remain interspersed. Since the variability of position in this PCA space is dependent on noise and variability of the sample, the ability to clearly identify materials relies on high SNR; decreasing this would lead to increased false positives.

In both the fused fluorescence/Raman and the Raman-only analyses, the deep-UV Raman resolution was ~45 cm^{-1}. The benefit of the fluorescence/Raman combined method is the ability to detect (fluorescence), decrease false positives (fluorescence+Raman), and identify the sample (Raman). An added advantage to this is that the requirements on spectral resolution are decreased from the traditional 5–10 cm^{-1} down to 45 cm^{-1} and are not reliant on high SNRs.

4.3 CURRENT CAPABILITIES AND NEW AND EMERGING EFFORTS

As presented in the previous sections, many of the instruments used to collect the data were laboratory instruments. However, these were configured with the perspective of field deployments in mind. For example, the lasers that were used were the NeCu and HeAg deep-UV sources from Photon Systems that are rugged and compact, and they have low power requirements (<10 W). In other respects, these lab instruments are the "gold standard" and included low-stray light spectrometers with liquid nitrogen (LN2)-cooled detectors and a stable platform. As such, they provided the ability to understand the potential of the compact deep-UV lasers and guided many of the resulting field instruments. Figure 4.10 shows the deep-UV instrument range from the laboratory gold standards (Figure 4.10a) to compact TE cooled versions for planetary science (Figure 4.10b), to fluorescence and fluorescence/Raman scanners, standoff instruments, underwater systems, and microscopic systems.

As stated in the beginning, the following questions drive instrument design: What are you looking for? What level of information is required? How do you intend on looking for it? As such, many of the deep-UV field instrument developments were driven by questions in planetary science. However, more recently, field instruments such as the targeted ultraviolet chemical biological explosive (TUCBE) sensor have been developed for detection of chemical/biological/explosives from a standoff distance. While the details of the instrument design are outside the scope of this chapter, the TUCBE uses a 248.6 nm laser to illuminate targets and collects both fluorescence and Raman emission with a Dall-Kirkham reflective objective lens. The detection system splits the deep-UV Raman and fluorescence paths into a dichroic stack with photomultiplier tubes (PMTs) for the fluorescence and a grating spectrometer path with a 32-channel PMT array for Raman detection. This is considered a low-resolution spectral instrument, but it provides a rapid solution to detect materials using the combined fluorescence/Raman method with high levels of success for CBE materials.

The next generation of instruments are rapidly evolving beyond the ability to simply demonstrate detection capability but are designed to operate within a use-scenario of the warfighter, checkpoint system, or alternative ConOps. However, in all of these scenarios, it is unclear how the operator searches for a possible explosive. While in some cases there is a need to analyze a suspicious "white" powder, trace detection requires a means to find potential areas of interest. As such, in addition to enabling point detection, the next generation of instruments includes the ability to map an area to provide the operator a means to find "hot spots." This is not possible with the current Raman methods as the integration time/point is too lengthy

FIGURE 4.10 Examples of the deep-UV fluorescence/Raman instruments that have been developed since 1996. (a) Deep-UV Raman fluorescence gold-standard Mineralogy and Organic Based Investigations with UV Spectroscopy (MOBIUS). (b) Custom miniature deep-UV Raman/fluorescence prototype (SHERLOC). The deep-UV laser is the silver cylinder. (c) Microbial and Organics Surface Analyzer/Image Constructor (MOSAIC), a macroscale mapping fluorescence/Raman instrument. (d) 1–2 m standoff deep-UV fluorescence instrument, Targeted Ultraviolet Chemical, Biological, Explosive (TUCBE) Gen 2 sensor. (e) 1–2 m standoff fluorescence/Raman instrument, TUCBE Gen 4.5. (f) 100 m submersible deep-UV fluorescence/Raman instrument for divers: Submersible Chemical, Biological, Explosives sensor (SubCBE). (g) 5 km submersible deep-UV fluorescence instrument for hydrothermal vent analysis: Dark Energy Biosphere Investigations Tool (DEBI-t). (h) Deep-UV LED-based fluorescence instruments for vapor and surface analysis (NAphthalene DOSimeter, NaDos). (i) Deep-UV fluorescence imaging microscope. microMOSIAC. Deep-UV fluorescence/Raman instruments for scientific and DoD related applications have been commercially available since 2001.

for any implementation other than on a stable platform/microscope stage. Native fluorescence, however, can operate at the 60–100 µs/point to rapidly create a map of larger areas to find potential "hot spots." Those areas can then be further investigated using the deep-UV Raman point analysis. Fortunately, both these capabilities can be incorporated into a single instrument; an example of this in a compact functional form is the scanning habitable environments with Raman and luminescence for organics and chemicals (SHERLOC) prototype shown in Figure 3.1.B [68].

REFERENCES

1. S. Wallin, A. Pettersson, H. Östmark, and A. Hobro, Laser-based standoff detection of explosives: A critical review, *Anal Bioanal Chem*, 395(2):259–274, 2009.
2. M. K. McPherson, Explosive detection equipment program, *2010 IEEE International Carnahan Conference on Security Technology (ICCST)*, pp. 403–406, San Jose, CA, 5–8 October, 2010.
3. D. S. Moore and R. J. Scharff, Portable Raman explosives detection, *Anal Bioanal Chem*, 393(6):1571–1578, 2009.
4. D. D. Tuschel, A. V. Mikhonin, B. E. Lemoff, and S. A. Asher, Deep ultraviolet resonance Raman excitation enables explosives detection, *Appl Spectrosc*, 64(4):425–432, 2010.
5. A. Blanco, L. C. Pacheco-Londoño, A. J. Peña-Quevedo, and S. P. Hernández-Rivera, UV Raman detection of 2,4-DNT in contact with sand particles, *Proc SPIE*, 6217:621737, 2006.
6. W. F. Hug, R. D. Reid, R. Bhartia, and A. L. Lane, A new miniature hand-held solar-blind reagentless standoff chemical, biological, and explosives (CBE) sensor, *Proc SPIE*, 6954:695401, 2008.
7. W. F. Hug, R. D. Reid, R. Bhartia, and A. L. Lane, Performance status of a small robot-mounted or hand-held, solar-blind, standoff chemical, biological, and explosives (CBE) sensor, *Proc SPIE*, 7304:73040Z, 2009.
8. R. Bhartia, W. F. Hug, and R. D. Reid, Improved sensing using simultaneous deep UV Raman and fluorescence detection, *Proc SPIE*, 8358:83581A, 2012.
9. D. S. Moore, Recent advances in trace explosives detection instrumentation, *Sens Imaging*, 8(1):9–38, 2007.
10. M. Wu, M. Ray, K. Fung, M. Ruckman, D. Harder, and A. Sedlecek, Stand-off detection of chemicals by UV Raman spectroscopy, *Appl Spectrosc*, 54, 800–806, 2000.
11. D. D. Tuschel, A. V. Mikhonin, B. E. Lemoff, and S. A. Asher, Deep ultraviolet resonance Raman spectroscopy of explosives. *AIP Conference Proc*, 1267:869–870, 2010.
12. C. C. Phifer, R. L. Schmitt, L. R. Thorne, P. Hargis Jr, and J. E. Parmeter, Studies of the laser-induced fluorescence of explosives and explosive compositions. SAND2006-6697. Sandia National Laboratories, Albuquerque, NM, 2006.
13. M. Gaft and L. Nagli, Standoff laser-based spectroscopy for explosives detection, *Proc SPIE*, 6739:673903, 2007.
14. R. Rumelfanger, S. A. Asher, and M. B. Perry, UV resonance Raman characterization of polycyclic aromatic hydrocarbons in coal liquid distillates, *Appl Spectrosc*, 42(2):267–272, 1988.
15. C. M. Jones and S. A. Asher, Ultraviolet resonance Raman study of the pyrene S4, S3, and S2 excited electronic states, *J Chem Phys*, 89(5):2649, 1988.
16. S. A. Asher and C. R. Johnson, Raman spectroscopy of a coal liquid shows that fluorescence interference is minimized with ultraviolet excitation, *Science*, 225(4659):311–313, 1984.
17. S. A. Asher, R. W. Bormett, X. G. Chen, D. H. Lemmon, N. Cho, P. Peterson, M. Arrigoni, L. Spinelli, and J. Cannon, UV resonance Raman spectroscopy using a new cw laser source: Convenience and experimental simplicity, *Appl Spectrosc*, 47(5):628–633, 1993.
18. C. R. Johnson and S. A. Asher, UV resonance Raman excitation profiles of l-cystine, *J Raman Spectrosc*, 18:345–349, 1987.
19. S. Asher, M. Ludwig, and C. Johnson, UV resonance Raman excitation profiles of the aromatic amino acids, *J Am Chem Soc*, 108(12):3186–3197, 1986.
20. S. A. Asher, UV resonance Raman spectroscopy for analytical, physical, and biophysical chemistry, *Anal Chem*, 65(4):201A–210A, 1993.

21. S. A. Asher, UV Resonance Raman Studies of Molecular Structure and Dynamics: Applications in Physical and Biophysical Chemistry, *Annu Rev Phys Chem*, 39:537–588, 1988.
22. S. A. Asher and J. L. Murtaugh, UV Raman excitation profiles of imidazole, imidazolium, and water, *Appl Spectrosc*, 42(1):83–90, 1988.
23. J. M. Dudik, C. R. Johnson, and S. A. Asher, UV resonance Raman studies of acetone, acetamide, and N-methylacetamide: Models for the peptide bond, *J Phys Chem*, 89(18):3805–3814, 1985.
24. C. M. Jones, V. L. Devito, P. A. Harmon, and S. A. Asher, High-repetition-rate excimer-based UV laser excitation source avoids saturation in resonance Raman measurements of tyrosinate and pyrene, *Appl Spectrosc*, 41(8):1268–1275, 1987.
25. J. M. Dudik, C. R. Johnson, and S. A. Asher, Wavelength dependence of the preresonance Raman cross sections of CH_3CN, SO_4^{2-}, ClO_4^-, and NO_3^-, *J Chem Phys*, 82(4):1732, 1985.
26. W. F. Hug and R. D. Reid, Apparatus for the efficient and accurate analysis of preferential compounds in sample, US Patent 6,287,869, 11 September 2001.
27. W. F. Hug and R. D. Reid, Sputtering metal ion laser, US Patent 6,693,944, 17 February 2004.
28. W. F. Hug and R. D. Reid, Spectroscopic chemical analysis methods and apparatus, US Patent 7,800,753, 21 September 2010.
29. W. F. Hug, R. D. Reid, and R. Bhartia, Spectroscopic chemical analysis methods and apparatus, US Patent 8,395,770, 12 March 2013.
30. M. Storrie-Lombardi, W. Hug, G. McDonald, A. Tsapin, and K. Nealson, Hollow cathode ion lasers for deep ultraviolet Raman spectroscopy and fluorescence imaging, *Rev Sci Instrum*, 72(12):4452, 2001.
31. X. Zhang and J. V. Sweedler, Ultraviolet native fluorescence detection in capillary electrophoresis using a metal vapor NeCu laser, *Anal Chem*, 73 (22):5620–5624, 2001.
32. M. Sparrow, J. Jackovitz, C. Munro, W. Hug, and S. Asher, New 224 nm hollow cathode laser–UV Raman spectrometer, *Appl Spectrosc*, 55:66, 2001.
33. R. Bhartia, W. F. Hug, E. C. Salas, K. Sijapati, A. L. Lane, R. D. Reid, and P. G. Conrad, Biochemical detection and identification false alarm rate dependence on wavelength using laser induced native fluorescence, *Proc SPIE*, 6218:62180J, 2006.
34. R. Bhartia, W. F. Hug, E. C. Salas, R. D. Reid, K. K. Sijapati, A. Tsapin, W. Abbey, K. H. Nealson, A. L. Lane, and P. G. Conrad, Classification of organic and biological materials with deep ultraviolet excitation, *Appl Spectrosc*, 62,(10):1070–1077, 2008.
35. P. V. Johnson, R. Hodyss, D. K. Bolser, R. Bhartia, A. L. Lane, and I. Kanik, Ultraviolet-stimulated fluorescence and phosphorescence of aromatic hydrocarbons in water ice, *Astrobiology*, 11(2):151–156, 2011.
36. M. D. Fries, R. Bhartia, L. W. Beegle, Y. Gursel, and G. S. Mungas, Microscopic sample interrogation through multi-wavelength spectroscopy coupled with variable magnification imaging, *Astrobiology Science Conference 2010: Evolution and Life: Surviving Catastrophes and Extremes on Earth and Beyond*, 1538:5214, 2010.
37. R. Bhartia, E. C. Salas, W. F. Hug, R. D. Reid, A. L. Lane, K. J. Edwards, and K. H. Nealson, Label-free bacterial imaging with deep-UV-laser-induced native fluorescence, *Appl Environ Microbiol*, 76(21):7321–7237, 2010.
38. M. Fries and R. Bhartia, UV fluorescence/Raman imaging of Allende, *Meteorit Planet Sci Suppl* 73:5441, 2010.
39. Lane 201040. A. U. Acuña, F. Amat-Guerri, P. Morcillo, M. Liras, and B. Rodríguez, Structure and formation of the fluorescent compound of *Lignum nephriticum*, *Org Lett*, 11(14):3020–3023, 2009.
41. R. Borkman and D. Kearns, Electronic relaxation processes in acetone, *J Chem Phys*, 44:945, 1966.
42. J. Kim, F. Kim, and J. Huang, Seeing graphene-based sheets, *Mater Today*, 13(3):28–38, 2010.

43. K. Loh, Q. Bao, and G. Eda, Graphene oxide as a chemically tunable platform for optical applications, *Nat Chem*, 2:1015–1024, 2010.
44. N. Tarcea, T. Frosch, P. Rosch, M. Hilchenbach, T. Stuffler, S. Hofer, H. Thiele, R. Hochleitner, and J. Popp, Raman spectroscopy: A powerful tool for *in situ* planetary science, *Space Sci Rev*, 135(1):281–292, 2008.
45. T. Frosch, N. Tarcea, M. Schmitt, H. Thiele, F. Langenhorst, and J. Popp, UV Raman imaging—A promising tool for astrobiology: comparative Raman studies with different excitation wavelengths on SNC Martian meteorites, *Anal Chem*, 79(3):1101–1108, 2007.
46. J. L. Anderson, A. A. Cantu, A. W. Chow, P. S. Fussell, R. G. Nuzzo, J. E. Parmeter, G. S. Sayler, J. M. Shreeve, R. E. Slusher, and M. Story, *Existing and Potential Standoff Explosives Detection Techniques*. The National Academies Press, Washington, DC, 2004.
47. P. L. Marinkas, Luminescence properties of RDX and HMX. Report No. AD-A015538, 1975.
48. C. C. Phifer, R. L. Schmitt, L. R. Thorne, and P. Hargis Jr, Studies of the laser-induced fluorescence of explosives and explosive compositions. SAND2006-6697. Sandia National Laboratories, Albuquerque, NM, 2006.
49. C. Capellos, P. Papagiannakopoulos, and Y. Liang, The 248 nm photodecomposition of hexahydro-1,3,5-trinitro-1,3,5-triazine, *Chem Phys Lett*, 164(5):533538,1989.
50. P. L. Marinkas, Luminescence of solid cyclic polynitramines, *J Lumin*, 15(1):57–67, 1977.
51. K. Katoh, S. Yoshino, S. Kubota, Y. Wada, Y. Ogata, M. Nakahama, S. Kawaguchi, and M. Arai, The effects of conventional stabilizers and phenol compounds used as antioxidants on the stabilization of nitrocellulose, *Prop Explos Pyrotech*, 32(4):314–321, 2007.
52. L. E. Brus and J. R. McDonald, Collision free, time resolved fluorescence of SO_2 excited near 2900 Å, *Chem Phys Lett* 21(2):283–288, 1973.
53. M. C. Thurber and R. K. Hanson, Pressure and composition dependences of acetone laser-induced fluorescence with excitation at 248, 266, and 308 nm, *Appl Phys B*, 69(3):229–240, 1999.
54. D. A. Skoog, F. J. Holler, and S. R. Crouch, *Principles of Instrumental Analysis*. Brooks/Cole, Stamford, CT, 2007.
55. C. V. Raman and K. S. Krishnan, molecular spectra in the extreme infra-red, *Nature*, 122(3069):278–278, 1928.
56. M. J. Pelletier, *Analytical Applications of Raman Spectroscopy*. Wiley-Blackwell, Hoboken, NJ, 1999.
57. S. Asher and C. Johnson, UV resonance Raman excitation profile through the 1B2u state of benzene, *J Phys Chem*, 89:1375–1379, 1985.
58. E. Ghiamati, R. Manoharan, W. Nelson, and J. Sperry, UV resonance Raman spectra of bacillus spores, *Appl Spectrosc*, 46:357364, 1992.
59. N. Tarcea, M. Harz, P. Rösch, T. Frosch, M. Schmitt, H. Thiele, R. Hochleitner, and J. Popp, UV Raman spectroscopy: A technique for biological and mineralogical *in situ* planetary studies. *Spectrochim Acta A*, 68(4):1029–1035, 2007.
60. D.-Y. Wu, X.-M. Liu, S. Duan, X. Xu, B. Ren, S.-H. Lin, and Z.-Q. Tian, Chemical enhancement effects in SERS spectra: A quantum chemical study of pyridine interacting with copper, silver, gold and platinum metals, *J Phys Chem C*, 112(11):4195–4204, 2008.
61. M. Ghosh, L. Wang, and S. A. Asher, Deep-ultraviolet resonance Raman excitation profiles of NH_4NO_3, PETN, TNT, HMX, and RDX, *Appl Spectrosc*, 66(9):1013–1021, 2012.
62. M. Gaft and L. Nagli, UV gated Raman spectroscopy for standoff detection of explosives, *Opt Mater (Amst)*, 30(11):1739–1746, 2008.

63. A. Ehlerding, I. Johansson, S. Wallin, and H. Östmark, Resonance-enhanced Raman spectroscopy on explosives vapor at standoff distances, *Int J Spectrosc*, 2012(3):1–9, 2012.

64. A. Tripathi, E. D. Emmons, P. G. Wilcox, J. A. Guicheteau, D. K. Emge, S. D. Christesen, and A. W. Fountain III, Semi-automated detection of trace explosives in fingerprints on strongly interfering surfaces with Raman chemical imaging, *Appl Spectrosc*, 65, (6):611–619, 2011.

65. G. H. Peters, W. Abbey, G. H. Bearman, G. S. Mungas, J. A. Smith, R. C. Anderson, S. Douglas, and L. W. Beegle, Mojave mars simulant—Characterization of a new geologic mars analog, *Icarus*, 197(2):470–479, 2008.

66. G. Malloci, G. Mulas, and C. Joblin, Electronic absorption spectra of PAHs up to vacuum UV. Towards a detailed model of interstellar PAH photophysics, *A&A*, 426(1):105–117, 2004.

67. N. Gomer, C. Gordon, P. Lucey, S. Sharma, J. Carter, and S. Angel, Raman spectroscopy using a spatial heterodyne spectrometer: Proof of concept, *Appl Spectrosc*, 65(8):849857, 2011.

68. L. W. Beegle, R. Bhartia, L. DeFlores, M. Darrach, R. D. Kidd, W. Abbey, S. Asher, et al. SHERLOC: Scanning habitable environments with raman and luminescence for organics and chemicals, an investigation for 2020, *45th Lunar and Planetary Science Conference Abstracts*, 2835, 2014.

5 Raman Detection of Explosives

Steven D. Christesen, Augustus W. Fountain III, Erik D. Emmons, and Jason A. Guicheteau

CONTENTS

5.1 INTRODUCTION

The ability to detect explosives both at close range and from a distance is important for providing a capability to warn potential victims. A wide variety of optical techniques are being explored in order to obtain this capability, but the problem is proving to be a significant technical and scientific challenge. These challenges were described in detail in a review by Steinfeld and Wormhoudt, who categorized them into a number of physical constraints, including low vapor pressures, limited sample size, deliberate concealment, and interferences.[1] Each of these constraints impacts the ability of optical techniques to detect and identify explosive material, especially in the vapor phase. Optical techniques are attractive, however, because they can potentially provide a noncontact and standoff detection capability requiring no sample preparation. Because of the aforementioned vapor pressure and concealment constraints, the application most suited to detection and identification of bulk or trace material on surfaces is optical spectroscopy. Raman spectroscopy is one such optical technique that has shown promise for explosives detection.[2–16]

As a vibrational spectroscopy, Raman spectra contain information on the specific arrangement and interaction of the atoms constituting the molecule.[17,18] This information is highly molecule-specific and allows for differentiation between types of explosives. Whereas infrared (IR) spectroscopy is also a vibrational spectroscopy and provides the same ability to differentiate molecular species, IR spectra are much more dependent on the particle size and nature of the background surface than Raman spectra. Sample thickness and surface reflectivity and roughness affect the IR peak positions, as well as their peak intensities.

5.2 RAMAN SPECTROSCOPY

5.2.1 CLASSICAL DESCRIPTION

In order to understand how the technique works, we will provide a very brief introduction to the theory of Raman scattering. A detailed treatment of Raman spectroscopy can be found in a number of good references in the literature.[19–25] In particular, the Long monograph covers both the classical and quantum mechanical theory of Raman scattering in detail.[17] Our description here will rely solely on the classical theory to describe the Raman effect.

The Raman scattering effect results from the interaction of the electric field of an incident photon (E) with a molecule's polarizability (α), resulting in an induced dipole moment μ:

$$\mu = \alpha E \tag{5.1}$$

Both the polarizability and the incident electric field vary with characteristic frequencies:

$$E = E_0 \cos(2\pi c \bar{\nu}_0 t) \tag{5.2}$$

$$\alpha = \alpha_0 + \alpha' \cos(2\pi c \bar{\nu}_j t) \tag{5.3}$$

where:
$\bar{\nu}_0$ and $\bar{\nu}_j$ = frequencies ($1/\lambda$ in units of cm^{-1}, where λ is wavelength) of the photon and molecular vibration, respectively
α_0 = equilibrium polarizability
α' = derivative of the polarizability with respect to a normal coordinate corresponding to a molecular vibration

The combination of an oscillating electric field with a polarizability modulated by the molecular vibration produces what can be considered beat frequencies in the induced dipole moment. Substituting Equations 5.2 and 5.3 into Equation 5.1 yields Equation 5.4:

$$\mu = \alpha_0 E_0 \cos\left(2\pi c \bar{v}_0 t\right) + \frac{1}{2}\alpha' E_0 \left\{\cos(2\pi c [\bar{v}_0 + \bar{v}_j]t + \cos(2\pi c [\bar{v}_0 - \bar{v}_j]t)\right\} \quad (5.4)$$

The first term in Equation 5.4 corresponds to Rayleigh scattering where there is no change in the incident photon frequency, while the second and third terms describe anti-Stokes Raman $[\bar{v}_0 + \bar{v}_j]$ and Stokes Raman $[\bar{v}_0 - \bar{v}_j]$ scattering (Figure 5.1). It is important to note that Raman scattering can only occur when the molecular vibration causes a change in polarizability, that is, $\alpha' \neq 0$

The polarizability is actually a 3×3 tensor (Equation 5.1) whose elements connect the polarization of the Raman scattered radiation to that of the incident radiation.[18]

$$\begin{bmatrix} \mu_x \\ \mu_y \\ \mu_z \end{bmatrix} = \begin{bmatrix} \alpha'_{xx} & \alpha'_{xy} & \alpha'_{xz} \\ \alpha'_{yx} & \alpha'_{yy} & \alpha'_{yz} \\ \alpha'_{zx} & \alpha'_{zy} & \alpha'_{zz} \end{bmatrix} \begin{bmatrix} E_x \\ E_y \\ E_z \end{bmatrix} \quad (5.5)$$

Some polarization information is even available from freely rotating molecules in the liquid or gas phase. In this case, it is useful to deal with average polarizabilities expressed in terms of quantities that are invariant to rotation, such as the trace, $\bar{\alpha}'$, and the anisotropy, γ'^2.

$$\bar{\alpha}' = \frac{1}{3}\left(\alpha'_{xx} + \alpha'_{yy} + \alpha'_{zz}\right) \quad (5.6)$$

$$\gamma'^2 = \frac{1}{2}\left[\left(\alpha'_{xx} - \alpha'_{yy}\right)^2 + \left(\alpha'_{yy} - \alpha'_{zz}\right)^2 + \left(\alpha'_{zz} - \alpha'_{xx}\right)^2 + 6\left(\alpha'^2_{xy} + \alpha'^2_{xz} + \alpha'^2_{yz}\right)\right] \quad (5.7)$$

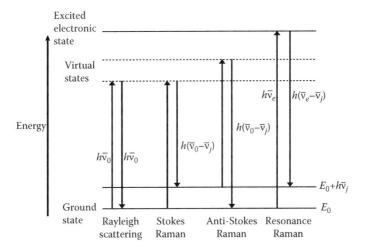

FIGURE 5.1 Modified energy diagram for Rayleigh and Raman scattering.

For linearly polarized incident laser light, the depolarization ratio ρ for any given Raman transition is defined as the ratio of the intensity of the Raman scattered light polarized perpendicular to the incident radiation to the intensity polarized parallel to the incident light.

$$\rho = \frac{I_\perp}{I_\parallel} = \frac{3\gamma'^2}{45\bar{\alpha}'^2 + 4\gamma'^2} \tag{5.8}$$

Molecular symmetry considerations dictate that $\rho < 3/4$ for totally symmetric vibrations and $\rho = 3/4$ for all other fundamental Raman active vibrations. For some totally symmetric vibrational modes, the depolarization ratio can be much smaller than 3/4. It has been shown that, in some cases, the polarization characteristics of Raman scattering from chemicals on surfaces can be used to discriminate against the broad, largely unpolarized fluorescence and Raman scattering from the surface to improve detection.[19]

5.2.2 RAMAN CROSS SECTIONS

The Raman cross section is the fundamental quantity relating the signal produced to the incident laser power, and it is given by Equation 5.9:

$$P = \sigma_R S_{inc} \tag{5.9}$$

where:
  P = power in the Raman scattered radiation
  σ_R = differential Raman cross section
  S_{inc} = incident laser irradiance

The differential Raman cross section is that fraction of incident laser radiation that is scattered into a particular Raman line per molecule irradiated and per unit solid scattering angle. It is also a function of the incident laser frequency $\bar{\nu}_0$ (in wavenumbers) and the frequency of the scattered light $\bar{\nu}_0 - \bar{\nu}_j$:

$$\sigma_R = \frac{(2\pi)^4 \, b_j^2 g_j \left(45\bar{\alpha}'^2_j + 7\gamma'^2_j\right) \bar{\nu}_0 (\bar{\nu}_0 - \bar{\nu}_j)^3}{45[1 - \exp(-hc\bar{\nu}_j / kT)]} \tag{5.10}$$

where:
  b_j^2 = zero point amplitude of the vibration
  g_j = vibrational degeneracy
  $\bar{\alpha}'_j$ and γ'_j = the trace and anisotropy, respectively, of the derived polarizability tensor as described previously

For normal Raman scattering, σ_R increases by the factor $\bar{\nu}_0(\bar{\nu}_0 - \bar{\nu}_j)^3$ as the excitation frequency increases for P expressed in units of photons per second per centimeter squared. The frequency dependence is $(\bar{\nu}_0 - \bar{\nu}_j)^4$ if P is expressed in watts per centimeter squared.[26] The polarizability tensor elements also become frequency dependent as the excitation wavelength approaches resonance with an allowed electronic transition. For this preresonance enhancement, the frequency dependence can be described by the modified Albrecht A-term:[26]

$$\sigma_R = K_1 \bar{\nu}_0 (\bar{\nu}_0 - \bar{\nu}_j)^3 \left[\frac{\bar{\nu}_e^2 + \bar{\nu}_0^2}{(\bar{\nu}_e^2 - \bar{\nu}_0^2)^2} + K_2 \right]^2 \tag{5.11}$$

where:

K_1 and K_2 = fitting constants

$\bar{\nu}_e$ = frequency in wavenumbers of the relevant electronic excited state

Both K_1 and K_2 are independent of $\bar{\nu}_0$, and Equation 5.11 simplifies to the standard A-term expression when $K_2 = 0$. According to Harmon and Asher,[27] the K_2 term models contributions from higher-frequency electronic states whose contribution can be considered to be frequency independent. Equation 5.11 tends to underestimate the energy of the resonant transition, while the standard A-term equation with $K_2 = 0$ overestimates the energy.[28] Equation 5.11 is valid only for preresonance Raman scattering as σ_R goes to infinity at resonance ($\bar{\nu}_0 = \bar{\nu}_e$).

5.3   RAMAN SPECTRA OF EXPLOSIVES

While some groups have worked on instrument development and the application of Raman spectroscopy to explosives detection, others have concentrated on the measurement of Raman spectra and cross sections.[4,11,29–33] These measurements provide basic information that can be used to determine both selectivity and sensitivity of actual as well as notional Raman-based sensor systems and to benchmark the development of instrumentation.

Early work focused on the assignments of the observed vibrational bands from explosives and was performed using visible and near-infrared (NIR) excitation.[29–31,34] More recent work has focused on ultraviolet (UV) excitation, as a significant enhancement of the cross section can be observed via the resonance Raman effect,[17] which occurs if the excitation wavelength corresponds to an electronic absorption band of the explosive (see Figure 5.1). Asher et al. measured UV resonance Raman cross sections of several explosives in acetonitrile solution using 204, 229, 244, and 257 nm excitation, and they observed significant resonance enhancements.[11,35] Comanescu et al. obtained two-dimensional UV resonance Raman spectra of explosives, again in acetonitrile solution, by measuring the Raman spectra as a function of the excitation wavelength, but they did not obtain cross sections.[4] Nagli et al. obtained cross sections for explosives in the solid state under UV excitation using an external standard technique, but they failed to account for the strong absorption by the explosives.[32]

5.3.1   RAMAN CROSS SECTIONS OF EXPLOSIVES

Experimentally, the cross sections are typically obtained by using either an internal or external standard technique. In each case, the band areas of the analyte are compared to those of molecules, such as acetonitrile, that have well-established cross section values.[11,20,28,36,37] Cross sections are usually obtained in this fashion since it is exceedingly difficult to calibrate all the experimental parameters necessary (laser beam profile, collection optics efficiency, detector efficiency, etc.) to independently and directly measure cross sections. When internal standard techniques are used, cross sections for only one analyte need to have been previously obtained in this manner. The formula for obtaining cross sections via the internal standard method is

$$\sigma_R^s = \sigma_R^r \left(\frac{I^s}{I^r} \right) \left[\frac{E\left(\overline{v}_0 - \overline{v}_j^r\right)}{E\left(\overline{v}_0 - \overline{v}_j^s\right)} \right] \left(\frac{C^r}{C^s} \right) \tag{5.12}$$

where:

σ_R	= differential cross section in units of cm^2/sr/molecule
superscripts r and s	= reference and sample, respectively[37]
I	= integrated area of a Raman band
E	= spectrometer efficiency at the Raman frequency
C	= sample concentration in molecules per unit volume

5.3.2 SOLUTION VERSUS SOLID PHASE

There are many examples in the literature of Raman cross sections of molecules in the gaseous, liquid, and solution phases, but there are a limited number of absolute cross-section measurements reported in the solid state.[20,33,38,39] For explosives, it is particularly important to measure the cross sections in the solid state, since explosives such as hexahydro-1,3,5-trinitro-1,3,5-triazine (RDX), octahydro-1,3,5,7-tetra-nitro-1,3,5,7-tetrazocine (HMX), pentaerythritol tetratnitrate (PETN), ammonium nitrate (NH$_4$NO$_3$), 2,4,6-trinitrotoluene (TNT), 2,4-dinitrotoluene (2,4-DNT), and 2,6-dinitrotoluene (2,6-DNT) are most commonly encountered in this form. The visible and NIR solid-state Raman cross sections of these explosives using sodium nitrate (NaNO$_3$) as an internal standard have been published by Emmons et al.[40] In the procedure used to make those measurements, the explosive and standard were finely ground and well mixed using a ball mill to ensure a uniform relative concentration of each in the sampling volume.

Solid-state measurements using UV excitation are more difficult because the UV penetration depths of solids in the vicinity of strong absorption bands can be on the order of only tens of nanometers.[41] Because of this, it is necessary to use nanoparticles with dimensions smaller than the penetration depth at the incident and Raman scattered wavelengths in order to reduce bias from absorption. The group of Asher et al. have measured the deep-UV resonance Raman cross sections of explosive nanoparticles in the solid state using a colloidal crystal technique to produce explosive nanoparticles and eliminate bias due to absorption.[33] In their approach, a stoichiometrically defined mixture of NaNO$_3$ and sodium sulfate (Na$_2$SO$_4$) was deposited from solution into the interstices of a close-packed photonic crystal. The Na$_2$SO$_4$ served as a nonresonant internal standard, while the interstices yielded particles significantly smaller than the penetration depth of the 229 nm radiation in

NaNO$_3$. They have also examined the photochemistry of sodium nitrate in solution and the solid state.[42]

Emmons et al. used a different technique to produce nanosized mixtures of explosive and an internal standard.[41] In their work, an aerosol jet of a solution containing known concentrations of the explosive material and the internal standard (sodium sulfate or ammonium sulfate) was sprayed onto a heated aluminum-coated microscope slide. The slide temperature was maintained at ~100°C in order to rapidly evaporate the solvent. Several factors contributed to allow for a nanostructured mixture of the analyte and internal standard to be formed. Since the solvent evaporation is rapid, there is little time for large crystals to grow. Also, the concentration of the resonant analyte was kept at least an order of magnitude less than that of the nonresonant internal standard. This made it even more difficult for large particles of the resonant analyte of interest to form, as the presence of the internal standard at a higher concentration would interfere with their growth. A summary table of the Raman cross sections of explosives is given in Table 5.1, including both solution and solid-phase cross sections (in parentheses) as a function of wavelength. The values listed in this table were taken from Ghosh et al.[35] and Emmons et al.[40,41]

It is expected that the Raman spectra and cross sections of the solid-phase explosives will differ from those obtained in solution due to differences in molecular conformation. For example, the most stable conformation for gas-phase RDX molecules is believed to be the chair conformation, with all three nitro groups pointing parallel to the axis of the ring, or the so-called AAA conformation (the conformation adopted in crystalline β-RDX).[43,44] In the solid state, RDX adopts a crystalline α-phase with an AAE chair conformation in which two of the nitro groups point along the axis of the ring and the third points in a nearly equatorial direction.[43,44] The Raman spectra and crystal structure have been studied in detail and vibrational mode assignments suggested.[45–47] HMX is similar to RDX except that it contains an eight-membered ring instead of a six-membered ring. The most stable phase is denoted as β-HMX, and this is the one present in bulk explosives.[48] Vibrational mode identifications for HMX have also been carried out.[48–51] For RDX and HMX, it has been shown that the appearance of the vibrational spectra is dominantly determined by the molecular conformation, with the crystalline structure playing a secondary role, indicating relatively weak intermolecular interactions in the solid state. The published spectra of RDX (Figure 5.2) and HMX show that the solution-phase normal Raman spectrum does not match the spectra of either of the dominant solid-phase spectra.[40] It is likely that the interaction with the solvent (acetonitrile) shifts the Raman bands of RDX and HMX. It is also possible that the differences are caused by the presence of multiple conformations in the solution.

There are also differences in the solid- and solution-phase normal Raman spectra of ammonium nitrate resulting from the different symmetries of the nitrate ion in the two environments.[52,53] Only minor differences in spectra are seen for solid- and solution-phase TNT, 2,4-DNT, and 2,6-DNT, indicating that the molecular conformations are similar for solid and solution.[40]

For the limited number of explosives where both solution and solid cross-section data are available, there is no clear trend in the ratio of the two values. The solid- and solution-phase cross sections for PETN in the UV are similar, while the UV

TABLE 5.1
Raman Cross Sections

Explosive	Mode Frequency (cm⁻¹)	Cross Section (cm²/molecule/sr)								
		204 nm	229 nm	238 nm	244 nm	257 nm	262 nm	532 nm	633 nm	785 nm
PETN	1279 + 1295		2.4×10^{-26} (3.2×10^{-26})	9.7×10^{-27}	6.1×10^{-27} (5.5×10^{-27})		1.4×10^{-27} (1.5×10^{-27})	9.9×10^{-30} (1.2×10^{-29})	4.5×10^{-30}	1.5×10^{-30} (1.8×10^{-30})
Ammonium Nitrate	1044	3.0×10^{-25}	1.8×10^{-26}	1.9×10^{-26}	3.0×10^{-27}	2.6×10^{-27}				
			6.4×10^{-26} (3.4×10^{-26})		1.0×10^{-26} (2.6×10^{-27})		2.6×10^{-27} (1.3×10^{-27})	1.0×10^{-29}	(7.9×10^{-30})	1.1×10^{-30} (3.8×10^{-30})
HMX	833	1.4×10^{-25}	7.3×10^{-26}		4.0×10^{-27}	3.6×10^{-27}				
	1260	1.1×10^{-26}	5.3×10^{-26}	4.8×10^{-26}	3.6×10^{-26}	1.2×10^{-26}	2.3×10^{-27}			
		1.1×10^{-25}	6.2×10^{-26}		3.2×10^{-26}					
RDX	885	1.5×10^{-26}	1.5×10^{-25}		1.7×10^{-26}	1.4×10^{-26}		1.8×10^{-29}	5.8×10^{-30} (4.7×10^{-30})	2.1×10^{-30} (1.9×10^{-30})
	1024	9.0×10^{-27}	2.5×10^{-26}	5.0×10^{-26}	3.6×10^{-26}	2.0×10^{-26}	1.3×10^{-26}			
			5.1×10^{-26}			3.6×10^{-26}				
	1273	1.3×10^{-25}	6.8×10^{-26}		3.5×10^{-26}	2.1×10^{-26}		1.8×10^{-30}	1.3×10^{-30}	2.9×10^{-31}
	1310	1.4×10^{-25}	3.0×10^{-26}		2.7×10^{-26}	1.3×10^{-26}		5.6×10^{-30}	3.9×10^{-30}	1.1×10^{-30}
					7.0×10^{-28}					

Compound	cm⁻¹									
TNT	794	**1.6×10^{-26}**						2.4×10^{-30} (6.4×10^{-30})	1.0×10^{-30} (2.2×10^{-30})	*5.1×10^{-31}* (*9.8×10^{-31}*)
	827							5.0×10^{-30} (1.2×10^{-29})	2.1×10^{-30} (3.6×10^{-30})	*8.6×10^{-31}* (*1.5×10^{-30}*)
	1360		**4.0×10^{-26}**		1.7×10^{-26}	**1.6×10^{-26}**		6.7×10^{-29}	2.0×10^{-29}	*5.7×10^{-30}*
	1548	**9.8×10^{-26}**	**6.5×10^{-25}**	2.6×10^{-25}	2.9×10^{-25}	**4.2×10^{-25}**		9.3×10^{-30}	4.7×10^{-30} (1.2×10^{-29})	*1.4×10^{-30}* (*3.3×10^{-30}*)
	1620		**3.3×10^{-25}**	3.8×10^{-25}	4.3×10^{-25}	**2.1×10^{-25}**	3.2×10^{-25}	5.9×10^{-30}	2.5×10^{-30} (7.0×10^{-30})	*6.4×10^{-31}* (*1.8×10^{-30}*)
2,4-DNT	792	**4.9×10^{-26}**						2.0×10^{-30} (4.3×10^{-30})	8.6×10^{-31} (1.5×10^{-30})	*3.8×10^{-31}* (*7.3×10^{-31}*)
	836							6.4×10^{-30} (1.8×10^{-29})	2.7×10^{-30} (4.7×10^{-30})	*1.0×10^{-30}* (*1.9×10^{-30}*)
	1537							9.6×10^{-30}	4.0×10^{-30}	*1.7×10^{-30}*
	1611		**8.7×10^{-26}**	1.8×10^{-25}	2.9×10^{-25}	**1.3×10^{-25}**	3.1×10^{-25}	1.2×10^{-29}	4.5×10^{-30}	*1.0×10^{-30}*
2,6-DNT	793							6.2×10^{-30} (1.3×10^{-29})	2.6×10^{-30} (5.1×10^{-30})	*1.1×10^{-30}* (*2.2×10^{-30}*)
	843							2.0×10^{-30} (4.4×10^{-30})	7.9×10^{-31} (1.9×10^{-30})	*3.4×10^{-31}* (*8.3×10^{-31}*)
	1538							6.7×10^{-30}	2.5×10^{-30}	*9.6×10^{-31}*
	1616		**4.6×10^{-26}**	4.7×10^{-26}	4.2×10^{-26}		1.6×10^{-26}			

Note: Values in bold are from Ghosh, M., Wang, L., Asher, S. A., *Applied Spectroscopy* **2012**, *66* (9), 1013–1021; values in normal font are from Emmons, E. D., Tripathi, A., Guicheteau, J. A., Fountain, A. W., Christesen, S. D., *Journal of Physical Chemistry A* **2013**, *117* (20), 4158–4166; values in italics are from Emmons, E. D., Guicheteau, J. A., Fountain, A. W., Christesen, S. D., *Applied Spectroscopy* **2012**, *66* (6), 636–643. Values in parentheses are measurements of solid explosives.

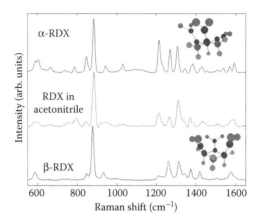

FIGURE 5.2 Raman spectra of RDX in different solid phases (top and bottom traces) and at a concentration of 40 mg/mL in acetonitrile solution (middle trace). The acetonitrile contribution was numerically subtracted from the solution-phase spectrum. (From Emmons, E. D., Guicheteau, J. A., Fountain, A. W., Christesen, S. D., *Applied Spectroscopy* **2012**, *66* (6), 636–643.)

ammonium nitrate solution-phase cross sections are two to four times larger than those of solid phase. The visible/NIR ammonium nitrate cross sections are approximately equal for solution and solid. Except for the 885 cm^{-1} carbon–nitrogen (C–N) ring-breathing mode of RDX, the solid-phase visible/NIR cross sections are all roughly double the solution-phase cross sections. Because of the scarcity of relevant data, it is difficult to draw any conclusions on the relative strengths of the Raman spectra of solid explosives compared to those in solution. It is clear for the analytes studied, however, that those differences are relatively minor and that the solution-phase cross sections (which are much easier to measure) provide a reasonable approximation to the solid-phase values.

5.4 RAMAN DETECTION SYSTEMS

Explosives detection broadly falls into three scenarios: identification of bulk material (close-range and short-range standoff), detection and identification of trace contamination on surfaces (short-range standoff), and identification of homemade explosives (HME) in glass or plastic containers (airport or checkpoint screening). Whereas commercial handheld Raman instruments such as the FirstDefender™ from Thermo Scientific or the CBEx from Snowy Range Instruments are generally capable of identifying bulk explosives, none of the commercial instruments can provide standoff detection of bulk or trace material. The checkpoint screening scenario is being addressed by the Cobalt Light Systems INSIGHT 100, using spatially offset Raman spectroscopy (SORS). There are no commercially available standoff explosives sensors, but a number of systems are in the development stage. Some of these systems are described in the following sections. However, data on their actual operational utility as determined by field-testing to measure probabilities of detection and false-alarm rates are very limited or nonexistent.

5.4.1   RAMAN SIGNAL

All commercial standoff Raman detectors operate in a backscattering configuration. This is true whether they are large Raman light detection and ranging (lidar) instruments or handheld instruments like the FirstDefender, operating at centimeter standoff distances. As such, the amount of signal observed for these Raman systems can be described by the basic Raman scattering lidar equation:[54]

$$S = I \times \frac{A}{r^2} \times \sigma_R \times \eta \times D_{\text{effective}} \times \varepsilon_{\text{spectrograph}} \times \varepsilon_{\text{detector}} \times e^{-(\kappa_0 + \kappa_R)r} \times f \qquad (5.13)$$

The intensity of Raman-scattered photons S in photons per second is a function of fundamental molecular parameters (Raman cross section in $cm^2/sr/molecule$, σ_R, and molecular density in $molecules/cm^3$, η), sample properties (sample thickness, $D_{\text{effective}}$), instrument response (total spectrograph, $\varepsilon_{\text{spectrograph}}$, and detector, $\varepsilon_{\text{detector}}$, efficiencies), the laser intensity at the sample in photons per second (I), the area of the receiver telescope primary collection optic (A), the atmospheric absorption coefficients at the laser (κ_0) and Raman (κ_R) wavelengths, and the range to the target (r).[54] In many standoff trace detection scenarios, the explosive residue will not be uniform, and the signal will also be dependent on the fractional surface coverage (f). $D_{\text{effective}}$ is the effective sample thickness when accounting for absorption of the incident and Raman-scattered radiation by the sample and is equal to

$$D_{\text{effective}} = \int_0^D 10^{-(\alpha_0 + \alpha_R)r} \, dr \qquad (5.14)$$

The explosives HMX, RDX, TNT, 2,4-DNT, 2,6-DNT, ammonium nitrate, and PETN are essentially transparent in the visible and NIR (ignoring attenuation due to scattering), and for laser excitation at those wavelengths, $D_{\text{effective}} \cong D$. Equation 5.13 is useful in that it allows one to determine the relative signal strengths for different explosives given a specific lidar configuration or to evaluate sensitivity to a specific explosive for different detector configurations. An example of the latter is described next where only the excitation wavelength is changed.

5.4.2   EXCITATION WAVELENGTH DEPENDENCE

In terms of sensor applications, it would generally appear advantageous to use as short an excitation wavelength as possible to obtain the greatest amount of signal. In addition to the signal increase due to the standard $1/\lambda^4$ dependence, many analytes of interest are also more likely to have resonance enhancement at shorter excitation wavelengths. This is seen for TNT and ammonium nitrate (Figure 5.3) as the Raman cross sections increase with increasing absorbance at shorter wavelengths and generally track the wavelength dependence of the molar absorptivity. It is also the case, however, that fluorescence interference from both the analyte of interest, as well as interferents (impurities), becomes stronger as the wavelength decreases.

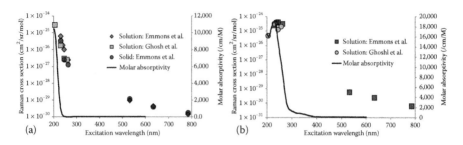

FIGURE 5.3 Wavelength dependence of the Raman cross section for 1620 cm⁻¹ line of TNT (a) and the 1044 cm⁻¹ line of ammonium nitrate (b) as well as their molar absorptivities (right axis). (Data from Ghosh, M., Wang, L., Asher, S. A., *Applied Spectroscopy* **2012**, *66* (9), 1013–1021; Emmons, E. D., Guicheteau, J. A., Fountain, A. W., Christesen, S. D., *Applied Spectroscopy* **2012**, *66* (6), 636–643; Emmons, E. D., Tripathi, A., Guicheteau, J. A., Fountain, A. W., Christesen, S. D., *Journal of Physical Chemistry A* **2013**, *117* (20), 4158–4166.)

This can sometimes be overcome by exciting with light of a wavelength less than 250 nm, which positions the Raman-scattered bands at a sufficiently short wavelength so they do not overlap with the fluorescence bands at longer wavelengths.[55] The possibility of photoinduced damage to the sample also increases with the increased absorbance at shorter wavelengths.[42] The increase in Raman cross sections associated with resonance enhancement is mitigated by the absorption of the incident laser radiation and Raman-scattered light, which limits the penetration depth and reduces the sampling volume. If fluorescence is a significant problem, as it is for many realistic samples of interest, it may instead be useful to shift to longer-wavelength excitation, such as 785 nm or even 1064 nm. The reduced cross section at these wavelengths may be more than offset by the ability to use higher-intensity laser excitation.

As mentioned before, the sensitivity of a Raman surface contamination detector is dependent on both the Raman scattering cross section and the molecular absorbance. The observed intensity of a given Raman line is proportional to the Raman-scattering cross section and the number of molecules that contribute to that scattering.[41] Ignoring the instrument-dependent parameters in Equation 5.13, we get

$$S \propto \sigma_R \times \eta \times \int_0^D 10^{-(\alpha_0 + \alpha_R)r}\, dr \qquad (5.15)$$

For nonabsorbing samples, the integral in Equation 5.14 ($D_{effective}$) is approximately equal to D, but as the sample absorbance increases; the penetration depth and therefore the Raman signal decrease. Thus, you can have a situation where the increase in Raman return due to the increase in Raman cross section resulting from resonance or preresonance enhancement is more than offset by the larger absorption cross section and reduction in penetration depth. Equation 5.14 demonstrates that

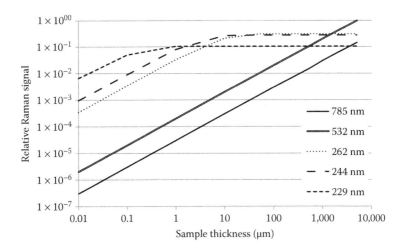

FIGURE 5.4   Relative Raman signal vs. sample thickness for the 1040 cm⁻¹ line of ammonium nitrate at different excitation wavelengths.

in assessing the relative sensitivities of visible or NIR excitation versus UV excitation for detecting surface contamination, the expected sample thickness D must be considered.

If we are only interested in the wavelength dependence of the Raman signal, it is useful to define a relative Raman signal as S/S_{max}, where S_{max} is the maximum signal calculated. The relative Raman signal as a function of sample thickness and excitation wavelength is shown for ammonium nitrate in Figure 5.4. In this case, S_{max} is achieved for a sample thickness of 5 mm (largest thickness used in the calculations) and 532 nm excitation. Based on the UV-visible (UV–Vis) absorption spectra shown in Emmons et al.,[41] the penetration depth of the excitation light as a function of wavelength can be calculated for ammonium nitrate if we assume that the molar absorptivity is the same in solution and the solid state. The penetration depths for the excitation wavelengths used are as follows: 262 nm, 18.0 μm; 244 nm, 4.0 μm; 238 nm, 1.2 μm; 229 nm, 0.2 μm. The samples are assumed to be transparent at 532, 633, and 785 nm. The Raman cross sections used for this analysis are the average values of the solution and solid measurements (where available) listed in Table 5.1. It is clear from Figure 5.4 that UV excitation is favored for thin samples (or small particles), as might be encountered when trying to detect trace contamination. As the sample thickness increases, however, absorption becomes an important effect and the relative Raman signal plateaus. At sample thicknesses of approximately 1 mm or greater, visible excitation yields the largest Raman signal, as a significantly larger sample volume is probed using longer-wavelength excitation. Operational considerations, such as eye safety and solar interference, may require the use of a wavelength other than the true optimum excitation wavelength. Whereas the results for TNT (Figure 5.5) are qualitatively similar to ammonium nitrate, different chemicals in general will produce different relative Raman signals and sample-thickness-dependent optimum excitation wavelengths.

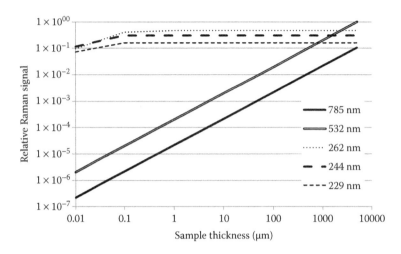

FIGURE 5.5 Relative Raman signal vs. sample thickness for the 1620 cm^{-1} line of TNT at different excitation wavelengths.

5.4.3 HANDHELD RAMAN INSTRUMENTS

A number of companies are marketing Raman systems for detection applications to the military, homeland defense, law enforcement, and other communities. Many of these are small handheld instruments designed for examining unknown bulk samples, such as explosives and drugs, at close range. Usually these systems use visible or NIR excitation. More recently, dual wavelength instruments containing two excitation lasers have been developed, including the Rigaku[56] Xantus-2™ (785 and 1064 nm) and the Enwave Optronics[57] EZRaman-I-Dual-G (532 and 785 nm). The longer wavelengths can provide improved sensitivity when the shorter wavelengths induce sample fluorescence that obscures the Raman spectrum. See Table 5.2 for examples of some commercially available handheld systems.

5.4.4 STANDOFF RAMAN DETECTION

In addition to the handheld instruments for very close-range detection, longer-range standoff instruments designed for detection at tens to hundreds of meters are being developed. Some of these systems use ultraviolet excitation, which yields larger Raman cross sections via the normal $1/\lambda^4$ dependence or, more significantly, through resonance enhancement. These shorter wavelength advantages yield the potential for detecting smaller quantities of materials at standoff distances. The group of Sedlacek et al. developed a standoff detection system based on 266 nm excitation.[58,59] Nagli and Gaft also worked on development of standoff detection systems and compared a variety of excitation wavelengths ranging from 248 to 785 nm.[6,60] Sharma and colleagues developed a 532 nm compact Raman system for use in mineral detection and demonstrated this system could be combined with the simultaneous use of laser-induced breakdown spectroscopy (LIBS) (refer to Chapter 7) to add an additional capability for

TABLE 5.2
Examples of Commercially Available Handheld Raman Instruments

System	Company	Description	Laser Wavelength	Spectral Range (cm⁻¹)	Spectral Resolution (cm⁻¹)
Fido-verdict	FLIR	Handheld system	785 nm	300–2000	12
Avalon	FLIR	Handheld system (standoff up to 3 m)	785 nm	300–2000	12–15
CBEX & CBEX 1064	Snowy range instruments	Handheld system	785 nm, 1064 nm	400–2300	12
First defender RM & RMX	Thermo scientific	Handheld system, RMX has fiber optic attachment	785 nm	250–2875	7–10.5
ReporteR	SciAps	Handheld system	785 nm	300–2500	10
RespondeR RCI	Smiths detection	Suitcase kit	785 nm	225–2400	12
StreetLab mobile	Morpho detection	Handheld system	785 nm	300–1800	10
TacticID	B&Wtek	Handheld system	785 nm	176–2900	9
First guard	Rigaku	Handheld systems	532 nm, 785 nm, 1064 nm	400–1850	10–15
EZRaman-I-Dual-G	Enwave optronics	Handheld dual-wavelength system	532 nm and 785 nm	250–2350 (785 nm laser), 250–3250 (532 nm laser)	6–7
Portable Raman analyzer	Real time analyzers	Suitcase kit	1064 nm	150–3500	Selectable between 4, 8, 16, and 32

elemental analysis.[61,62] Other groups have also built standoff Raman systems for explosives analysis, which have undergone some limited but realistic outdoor testing.[10,16]

5.4.4.1 Visible Raman

A number of researchers have demonstrated standoff detection of explosives using visible laser excitation.[3,7–9] Pacheco-Londoño et al. were able to detect TNT, DNT, triacetone triperoxide (TATP), RDX, and the plastic explosive C-4 at a range of 7 m using the 488 nm (C-4) and 514.5 nm (all others) lines from a continuous-wave (CW) argonion laser.[9] Their detection limits were generally less than 10 mg of explosive material, using an acquisition time of 10 s, although their use of a CW laser required that the system be operated in the dark to minimize the background radiation interference. Sharma and colleagues at the University of Hawaii have been developing standoff Raman systems for more than 10 years, and they have recently demonstrated daylight detection of ammonium nitrate, potassium chlorate, and nitrobenzene at a range of 120 m with a single 100 mJ pulse from a frequency-doubled neodymium-doped yttrium aluminum garnet (Nd:YAG) laser operating at 532 nm.[7] In this study, the samples were contained in 1 in. diameter, 20 mL glass vials and represented a scenario for detection of bulk material. The authors see this technology being applicable to identifying contaminants on moving targets such as items on a conveyer belt at airports. Eye-safety issues, as well as limited sensitivity to trace amounts of explosives, would likely make this application problematic.

Laserna and coauthors have recently performed an evaluation of the factors governing the ultimate sensitivity of a visible standoff Raman system.[8] Their experiments were performed using 2,6-DNT, RDX, PETN, C-4, sodium chlorate, potassium chlorate, and Goma-2 ECO (an ammonium nitrate–based explosive used primarily in mining) at ranges up to 45 m. In this study, the authors measured the Raman response as a function of pulse energy, total acquisition time, and range to target and demonstrated that the spectral quality is a complex function of these parameters. Significantly, they concluded that the pulse energy is of primary importance in improving limits of detection, and accumulating multiple pulses does not make up for lower pulse energy if that energy is not sufficient to produce a reasonable signal. They concluded that the limiting factor on pulse energy should be that level at which it starts to produce photoinduced damage.[8]

5.4.4.2 UV Raman

Although there are a number of significant advantages to using deep-UV laser excitation below 250 nm, including operating in the solar-blind region and increased signal due to both the $1/\lambda^4$ dependence of the Raman cross section and resonance or preresonance enhancement, development of UV Raman chemical or explosives sensors suffers from a dearth of suitable laser sources. ITT (now Exelis) developed a system (Joint Contamination Surface Detector or JCSD) for on-the-move detection of chemical contamination on the ground using a modified compact KrF (krypton fluoride) excimer laser (248 nm) manufactured by GAM Laser, Inc.[63] This is also the laser used by Alakai Defense Systems in their standoff explosive sensor (see Section 5.4.4.2.1). Whereas excimer lasers are relatively simple and require no frequency doubling or mixing, they must be refilled on a regular basis, and the associated gas cylinders and recharging stations are logistical burdens that have to be considered before fielding such a system. The high peak powers associated with the short 15–20 ns pulse widths are also conducive to sample photodegradation, and photoinduced sample damage becomes more likely with multiple laser pulses hitting the same spot. The usable laser power per pulse is also limited by plasma formation that can occur when the laser is focused onto a surface. The helium silver (HeAg) laser at 224.3 nm and neon copper (NeCu) laser at 248.6 nm produced by Photon Systems providing quasi-CW output, with average powers ranging from approximately 10 mW to 250 mW, are also suitable for some short-range applications (see Section 5.4.4.2.2). The development of a compact UV solid-state laser would significantly change the prospects for

TABLE 5.3
Desired Solid-State UV Laser Characteristics

	Pulsed	CW or Quasi-CW
Wavelength	220–250 nm	220–250 nm
Linewidth	<1 cm^{-1}	<1 cm^{-1}
Pulse energy/power	>10 mJ	>50 mW
Repetition rate	75 Hz	NA
Pulse width	>100 ns	NA

fielding a UV lidar system. Some of the desired characteristics for both a pulsed and CW or quasi-CW laser are listed in Table 5.3.

In the following sections, we discuss two UV Raman systems specifically designed for detection of explosives on surfaces. The JCSD is similar to the Alakai system, although it was not designed for or tested for detection of explosives. The following UV systems are not the only ones in operation, but they help to illustrate the current state of the art in UV Raman detection of explosives and the possible advantages of combining Raman with another optical technique.

5.4.4.2.1 Alakai Defense Systems: CPEDS

Alakai Defense Systems (ADS) has fabricated a standoff checkpoint explosives detection system (CPEDS) for trace detection of explosives at vehicular checkpoints.[64] The system is designed to incorporate multiwavelength Raman spectroscopy and LIBS, combined with a LIBS enhancement technique called Townsend effect plasma spectroscopy (TEPS). The CPEDS platform uses a frequency-tripled Nd:YAG at 355 nm and an excimer laser at 248 nm for the Raman spectroscopy. The 355 nm beam is combined with a carbon dioxide (CO_2) laser at 10.6 μm to generate the TEPS signal. While ADS has demonstrated UV Raman and TEPS detection of explosives at ranges of 25 m,[14,15] a more recent technology assessment conducted by the Edgewood Chemical Biological Center (ECBC) and the Army Research Laboratory (ARL) at ECBC in 2012 focused exclusively on UV (248 and 355 nm) Raman detection of bulk explosive material and explosive residues on automobile panels (Figure 5.6a). While the specifics of the technology assessment conducted by ECBC and ARL at ECBC are not releasable to the public, the results confirmed much of the previous ADS testing and its ability to identify bulk (>100 μg) material at ranges up to 25 m. ADS subsequently designed and built a more ruggedized CPEDS system utilizing only 248 nm Raman spectroscopy (Figure 5.6b).

5.4.4.2.2 Photon Systems: Targeted Ultraviolet Chemical Biological Explosive

Photon Systems is developing a combined laser-induced fluorescence (LIF) and deep-UV Raman system (refer to Chapters 6 and 4, respectively) for the noninvasive short-range (up to 10 m) standoff detection of chemical, explosive, and biological contamination on surfaces (Figure 5.7).[65] The targeted ultraviolet chemical biological explosive (TUCBE) sensor employs a Photon Systems deep-UV laser operating at either

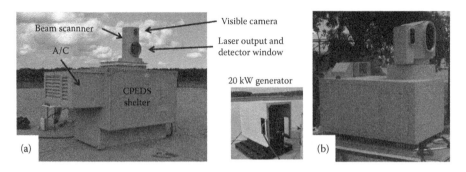

FIGURE 5.6   (a) CPEDS system components and (b) modified CPEDS with 248 nm Raman detection only.

FIGURE 5.7   Photon Systems TUCBE GEN-4 combination fluorescence and UV Raman instrument.

224.3 nm (HeAg) or 248.6 nm (NeCu). The system fires a laser pulse onto an unknown target and simultaneously collects both Raman and fluorescence spectral emissions from the target. Instead of a spectrograph and CCD detector, the TUCBE divides the spectral emissions into the Raman (below 270 nm) and fluorescence (above 270 nm) spectral regions. The Raman scattering is measured using a grating spectrograph with a 32-element PMT array detector, while a dichroic filter stack with individual PMT detectors is used to measure the fluorescence spectrum. The Raman resolution is about 80 cm^{-1} in 32 channels and the fluorescence resolution is about 10–16 nm in 5 channels to produce the simultaneous Raman and fluorescence spectra. As discussed previously, operating in the deep UV below 250 nm enables spectral separation of Raman and fluorescence spectra, allowing for detection of weak Raman emissions without obscuration by fluorescence or alteration of fluorescence emissions by strong Raman bands.

The combination of orthogonal spectroscopic techniques such as fluorescence and Raman is expected to enhance detection and reduce false positives and negatives. Some recently published results on the TUCBE hint to some advantages to this approach, although the exact nature of the data fusion still needs to be determined.[66] Bhartia et al. advocate a two-step method where the fluorescence spectrum, which is much more intense than the Raman spectrum, is used to further differentiate samples that have similar low-resolution Raman spectra.[66] As an example, they use the fluorescence to differentiate samples that have aromatic structures and cluster in the principal components analysis (PCA) of their Raman spectra.

5.4.5 COHERENT RAMAN DETECTION

Nonlinear Raman spectroscopy techniques such as coherent anti-Stokes Raman spectroscopy (CARS) and stimulated Raman spectroscopy (SRS) offer the promise of extremely low-level explosives detection and a signal improvement over normal Raman of as much as a factor of 50,000 (refer to Chapter 13).[5] In addition, these techniques also offer the promise of a simultaneous increase in selectivity between different analytes.[67] While these nonlinear Raman effects have been known since the early 1960s,[68] early applications required the use of multiple pulsed or, in the case of stimulated Raman gain or loss, multiple narrowband CW lasers.[69] The advent of femtosecond pulsed lasers and the associated pulse shaping have led to single-laser techniques such as single-beam SRS (SB-SRS)[2] and multiplexed CARS (MCARS).[70–72] Individually adjusting the intensity and phase of the different wavelength components of the ultrafast pulses (i.e., the pulse shape) allows for optimization of the detector system.[67] Different control algorithms such as genetic algorithms have been used to search for nonintuitive optimal pulse shapes.[67,73,74]

Of particular relevance to us are the applications that deal with standoff detection of explosive materials, especially on surfaces.[2,5,12,70,71] Bremer et al. have demonstrated detection of 100 µg/cm² of ammonium nitrate and TNT on cotton at a range of over 7 m with their SB-SRS system,[2] and Dogariu and coauthors reported detection of 100 ng of RDX and PETN with 100 ms integration times, but only at a range of 20 cm.[5] Bremer et al. also used MCARS to perform standoff chemical imaging of trace chemicals in background matrices.[72] In related measurements, Brady et al. have used MCARS to detect chemical warfare agent simulants at a 1 m standoff distance using integration times on the order of milliseconds with a compact, low-sensitivity USB spectrometer.[70,71] While these laboratory results are encouraging, the nonlinear spectroscopy systems clearly lag behind their normal Raman spectroscopy counterparts in testing with highly controlled samples on relevant surfaces; these types of tests will be required to truly determine detection limits.

5.5 CONCLUSIONS

Much progress has been made on the measurement of the Raman spectra and cross sections of explosive material, including the dependence of the spectra on crystalline phase and the differences between solution and solid-state spectra. Armed with the Raman scattering and UV/Vis absorption spectra, we are able to predict both selectivity and sensitivity as a function of sensor parameters (e.g., excitation wavelength) and sample characteristics (e.g., particle size and area coverage). This can help guide future Raman sensor design for the specific application of explosives detection, whether for the detection of bulk material on the back of a truck or detection of trace residue deposited as a result of manufacturing or transportation.

A number of standoff Raman systems have been developed that have demonstrated some capability to detect explosive contamination on surfaces, albeit at levels above what could be considered trace. While the analysis of the wavelength dependence of the Raman signal shows the advantage of using UV excitation for detecting small particles, the lack of a suitable solid-state laser operating below 250 nm limits its employment operationally. Another issue facing any standoff optical detection technique has to do with targeting or sampling. While Raman spectroscopy may be capable of detecting and identifying relatively small (millimeter-sized) particles at standoff distances, the time required to scan any reasonably sized surface would likely be prohibitive. Because of this, standoff trace detection will likely require imaging of some kind, either full Raman[13,75] or coherent Raman[2] imaging of the contaminated surface; or normal point Raman spectroscopy coupled with a more sensitive but less specific imaging technique such as short-wave IR (SWIR) or fluorescence to provide positional targeting and identification of regions of interest.

Regardless of the challenges, the ability of Raman spectroscopy to provide chemical identification, with spectra that are qualitatively insensitive to sample morphology, make it very attractive for standoff detection of explosive material. Advances in deep-UV and femtosecond laser technologies promise to improve detection limits and yield more rugged and fieldable Raman-based sensors.

ACKNOWLEDGMENTS

The opinions, interpretations, conclusions, and recommendations are those of the authors and are not necessarily endorsed by the US government.

REFERENCES

1. Steinfeld, J. I., Wormhoudt, J., Explosives detection: A challenge for physical chemistry. *Annual Review of Physical Chemistry* **1998**, *49*, 203–232.
2. Bremer, M. T., Dantus, M., Standoff explosives trace detection and imaging by selective stimulated Raman scattering. *Applied Physics Letters* **2013**, *103*, 061119.
3. Carter, J. C., Angel, S. M., Lawrence-Snyder, M., Scaffidi, J., Whipple, R. E., Reynolds, J. G., Standoff detection of high explosive materials at 50 meters in ambient light conditions using a small Raman instrument. *Applied Spectroscopy* **2005**, *59* (6), 769–775.
4. Comanescu, G., Manka, C. K., Grun, J., Nikitin, S., Zabetakis, D., Identification of explosives with two-dimensional ultraviolet resonance Raman spectroscopy. *Applied Spectroscopy* **2008**, *62* (8), 833–839.
5. Dogariu, A., Pidwerbetsky, A., Coherent anti-stokes Raman spectroscopy for detecting explosives in real-time. *Proceedings of SPIE* **2012**, *8358*, 83580R.
6. Gaft, M., Nagli, L., UV gated Raman spectroscopy for standoff detection of explosives. *Optical Materials* **2008**, *30* (11), 1739–1746.
7. Misra, A. K., Sharma, S. K., Acosta, T. E., Porter, J. N., Bates, D. E., Single-pulse standoff Raman detection of chemicals from 120 m distance during daytime. *Applied Spectroscopy* **2012**, *66* (11), 1279–1285.
8. Moros, J., Lorenzo, J. A., Novotny, K., Laserna, J. J., Fundamentals of stand-off Raman scattering spectroscopy for explosive fingerprinting. *Journal of Raman Spectroscopy* **2013**, *44* (1), 121–130.
9. Pacheco-Londoño, L. C., Ortiz-Rivera, W., Primera-Pedrozo, O. M., Hernandez-Rivera, S. P., Vibrational spectroscopy standoff detection of explosives. *Analytical and Bioanalytical Chemistry* **2009**, *395* (2), 323–335.
10. Pettersson, A., Johansson, I., Wallin, S., Nordberg, M., Ostmark, H., Near real-time standoff detection of explosives in a realistic outdoor environment at 55 m distance. *Propellants Explosives Pyrotechnics* **2009**, *34* (4), 297–306.

11. Tuschel, D. D., Mikhonin, A. V., Lemoff, B. E., Asher, S. A., Deep ultraviolet resonance Raman excitation enables explosives detection. *Applied Spectroscopy* **2010**, *64* (4), 425–432.
12. Van Neste, C. W., Liu, X., Gupta, M., Kim, S., Tsui, Y., Thundat, T., Standoff detection of explosive residues on unknown surfaces. *Proceedings of SPIE*, **2012**, *8373*, 83732F.
13. Wallin, S., Pettersson, A., Onnerud, H., Ostmark, H., Nordberg, M., Ceco, E., Ehlerding, A., Johansson, I., Kack, P., Possibilities for standoff Raman detection applications for explosives. In Fountain, A. W., (ed.), *Proceedings of SPIE: Chemical, Biological, Radiological, Nuclear, and Explosives Sensing XIII*, 2012, 23 April, Baltimore, Maryland, Vol. 8358, pp. 83580P–83580P-9.
14. Waterbury, R., Rose, J., Vunck, D., Blank, T., Pohl, K., Ford, A., McVay, T., Dottery, E., Fabrication and testing of a standoff trace explosives detection system. *Proceedings of SPIE* **2012**, *8358*, 83580P.
15. Waterbury, R., Vunck, D., Hopkins, A. J., Pohl, K., Ford, A., Dottery, E., Recent improvements and testing of a check point explosives detection system. *Proceedings of SPIE* **2012**, *8358*, 83580N.
16. Zachhuber, B., Ramer, G., Hobro, A., Chrysostom, E. T. H., Lendl, B., Stand-off Raman spectroscopy: A powerful technique for qualitative and quantitative analysis of inorganic and organic compounds including explosives. *Analytical and Bioanalytical Chemistry* **2011**, *400* (8), 2439–2447.
17. Long, D. A., *The Raman Effect*. John Wiley and Sons: Chichester, UK, 2002.
18. Smith, E., Dent, G., *Modern Raman Spectroscopy: A Practical Approach*. John Wiley and Sons: Chichester, UK, 2005.
19. Christesen, S., Improved Raman sensitivity using polarization analysis. *Proceedings of SPIE* **2010**, *7665*, 76651B.
20. Eckhardt, G., Wagner, W. G., On calculation of absolute Raman scattering cross sections from raman scattering coefficients. *Journal of Molecular Spectroscopy* **1966**, *19* (4), 407.
21. McCreery, R. L., *Photometric Standards for Raman Spectroscopy*. John Wiley and Sons: West Sussex, UK, 2002, pp. 920–932.
22. Michalska, D., Wysokinski, R., The prediction of Raman spectra of platinum(II) anticancer drugs by density functional theory. *Chemical Physics Letters* **2005**, *403* (1–3), 211–217.
23. Polavarapu, P. L., *Ab initio* vibrational Raman and Raman optical-activity spectra. *Journal of Physical Chemistry* **1990**, *94* (21), 8106–8112.
24. Scott, A. P., Radom, L., Harmonic vibrational frequencies: An evaluation of Hartree-Fock, Moller-Plesset, quadratic configuration interaction, density functional theory, and semiempirical scale factors. *Journal of Physical Chemistry* **1996**, *100* (41), 16502–16513.
25. Williams, S. D., Johnson, T. J., Gibbons, T. P., Kitchens, C. L., Relative Raman intensities in C_6H_6, C_6D_6, and C_6F_6: A comparison of different computational methods. *Theoretical Chemistry Accounts* **2007**, *117* (2), 283–290.
26. Dudik, J. M., Johnson, C. R., Asher, S. A., UV resonance Raman studies of acetone, acetamide, and *N*-methylacetamide: Models for the peptide bond. *The Journal of Physical Chemistry* **1985**, *89* (18), 3805–3814.
27. Harmon, P. A., Asher, S. A., Environmental dependence of preresonance Raman cross-section dispersions: Benzene vapor-phase excitation profiles. *The Journal of Chemical Physics* **1990**, *93* (5), 3094–3100.
28. Dudik, J. M., Johnson, C. R., Asher, S. A., Wavelength dependence of the preresonance Raman cross-sections of CH_3CN, SO_4^{2-}, ClO_4^-, and NO_3^-. *Journal of Chemical Physics* **1985**, *82* (4), 1732–1740.

29. Fell, N. F., Widder, J. M., Medlin, S. V., Morris, J. B., PesceRodriguez, R. A., McNesby, K. L., Fourier transform Raman spectroscopy of some energetic materials and propellant formulations. *Journal of Raman Spectroscopy* **1996**, *27* (2), 97–104.

30. Lewis, I. R., Daniel, N. W., Griffiths, P. R., Interpretation of Raman spectra of nitro-containing explosive materials. Part I: Group frequency and structural class membership. *Applied Spectroscopy* **1997**, *51* (12), 1854–1867.

31. McNesby, K. L., Wolfe, J. E., Morris, J. B., Pescerodriguez, R. A., Fourier-transform Raman-spectroscopy of some energetic materials and propellant formulations. *Journal of Raman Spectroscopy* **1994**, *25* (1), 75–87.

32. Nagli, L., Gaft, M., Fleger, Y., Rosenbluh, M., Absolute Raman cross-sections of some explosives: Trend to UV. *Optical Materials* **2008**, *30* (11), 1747–1754.

33. Wang, L. L., Tuschel, D., Asher, S. A., Templated photonic crystal fabrication of stoichiometrically complex nanoparticles for resonance Raman solid cross section determinations. *Journal of Physical Chemistry C* **2011**, *115* (32), 15767–15771.

34. Akhavan, J., Analysis of high-explosive samples by Fourier-transform Raman-spectroscopy. *Spectrochimica Acta Part A: Molecular and Biomolecular Spectroscopy* **1991**, *47* (9–10), 1247–1250.

35. Ghosh, M., Wang, L., Asher, S. A., Deep-ultraviolet resonance Raman excitation profiles of NH_4NO_3, PETN, TNT, HMX, and RDX. *Applied Spectroscopy* **2012**, *66* (9), 1013–1021.

36. Christesen, S. D., Raman cross-sections of chemical-agents and simulants. *Applied Spectroscopy* **1988**, *42* (2), 318–321.

37. Christesen, S. D., Jones, J. P., Lochner, J. M., Hyre, A. M., Ultraviolet Raman spectra and cross-sections of the G-series nerve agents. *Applied Spectroscopy* **2008**, *62* (10), 1078–1083.

38. Aggarwal, R. L., Farrar, L. W., Polla, D. L., Measurement of the absolute Stokes Raman cross sections of the longitudinal optical (LO) phonons of room-temperature GaP. *Solid State Communications* **2009**, *149* (33–34), 1330–1332.

39. Aggarwal, R. L., Farrar, L. W., Polla, D. L., Measurement of the absolute Raman scattering cross sections of sulfur and the standoff Raman detection of a 6-mm-thick sulfur specimen at 1500 m. *Journal of Raman Spectroscopy* **2011**, *42* (3), 461–464.

40. Emmons, E. D., Guicheteau, J. A., Fountain, A. W., Christesen, S. D., Comparison of visible and near-infrared Raman cross-sections of explosives in solution and in the solid state. *Applied Spectroscopy* **2012**, *66* (6), 636–643.

41. Emmons, E. D., Tripathi, A., Guicheteau, J. A., Fountain, A. W., Christesen, S. D., Ultraviolet resonance Raman spectroscopy of explosives in solution and the solid state. *Journal of Physical Chemistry A* **2013**, *117* (20), 4158–4166.

42. Asher, S. A., Tuschel, D. D., Vargson, T. A., Wang, L., Geib, S. J., Solid state and solution nitrate photochemistry: Photochemical evolution of the solid state lattice. *Journal of Physical Chemistry A* **2011**, *115* (17), 4279–4287.

43. Infante-Castillo, R., Pacheco-Londoño, L. C., Hernandez-Rivera, S. P., Monitoring the alpha→beta solid-solid phase transition of RDX with Raman spectroscopy: A theoretical and experimental study. *Journal of Molecular Structure* **2004**, *970* (1–3), 51–58.

44. Torres, P., Mercado, L., Cotte, I., Hernandez, S. P., Mina, N., Santana, A., Chamberlain, R. T., Lareau, R., Castro, M. E., Vibrational spectroscopy study of beta and alpha RDX deposits. *Journal of Physical Chemistry B* **2004**, *108* (26), 8799–8805.

45. Dreger, Z. A., Gupta, Y. M., High pressure Raman spectroscopy of single crystals of hexahydro-1,3,5-trinitro-1,3,5-triazine (RDX). *Journal of Physical Chemistry B* **2007**, *111* (15), 3893–3903.

46. Haycraft, J. J., Stevens, L. L., Eckhardt, C. J., Single-crystal, polarized, Raman scattering study of the molecular and lattice vibrations for the energetic material cyclotrimethylene trinitramine. *Journal of Applied Physics* **2006**, *100* (5), 053508-1–053508-9.

47. Rice, B. M., Chabalowski, C. F., *Ab initio* and nonlocal density functional study of 1,3,5-trinitro-s-triazine (RDX) conformers. *Journal of Physical Chemistry A* **1997**, *101* (46), 8720–8726.
48. Goetz, F., Brill, T. B., Laser Raman-spectra of alpha-octahydro-1,3,5,7-tetra-nitro-1,3,5,7-tetrazocine, beta-octahydro-1,3,5,7-tetranito-1,3,5,7-tetrazocine, gamma-octahydro-1,3,5,7-tetranitro-1,3,5,7-tetrazocine, and delta-octahydro-1,3,5,7-tetranitro-1,3,5,7-tetrazocine and their temperature-dependence. *Journal of Physical Chemistry* **1979**, *83* (3), 340–346.
49. Brand, H. V., Rabie, R. L., Funk, D. J., Diaz-Acosta, I., Pulay, P., Lippert, T. K., Theoretical and experimental study of the vibrational spectra of the alpha, beta, and delta phases of octahydro-1,3,5,7-tetranitro-1,3,5,7-tetrazocine (HMX). *Journal of Physical Chemistry B* **2002**, *106* (41), 10594–10604.
50. Iqbal, Z., Bulusu, S., Autera, J. R., Vibrational-spectra of beta-cyclotetramethylene tetranitramine and some of its isotopic isomers. *Journal of Chemical Physics* **1974**, *60* (1), 221–230.
51. Zhu, W. H., Xiao, J. J., Ji, G. F., Zhao, F., Xiao, H. M., First-principles study of the four polymorphs of crystalline octahydro-1,3,5,7-tetranitro-1,3,5,7-tetrazocine. *Journal of Physical Chemistry B* **2007**, *111* (44), 12715–12722.
52. Waterland, M. R., Kelley, A. M., Far-ultraviolet resonance Raman spectroscopy of nitrate ion in solution. *Journal of Chemical Physics* **2000**, *113* (16), 6760–6773.
53. Waterland, M. R., Stockwell, D., Kelley, A. M., Symmetry breaking effects in NO_3^-: Raman spectra of nitrate salts and *ab initio* resonance Raman spectra of nitrate–water complexes. *Journal of Chemical Physics* **2001**, *114* (14), 6249–6258.
54. Measures, R. M., *Laser Remote Chemical Analysis*. John Wiley and Sons: New York, 1988.
55. Asher, S. A., Johnson, C. R., Raman-spectroscopy of a coal liquid shows that fluorescence interference is minimized with ultraviolet excitation. *Science* **1984**, *225* (4659), 311–313.
56. Rigaku, Xantus: Advanced portable Raman spectrometer. Rigaku. http://www.rigaku.com/products/raman/xantus.
57. Enwave Optronics, Inc. Spectroscopy **2012**, *27* (12), 47. Available online at http://www.spectroscopyonline.com/spectroscopy/Corporate+Profiles/Enwave-Optronics-Inc/ArticleStandard/Article/detail/806287.
58. Ray, M. D., Sedlacek, A. J., Wu, M., Ultraviolet mini-Raman lidar for stand-off, *in situ* identification of chemical surface contaminants. *Review of Scientific Instruments* **2000**, *71* (9), 3485–3489.
59. Wu, M., Ray, M., Fung, K. H., Ruckman, M. W., Harder, D., Sedlacek, A. J., Stand-off detection of chemicals by UV Raman spectroscopy. *Applied Spectroscopy* **2000**, *54* (6), 800–806.
60. Fleger, Y., Nagli, L., Gaft, M., Rosenbluh, M., Narrow gated Raman and luminescence of explosives. *Journal of Luminescence* **2009**, *129* (9), 979–983.
61. Sharma, S. K., Misra, A. K., Clegg, S. M., Barefield, J. E., Wiens, R. C., Acosta, T. E., Bates, D. E., Remote-Raman spectroscopic study of minerals under supercritical CO_2 relevant to Venus exploration. *Spectrochimica Acta Part A: Molecular and Biomolecular Spectroscopy* **2011**, *80* (1), 75–81.
62. Sharma, S. K., Misra, A. K., Lucey, P. G., Lentz, R. C. F., A combined remote Raman and LIBS instrument for characterizing minerals with 532 nm laser excitation. *Spectrochimica Acta Part A: Molecular and Biomolecular Spectroscopy* **2009**, *73* (3), 468–476.
63. Exelis, Joint contaminated surface detector (JCSD). Exelis. http://www.exelisinc.com/solutions/joint-contaminated-surface-detector-JCSD/Pages/default.aspx.
64. Alakai. http://alakaidefensesystems.com/.
65. Photon Systems. http://www.photonsystems.com/home.

66. Bhartia, R., Hug, W. F., Reid, R. D., Improved sensing using simultaneous deep UV Raman and fluorescence detection. *Proceedings of SPIE* **2012**, *8358*, 83581A.

67. Moore, D., Optimal coherent control of sensitivity and selectivity in spectrochemical analysis. *Analytical and Bioanalytical Chemistry* **2009**, *393* (1), 51–56.

68. Maker, P. D., Terhune, R. W., study of optical effects due to an induced polarization third order in electric field strength. *Physical Review* **1965**, *137* (3A), A801.

69. Owyoung, A., Jones, E. D., Stimulated Raman spectroscopy using low-power CW lasers. *Optics Letters* **1977**, *1* (5), 152–154.

70. Brady, J. J., Farrell, M. E., Pellegrino, P. M., Discrimination of chemical warfare simulants via multiplex coherent anti-Stokes Raman scattering and multivariate statistical analysis. *Optical Engineering* **2014**, *53* (2), 021105.

71. Brady, J. J., Pellegrino, P. M., Next generation hazard detection via ultrafast coherent anti-Stokes Raman spectroscopy. *Proceedings of SPIE* **2013**, *8710*, 87100Q.

72. Bremer, M. T., Wrzesinski, P. J., Butcher, N., Lozovoy, V. V., Dantus, M., Highly selective standoff detection and imaging of trace chemicals in a complex background using single-beam coherent anti-Stokes Raman scattering. *Applied Physics Letters* **2011**, *99* (10), 101109.

73. Moore, D. S., McGrane, S. D., Greenfield, M. T., Scharff, R. J., Chalmers, R. E., Use of the Gerchberg-Saxton algorithm in optimal coherent anti-Stokes Raman spectroscopy. *Analytical and Bioanalytical Chemistry* **2012**, *402* (1), 423–428.

74. Rabitz, H. A., Hsieh, M. M., Rosenthal, C. M., Quantum optimally controlled transition landscapes. *Science* **2004**, *303* (5666), 1998–2001.

75. McCain, S. T., Guenther, B. D., Brady, D. J., Krishnamurthy, K., Willett, R., Coded-aperture Raman imaging for standoff explosive detection. *Proceedings of SPIE* **2012**, *8358*, 83580Q.

6 Standoff Detection of Explosive Residue via Photodissociation Followed by Laser-Induced Fluorescence

Charles M. Wynn

CONTENTS

6.1 DETECTION OF TRACE EXPLOSIVES

Events in recent years have led to an increased need for improvements in our ability to detect explosives. In numerous situations, the capability to detect explosives sensitively, accurately, and rapidly could have great benefits to national security both at home and abroad. Some of the homeland situations that would benefit from improvements in explosives detection include screening passengers and luggage at airports and other sensitive locations and screening vehicles and people along the perimeters of high-value installations such as federal buildings. Abroad, the improvised explosive device (IED) problem is clearly in need of creative solutions to mitigate the devices' very damaging effects. Much research has been conducted to help solve these problems; however, it is unlikely that any one solution will suffice.

If we view the problem of explosive devices as a time line, from the initial planning to construction of the devices and ultimately to their detonation, we can see that it is preferable to detect the activities as early in the process as possible. The term "left of boom" derives from displaying the explosives creation-to-detonation time line from left to right and is sometimes used to refer to the concept of early explosives detection. A potential means of early identification is the detection of quantifiable residues left behind as a result of bomb-making activities. Such residues could be used as highly reliable cues of nearby explosive threats. Trace detection techniques could be used forensically, identifying devices and bomb-related activities, such as assembly, earlier in the process, thus preventing the need for more difficult countermeasures later on.

It is well known that the handling of explosives results in the spread of trace amounts of explosive particulates, yielding contamination levels from low to mid ng/cm^2 to in excess of 10 $\mu g/cm^2$.[1-3] A technique capable of detecting explosive residue is thus of great interest.[4,5] To be useful, a technique must not only display sensitivity to the levels noted above but must also have immunity to false positives from common nonexplosives, the ability to rapidly scan wide areas (since locations of explosives are unknown), and minimal spectral or algorithmic fine tuning requirements across a broad range of explosives. Despite much research,[6-9] a technique *simultaneously* demonstrating the three characteristics of (1) sensitivity, (2) rapid, remote detection, and (3) immunity to false positives has not been robustly demonstrated to date.

Photodissociation followed by laser-induced fluorescence (PD-LIF) is an optical means of explosive detection. With the proper choice of laser wavelength (deep ultraviolet [UV]), PD-LIF can be used to generate energetic nitric oxide fragments from certain nitro-bearing explosives such as 2,4,6-trinitrotoluene (TNT), which are used for detection of the parent explosive species. High-sensitivity (~ng/cm^2) detection has been demonstrated for numerous nitro-bearing explosive species. Detection occurs rapidly, generally within several laser pulses (~7 ns each). Since the photons utilized for detection have higher energy than the source laser, the technique has minimal response to nonexplosive background materials; ambient fluorescence also poses less of an issue than for other techniques using UV lasers.

6.2 PHOTODISSOCIATION FOLLOWED BY LASER-INDUCED FLUORESCENCE (PD-LIF) CONCEPT

When the electromagnetic energy of a photon is absorbed by a target molecule and converted into internal energy, there is a possibility that the target molecule will fragment or break apart into smaller atoms or molecules. This can occur if the transferred energy is greater than the binding energy of a relevant molecular bond.[10] This process, known as photodissociation (PD),[11] can involve a single photon or multiple photons, depending on their energies. UV photons are particularly useful for performing PD because of their energy (several electron volts).

Laser-induced fluorescence (LIF) involves the utilization of a laser at a particular wavelength to probe a target molecule.[12,13] Photons of a particular wavelength interrogate a target molecule, and the resultant fluorescent photons are detected, revealing information about the internal energy levels of the target molecule. Both the interrogation wavelength and the detected photon wavelength can be varied, each yielding information particular to the energy structure of the molecule of interest. LIF can be particularly effective when probing smaller molecules that possess strong and distinct optical transitions.

The combination of the above two processes, PD-LIF, is a promising explosives detection method that offers the potential to achieve the simultaneous requirements noted in Section 6.1.[14,15] This overall process is sometimes referred to by slightly different names, for example, photofragmentation followed by laser-induced fluorescence (PF-LIF). When applied to the explosives detection problem, PD-LIF utilizes a pulsed deep-UV laser. The choice of a deep-UV laser wavelength is dictated by the absorption spectra of the materials of interest. In order to be effective, PD-LIF must efficiently couple energy into the explosives via absorption of the incident photons. The absorption spectra of typical military-grade explosives are shown in Figure 6.1.[15] The strongest absorption generally occurs below ~250 nm; hence the need for a deep-UV laser when detecting military-grade explosives. If the wavelength, pulse width, and intensity of the laser are chosen properly, PD-LIF detection of trace explosives occurs very rapidly, potentially within a single (ns duration) laser pulse. The choice of these operating parameters is an area of active research in which signal-to-noise optimization must be balanced against other factors such as availability of reliable, fieldable technology solutions.

The majority of work using PD-LIF to detect explosives has focused on nitro-bearing species in which fragments of nitric oxide (NO) are used for detection.[16,17] While it is conceivable that PD-LIF could be used to detect other explosives (such as the peroxide-based family of homemade explosives) via the detection of oxygen–hydrogen (OH) fragments, no such measurements have been made to date.

As is evident from Figure 6.1, the absorption spectra of explosives of interest are fairly broad, with few distinct features easily amenable to a detection scheme. Instead of relying upon the broad, relatively featureless spectra of the polyatomic explosive solids, PD-LIF instead uses a dissociation product, NO, for detection. This yields several very important advantages. First, by operating on a gaseous diatomic molecule, one can take advantage of its very distinct spectrum

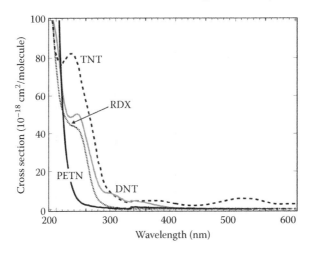

FIGURE 6.1 Absorption spectra of various military-grade explosives. TNT, 2,4,6-trinitrotoluene; DNT, 2,4-dinitrotoluene; RDX, hexahydro-1,3,5-trinitro-1,3,5-triazine; PETN, pentaerythritol tetranitrate.

for detection (see Figure 6.2). It is well known that most military-grade explosives have very low vapor pressures,[18] and thus detection techniques that rely upon that ambient vapor pressure suffer accordingly. The second advantage of the photo-dissociation process of PD-LIF is that it creates its own vapor instead of relying upon the ambient vapor pressure of the explosives themselves. Finally, as will be discussed below, many of the NO fragments are created with additional vibrational energy (see Figure 6.2). Exceedingly little ambient NO exists in this higher-energy state. This is perhaps the most important advantage of the dissociation process, in that it allows one to discriminate between NO resulting from explosives and NO created via other means. Also, as will be discussed below, this allows the detection to occur at wavelengths shorter than that of the illuminating laser, thus eliminating interference from traditional fluorescence processes.

PD-LIF, in the context of standoff explosives detection, requires the near-simultaneous occurrence of the following steps (Figure 6.3):[15] (1) vaporization of the explosives molecules residing at or near a surface; (2) photoinduced dissociation of the explosives molecules, generating appropriate photofragments; (3) resonant excitation of the vibrationally excited photofragments; and (4) fluorescence from the excited electronic state of the photofragments, to be detected by a suitable optical system. Typically, all four steps occur within the span of a single laser pulse, which lasts several nanoseconds.

When nitro-bearing explosives are illuminated by UV light of a sufficient intensity, the molecules both vaporize and dissociate (Figure 6.3), creating, among other things, fragments of NO. The process is very rapid, orders of magnitude faster than the typical nanosecond laser pulse used in PD-LIF detection systems.[19] Optimum dissociation occurs at the peak absorption of the material of interest, hence the use of deep-UV lasers. Note, however, that the absorption spectra are relatively broad and a highly specific laser wavelength is not required for dissociation to occur. In contrast, a very specific laser wavelength is required for resonant excitation. Overall

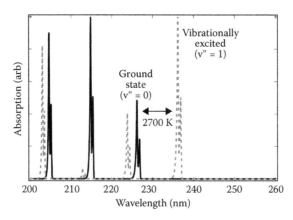

FIGURE 6.2 Absorption spectrum of NO. Both the ground state (v"=0; black) and first vibrationally excited state (v" = 1; gray dashes) are shown. The first vibrationally excited state can be created using a deep-UV laser to dissociate nitro-bearing explosives, as is done in PD-LIF.

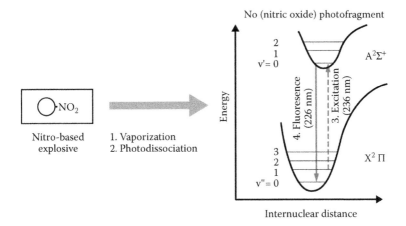

FIGURE 6.3 Schematic of the PD-LIF process. One possible excitation and emission path is shown. Others may also be viable.

system requirements may dictate which laser wavelength to use for the absorption step. If one is interested in a simpler system composed of only one laser, then the dissociation laser wavelength is chosen to match the resonant excitation wavelength. A schematic of a one-laser detection system is shown in Figure 6.4. Alternatively, it may be desirable to choose a more readily available laser (such as an excimer at 248 nm or a quadrupled neodymium:yttrium aluminum garnet [Nd:YAG] at 266 nm) to be used for the dissociation step and couple it with a smaller laser to be used for resonant excitation. Generally, the approach chosen to date has been to opt for a simpler one-laser solution in which the laser wavelength is dictated by the resonant excitation requirement. The resultant detection system is thus one in which a single source wavelength is used, and photons at a single wavelength are detected.

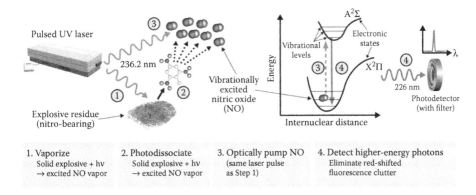

FIGURE 6.4 Schematic of PD-LIF detection setup including the two primary components: a pulsed UV laser and a single-photon sensitive detector with optical filter. Labeled steps correspond to those discussed in Figure 6.3.

As illustrated in Figure 6.4, resonant excitation (Step 3) requires the choice of a very specific wavelength. It is the use of this wavelength that helps give the technique its specificity. Step 3 can be accomplished using several wavelengths, depending on the choice of the lower and upper molecular energy levels. Transitions to the electronic excited state $A^2\Sigma$ are in the UV, and the resultant radiative decay back to the $X^2\Pi$ state, Step 4, would also be in the UV. In order to increase the sensitivity of detection, it would be preferable to excite the vibrational state with the highest nascent population following photodissociation. This is the vibrational ground state, $v'' = 0$. However, excitation of $v'' = 0$ leads to fluorescence at either the excitation wavelength itself or at wavelengths longer than the excitation wavelength, both scenarios leading to increased risk of false alarms. At the laser wavelength, scattered photons from the exciting laser will swamp any fluorescence from NO; false alarms may also be generated by atmospheric NO, which exists in the vibrational ground state under ambient conditions. At wavelengths longer than the excitation wavelength, many naturally occurring compounds will exhibit similar UV-induced fluorescence (including atmospheric NO and NO_x). One strategy to overcome this problem is to probe $v'' > 0$ states, since now some of the fluorescence will take place at wavelengths shorter than the excitation wavelength, and this is an unusual enough occurrence under normal circumstances to significantly reduce the probability of false alarms. Note that Boltzmann statistics predict that the $v'' = 1$ state of NO will have a relative population of $\sim 10^{-5}$ at room temperature. Excitation of $v'' > 0$ states is, of course, predicated on the assumption that these states are populated to some degree when the parent molecule is dissociated. As will be discussed, many of these states are indeed populated to some appreciable level during UV dissociation.

The final step in the detection process is that of detection of the fluorescent photons. The wavelength of the fluorescent photons depends on the resonant excitation wavelength chosen, or more particularly, the upper state that is accessed. The vibrational ground state of the upper electronic state has been the focus of most investigations to date such that the fluorescence emission is at 226 nm. There is no *a priori* reason (other than perhaps the availability of commercial laser sources) that excitation to other vibrational states cannot be used and an alternate fluorescence emission observed. Section 6.5 includes a brief discussion of alternate pathways that have been investigated. Regardless of which upper state is accessed, the resultant fluorescent photons will have more energy than the laser photons; they will be "blue-shifted".[20] This represents a significant advantage of the PD-LIF technique in that the technique is naturally much more immune to optical clutter (i.e., background interference) than many other optical detection techniques.

Laser-induced clutter can be subdivided into three broad spectral categories: laser scatter, that is, photons at the laser wavelength; red-shifted clutter (photons with less energy and longer wavelengths than the laser); and blue-shifted clutter (photons with more energy and shorter wavelengths than the laser). Generally, the scattered photons (at the laser wavelength) are the strongest source of clutter. Fluorescence processes (the dominant form of red-shifted clutter) can yield signals as strong as $\sim 10\%$ of the laser scatter, depending on the material and wavelengths chosen. Note that many materials fluoresce when irradiated with UV light. Photons that are created with more energy than the energy of incident photons are quite rare because they require the addition of excess energy. Thus, blue-shifted clutter is much weaker than either of the other two. Another potential detection technique that relies upon a blue-shifted signal (and shares a similar immunity as PD-LIF to the other forms of optical clutter) is known as coherent anti-Stokes Raman spectroscopy (CARS; refer to Chapters 5 and 13).[21]

In order to benefit from an immunity to photons generated out of band, it is necessary to reject all photons except those at the wavelength of interest (typically 226 nm for most experiments to date). The choice of the v" > 0 vibrationally excited state dictates the wavelength of the source laser. As shown in Figure 6.3, this wavelength can be as short as 236 nm if the v" = 1 state is chosen. The advantage of choosing the v" = 1 vibrational state is that it is the most populated state, and thus the overall signal is strongest at this wavelength.[22] The disadvantage is that an optical filter must now strongly reject laser photons (236 nm) that are only separated by 10 nm from the signal photons (226 nm). Given the various efficiencies of the process, detection is generally performed with a single-photon sensitive detector such as a photomultiplier tube. Given that the signal is so weak, it is imperative to have as high an optical throughput as possible at the signal wavelength. Additionally, the rejection ratio at the laser wavelength relative to the signal must be eight orders of magnitude or more given that the signal photons are small relative to the laser photons.[15] Filters meeting these requirements have been designed though generally with a lower transmission at the signal wavelength than desirable. Optimization of the filter and related choice of laser wavelength remains an engineering challenge, which, if improved appreciably, could significantly increase the performance and viability of the PD-LIF technique.

The fluorescence lifetime of the excited NO photofragment is much longer than the timescale of the dissociation/vaporization process that creates it. In practical situations under ambient conditions, the detectable PD-LIF signal will be limited not by the natural lifetime of excited NO but rather by collisional quenching of the NO primarily by oxygen, with a resultant timescale of ~1.5 ns under standard ambient conditions.[23] Ideally, the source laser pulse will be of this duration or shorter.

Noting the discussion in this section, a generalized PD-LIF detection system can now be defined. It will consist of three primary components: (1) an intense UV source producing light at a specific wavelength; (2) a sensitive photodetector (likely with ~ns time resolution); and (3) a narrowband UV filter. These components are all encapsulated within Figure 6.4. The photodetector is gated to respond only during the brief duration of the laser pulse (in order to minimize detection of ambient photons). The occurrence of photons at the photodetector during the laser pulse indicates explosives. Note that the background dark count rate of typical detectors will be exceedingly small on this timescale. Given the need for high-intensity UV light, it is likely that the light source will be a laser (see Section 6.5 for a discussion of laser options); however, depending on detection specifications and technology advances, it may be possible that light-emitting diodes or other sources are used in the future. The photodetector will most likely need near single-photon sensitivity, thus either a photomultiplier or an avalanche photodiode is used. Depending on the necessary working distances, focusing or collection optics may be required. A detailed description of a particular PD-LIF incarnation can be found in [15].

6.3 HISTORY AND CURRENT STATUS OF PD-LIF USED FOR EXPLOSIVES DETECTION

In the late 1990s, researchers from Mississippi State University reported experiments designed to detect TNT in soil and groundwater using PD-LIF.[24,25] These studies utilized a laser at 226 nm to probe fragmented NO created in its ground vibrational state (but electronically excited state) from explosives. The use of ground-vibrational-state NO for detection did not provide the immunity to background ambient NO or the ability to optically filter the source laser photons that come from using the excited vibrational states of the NO molecule for detection. In 1999, researchers from Ben-Gurion University utilized PD-LIF to probe gaseous nitro-bearing molecules including 2,4-dinitrotoluene (DNT). These experiments showed that the NO fragments were created with appreciable vibrational energy, and that at least the first three excited vibrational states ($v" = 1,2,3$) had significant populations.[22,26] In 2001, Arusi-Parpar et al., of Soreq NRC, performed PD-LIF measurements of TNT vapor showing clear evidence of vibrationally excited NO.[27] Excitation of its $v"=2$ state near 248 nm resulted in fluorescence at 226 nm.14 The formation of the vapor phase was accomplished not by the laser itself but rather by heating the solid or by other means. Cabalo and Sausa (Army Research Laboratory) addressed the very important question of how solid materials on substrates (as opposed to gases) respond to laser vaporization and dissociation.[28,29] Their 248 nm laser successfully formed NO photofragments from solids of energetic materials, but an alternate means of detection (not readily amenable to standoff detection) was used to detect them. Additionally, the fragments they detected were in their ground vibrational state ($v" = 0$).

In 2008, Wynn et al. (Massachusetts Institute of Technology, Lincoln Laboratory; MIT/LL) demonstrated all four steps of PD-LIF within a single laser pulse on solid samples of military-grade explosives.[15,20] They used an excitation source at 236.2 nm in order to access the $v" = 1$ vibrational state (highest excited state population) and maximize their signal strength. The ability to perform all four steps not only with a laser but within a single laser pulse indicated that a relatively simple system could be designed for measurements outside the laboratory (i.e., field measurements). Both the Soreq group[23] and the MIT/LL group[30] built and demonstrated systems suitable for remote field measurements. Estimates of detection ranges on the order of tens of meters were made for telescope-based systems. The MIT/LL group used their PD-LIF system to participate in a government-sponsored study of roughly 30 different remote trace-explosive detection techniques (most were Raman based). This double-blind study was coordinated by the Technical Support Working Group (TSWG) and executed by the Transportation Security Laboratory (TSL). The results of the study were limited to a range of government agencies.

Early PD-LIF studies focused primarily on military-grade nitro-bearing explosives. Successful detection of TNT, DNT, hexahydro-1,3,5-trinitro-1,3,5-triazine (RDX; the active component in C4 plastic explosives), and pentaerythritol tetranitrate (PETN) was observed.15 The PD-LIF signal for PETN was the weakest, due presumably to the fact that its absorption at the laser wavelength (236 nm) is weakest (see Figure 6.1). In 2010, the technique was extended to homemade explosives (HMEs) and their precursors.[31] Urea nitrate (UN) and the precursor materials nitromethane (NM) and ammonium nitrate (AN; a component of ammonium nitrate with fuel oil [ANFO] explosives) were successfully detected. In all cases, a vibrationally excited NO photofragment was detected (as was the case for the military-grade materials). The AN signal was much weaker (by ~2 orders of magnitude) than that of any other materials detected. This large difference was attributed to at least two factors related to the photochemistry of the compounds: differences in the photodissociation pathways leading to different yields of NO fragments; and (keeping in mind that the PD-LIF experiments do not sense NO in $v''=0$, even if the NO yield is the same) the dissociation pathways leading to vastly different vibrational temperatures. The relevant difference between AN and the other materials tested is that AN is a simple salt containing a nitrate (NO_3^-) ion. Presumably, the photodissociation of the nitrate ion into NO proceeds along very different pathways than the other nitro-bearing materials for which NO_2 groups are the relevant parent species. The fact that AN does generate a PD-LIF signal is of note, even if it is very weak. Presumably other simple nitrate salts will generate a PD-LIF signal, and these could be of concern as potential false alarms, though it should be noted that their signal levels are significantly lower than those of the nitro-bearing (NO_2-bearing) materials.

The UV dissociation of nitro-bearing explosives creates NO fragments not only with excess vibrational energy but also with excess rotational energy. These rovibrational states (with energy spacing less than those of the vibrational states) are not shown in Figure 6.3 for clarity; nonetheless, they can be accessed by choosing the proper laser wavelength. Using a tunable laser source and varying the excitation wavelength (which is the same as the dissociation wavelength), PD-LIF excitation spectra have been obtained. Typical excitation spectra are displayed in Figure 6.5. For these measurements, the excitation source was varied to span both the $v''=1$ (236.2 nm) and $v''=2$ (247 nm) vibrationally excited states. The detector was tuned to 226 nm (Step 4 of Figure 6.3 and 6.4). Overlaid upon the data (symbols) are fits (lines) yielding an effective rotational temperature for the NO. These fits were created using the modeling software LIFBASE.[32] They indicate that dissociated explosives yield NO fragments that are both vibrationally and rotationally hot. Rotational temperature estimates were well above room temperature, with values ranging between 500 and 2000 K.

In addition to the nitroaromatic ($C-NO_2$), nitramine ($N-NO_2$), and nitrate-based ($O-NO_2$) materials mentioned above, PD-LIF has successfully detected a class of furazan-based materials with different $N-O$ bond structure than the nitro-bearing or nitrate-based materials. Both the $v''=1$ and $v''=2$ vibrational states yielded strong PD-LIF signals. Excitation spectra are displayed in Figure 6.6.

Most common nonexplosive materials (especially those not containing NO) will not elicit any PD-LIF response at these wavelengths. This is because there is no ready mechanism for the creation of higher-energy photons at 226 nm.[20] Accordingly, the PD-LIF excitation spectra of common nonexplosive materials is near zero (being generally dark-count limited). An explosive detection system could thus utilize any of the excitation wavelengths (Figures 6.5 and 6.6) that access an excited NO rovibrational state and generate a 226 nm photon. As seen in Figures 6.5 and 6.6, the PD-LIF response is maximized at 236.2 nm, thus this wavelength has been the focus of considerable interest.

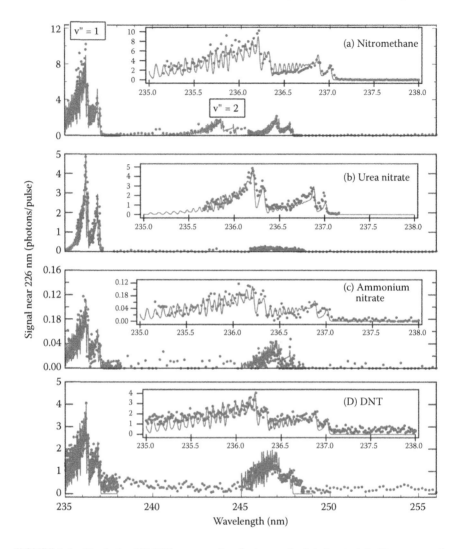

FIGURE 6.5 Excitation PD-LIF spectra of explosives and related materials. Data were collected by varying the source laser, while recording the signal at 226 nm. Both the first vibrationally excited ($v'' = 1$; near 236 nm) and second vibrationally excited ($v'' = 2$; near 247 nm) states of NO were probed. Lines are fits assuming a fixed rotational energy.

Since nonexplosive materials containing N and O could dissociate into NO fragments, and these fragments could cause a false-positive signal, it is of interest to determine the rotational and vibrational temperature of their NO dissociation by-products. Guo and others have performed gas-phase measurements on the dissociation products of both explosives and so-called "model compounds," by which they mean compounds containing N and O that are not explosives or fuels.[33,34] They observed that RDX, octahydro-1,3,5,7-tetranitro-1,3,5,7-tetrazocine (HMX), and 2,4,6,8,10,12-hexanitro-2,4,6,8,10,12-hexaazaisowurtzitane (CL20) (all explosives)

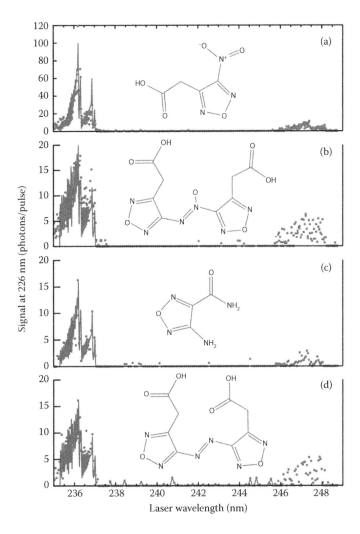

FIGURE 6.6   Excitation PD-LIF spectra of furazan-based materials. Data were collected by varying the source laser, while recording the signal at 226 nm. Both the first vibrationally excited (v″ = 1; near 236 nm) and second vibrationally excited (v″ = 2; near 247 nm) states of NO were probed. Chemical structures for materials are included as insets. (a) (4-Nitro-furazan-3-yl)-acetic acid; (b) (4-(N'-(4-carboxylmethyl-furazan-3-yl)-ONN-azoxy)-furazan-3-yl)-acetic acid; (c) 4-amino-furazan-3-carboxylic acid amide; (d) (4-(4-carboxymethyl-furazan-3-ylazo)-furazan-3-yl)-acetic acid. Lines are fits assuming a fixed rotational energy.

yielded vibrationally hot NO upon photodissociation (consistent with the PD-LIF results discussed above). In contrast, the "model compounds," including materials such as nitropyrrolidine and dimethylnitramine, yielded vibrationally cold NO. Thus, one would expect that at least this class of nonexplosive materials will not cause false positives for a PD-LIF explosives detection system.

The detection sensitivity of a PD-LIF system using a 236.2 nm excitation and 226 nm emission was estimated using drop-cast fingerprints of TNT solution.[35] Samples of a wide range of mass concentrations of TNT per unit area (C) were prepared via drop-casting from dilute solutions containing varying amounts of TNT in spectroscopic-grade acetone. Ten microliter aliquots were delivered onto clean silicon (Si) wafers and the solvent was allowed to evaporate. The residue deposited following solvent evaporation tended to form in a "coffee-ring" pattern in which the rim of the residue was denser than its interior. Individual residue areas measured 1–2 cm². For each residue pattern, six PD-LIF measurements spanning the residue diameter were made, and an average PD-LIF signal was obtained. Each PD-LIF measurement used an average of six laser pulses (0.2 s measurement time). For each C, three to four residue samples were created, and an average over these replicates was obtained. The results are displayed in Figure 6.7. Note that the background signal (acetone on bare Si) was 0.02 photons/pulse. These data indicate that for benign substrates such as Si, the sensitivity of PD-LIF to TNT is on the order of ng/cm² (signal of ~1 photon/pulse compared to background of only 0.02 photons/pulse). This work also deduced that the PD-LIF signal depends directly on the areal coverage (sometimes also referred to as fill factor) of the sample. It was noted that the fill factor for a drop-cast fingerprint differs from that of an actual fingerprint for the same mass loading. Drop-casting generally leads to a larger fill factor for a given mass loading than the fill factor of a corresponding actual fingerprint of the same mass load. Thus, one must be careful when estimating a detection sensitivity to specify not only mass loading but also means of preparation (refer to Chapter 3), which can significantly change the response of an optical-based technique such as PD-LIF.

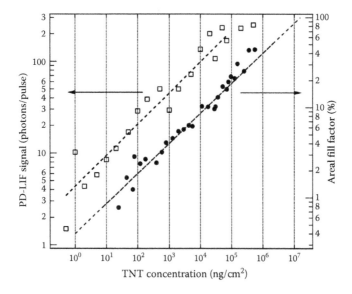

FIGURE 6.7 PD-LIF signal as a function of TNT mass concentration (left axis) for drop-cast TNT fingerprints. Areal fill factor of drop-cast TNT fingerprints as a function of TNT mass concentration (right axis).

6.4   ADVANTAGES/DISADVANTAGES OF PD-LIF AS APPLIED TO EXPLOSIVE DETECTION

A number of potentially related factors should be considered when comparing stand-off techniques, including speed, sensitivity, selectivity (immunity to false positives), and practicality of a field system (eye/skin safety; cost/availability/maturity of critical technology components; size, weight, and power (SWAP)/portability) (refer to Chapter 2).

The efficiency with which laser photons can be converted into signal photons affects the overall system detection speed and scan rate. The TNT cross section for PD-LIF was estimated by MIT/LL[15] and compared to Raman cross sections available at the time; the results are shown in Figure 6.8. Because PD-LIF is a multiphoton technique, its effective cross section will vary with laser intensity, whereas the Raman process is linear with a cross section independent of laser intensity. The analysis indicates an advantage in signal strength for PD-LIF as compared to traditional Raman techniques over the range of fluences tested. It should be noted, however, that the Raman cross sections would be expected to increase for alternative Raman-based techniques such as CARS or stimulated Raman;[36] however, no cross sections are presently available for these techniques. This relatively strong PD-LIF signal enables a relatively quick detection time. Researchers generally report data from only a handful of short pulses, translating into measurement times of a fraction of a second.

Because the PD-LIF signal is blue-shifted, its optical clutter is low, especially as compared to techniques that rely upon a signal at the laser wavelength or at a red-shifted wavelength (e.g., Raman). Techniques (such as Stokes-shifted Raman) that rely upon a red-shifted return can suffer from interference from the fluorescence of

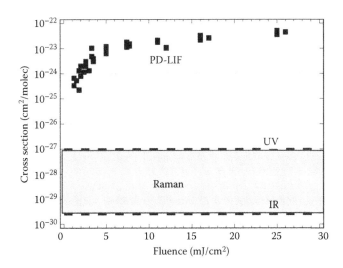

FIGURE 6.8   Effective PD-LIF cross section for TNT as a function of laser fluence. Also included are the Raman cross sections for both UV and infrared (IR) excitation wavelengths (dashed lines). Note that the Raman cross sections are constant with laser fluence.

many nonexplosive background materials, especially if a UV source laser is used. Generally, one would expect it to be similar to that observed by other techniques relying upon a blue-shifted signal such as an anti-Stokes Raman, which is used in CARS techniques. Researchers have observed that the PD-LIF background signal from a "clean" non-nitro-bearing substrate is not always detector-noise limited as might be expected but rather determined by a very weak, broadband spectrum of photons for some substrates. Similar continuum background signals have been observed in CARS experiments.[37]

Realistic signatures of explosives generally do not occur as uniform films, instead tending to be collections of sparse particulates (refer to Chapter 8).[38] Depending on the nature of the signature (e.g., footprint, fingerprint, spill), it is quite conceivable that the areal fill factor will be well below 1%. In other words, a large majority of the area interrogated will contain benign background material and not the explosive of interest. In order for a technique to successfully detect trace explosives while minimizing its false positives, it must have a very low response to background materials and/or have resolution sufficient to resolve the individual explosive particulates, which for fingerprints are on the order of 10 μm.[38] As already mentioned, PD-LIF has a very low background response to most materials due to its use of a blue-shifted signal for detection. Current incarnations of PD-LIF systems utilize collection optics that collect light from a wide field of view into a single-element photodetector (e.g., photomultiplier tube). The spatial resolution comes from scanning a laser within this field of view. Because there is essentially no background signal from any of the scene other than that directly interrogated by the laser, the system resolution is dictated by the laser spot size. This is in contrast to systems that rely upon a traditional camera-based solution with focal plane array detector to achieve resolution. PD-LIF imaging systems have successfully resolved features within actual fingerprints. One such example of a second-generation fingerprint imaged by a system using a 2 mm diameter laser spot is shown in Figure 6.9.[35]

Currently, there are no published results comparing the sensitivities of the various explosives detection techniques against the same set of realistic signatures with correspondingly low fill factors. The closest such study is that performed by TSL and mentioned in Section 6.3. The results of that study, however, are not openly distributed. Table 6.1 is an effort to make a comparison based on different experiments performed under different conditions. Relevant differences are noted directly in the table. The reported fill factors in the table are best-case results based on the experiments performed. These results imply a distinct detection advantage for PD-LIF against real-world sparse signatures in that it can detect the lowest fill factor (most sparse) signatures of any of the techniques discussed in the table. This has implications for remote detection of many signatures, such as fingerprints, which have a fill factor of <1%.

Because PD-LIF utilizes an indirect species (NO) to indicate the presence of explosives, it is subject to chemical clutter or false alarms. Data to date indicate that nonexplosive nitrates will likely contribute a relatively small background signal (see above) that could reduce system sensitivity or cause false alarms. The few other nonexplosive materials containing NO that have been tested show no PD-LIF signal; however, the possibility exists that there are other materials that could serve as a potential source of

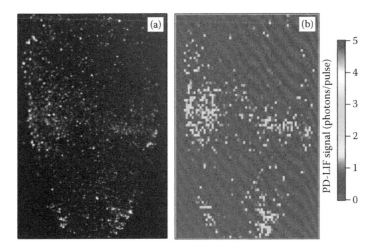

FIGURE 6.9 Images of a second-generation TNT fingerprint (areal coverage = 0.8%; estimated mass loading 2 $\mu g/cm^2$) on a Si wafer. (a) Optical image is an optical image obtained via a microscope using cross-polarizing filters to enhance the explosive contrast. (b) PD-LIF signal.

background signal. While ambient gaseous NO is not expected to generate a PD-LIF signal (since it exists primarily in its ground state), other ambient NO_x materials may dissociate into excited NO and could serve as a cause of background signal. In the case of these airborne contaminants, this signal could be mitigated by time-gating the return signal to collect photons only related to the range near the target.

The range of threats to which a technique can respond is important in many scenarios. To date, PD-LIF has only been demonstrated against materials containing both N and O. This represents a limitation, especially as compared to spectroscopic techniques such as those based on Raman or long-wave IR spectroscopy. Nonetheless it has been demonstrated against most military-grade threats of interest.

Eye safety is always a concern when considering laser-based techniques, especially if they will be used in the vicinity of people. PD-LIF utilizes a pulsed deep-UV

TABLE 6.1

Summary of Empirically Determined Threshold Fill Factors for Several Candidate Standoff Detection Approaches

Technology	Material	Substrate	Lowest Detectable Fill Factor (%)
PD-LIF[35]	TNT	Si	0.4
Active LWIR[39]	$KClO_3$	Car panel	3
Raman (UV, Vis)[40]	RDX	Polystyrene	34
Passive SWIR HSI[41]	NH_4NO_3	Car panel	> 50

Notes: LWIR, long-wave infrared; SWIR = shortwave infrared; HSI = hyperspectral imaging.

laser. At deep-UV wavelengths, the human safety levels are the same for the eye as for skin. Due to the nature of the damage mechanism (DNA damage), the risk is considered cumulative over the course of an 8 h day. The safe cumulative dose for an 8 h day is 3 mJ/cm^2.[42] Thus, to be truly safe, one would have to operate PD-LIF such that its single-pulse energy was only 3 mJ/cm^2 and ensure that only a single pulse (or less) interrogated any human bystander. To date, the pulse widths have been of the order of several ns (Q-switched laser system).[14,15,20,22,23,26,27,35,43] Noting the cross-section curves of Figure 6.8, this is feasible; however, signal strength would be weaker than it could be at higher fluence. Other techniques utilizing deep-UV lasers will be similarly restricted, while those utilizing visible, near, or long-wave IR will not have cumulative dose restrictions but will be limited in their peak intensity. Currently, commercial portable Raman systems (even in the visible/near IR) are not eye safe. These systems are not specifically designed for explosives detection nor do they perform well against trace quantities of materials.*

6.5 FUTURE

The successful implementation of a fieldable PD-LIF system depends largely on two critical technology components: a laser of suitable energy, repetition rate, pulse width, and wavelength; and a filter with sufficiently high throughput at the signal wavelength. These depend critically upon the choice of laser source wavelength and detection signal wavelength. PD-LIF (and related) experiments indicate that the $v'' = 1$ vibrational state has the highest excited-state population, suggesting that a source laser at 236.2 nm yields the strongest signal (see Figure 6.5). However, the use of a laser at 236.2 nm requires the need for an optical filter at the nearby wavelength of 226 nm, which can be challenging and may limit optical throughput. Other wavelength possibilities could include a system that promotes the electrons to the $v' = 1$ state instead of the $v' = 0$ state or a system that utilizes the $v'' = 2 \rightarrow v' = 0$ transition (these might increase filter throughput). Emission spectra demonstrating some possible alternative wavelengths are shown in Figure 6.10.[31] These demonstrate that both the $v' = 0$ and $v' = 1$ upper states can be utilized to generate a PD-LIF signal. A final system will have to trade the difficulty of achieving a particular laser wavelength against the throughput of its partner filter.

A 236.2 nm laser system was developed by John Zayhowski of MIT/LL for use in a portable system for the standoff detection of nitro-bearing explosive residues using PD-LIF.[43] The laser system is based on a frequency-quadrupled 944.8 nm regenerative amplifier that uses a compositionally tuned Nd-doped garnet as the gain medium. The output of the system is a 1 kHz train of near-diffraction-limited, 35 µJ, 236.2 nm pulses, with temporal and spectral characteristics tailored to the PD-LIF application. This is the only example, to date, of a laser system developed specifically for PD-LIF. Its kilohertz pulse repetition rate is significantly superior to commercial deep-UV systems, which are generally limited to 30 Hz repetition rates or below. A rapid pulse repetition rate

---

* Currently available Raman systems include Fido Verdict by ICX, FirstDefender by Ahura Scientific, and RespondeR RCI by Smiths. See product manuals for relevant information.

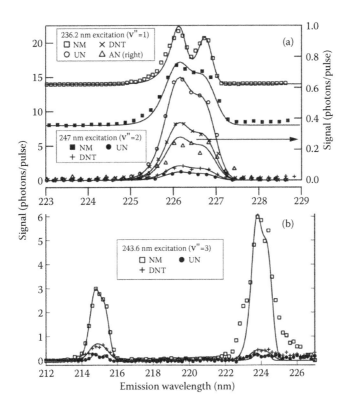

FIGURE 6.10 PD-LIF emission spectra for various explosives. Symbols are experimental data. NM data in (a) have been offset for clarity. Different excitation source wavelengths were used to access different vibrationally excited states of NO. Excitations in (a) are $(v''=1,2) \rightarrow (v'=0)$, whereas excitations in (b) are $(v''=3) \rightarrow (v'=1)$. Lines are simulated NO emission with $T_{Rot}=350$ K. In (a) lines correspond to $(v'=0) \rightarrow (v''=0)$ transition; in (b) lines correspond to $(v'=1) \rightarrow (v''=0,1)$ transitions.

directly impacts the effective areal scan rate of a detection system. Modifications to the system in order to improve its performance are being planned. Since the PD-LIF process involves multiple photons, the signal increases nonlinearly with laser intensity. Thus, it is presumed that a shorter laser pulse could significantly increase the PD-LIF signal strength.

The technology components required for PD-LIF share much with those required for a Raman-based detection system. A pulsed UV laser developed for PD-LIF could also be used for Raman-based detection. In fact, it would provide advantages over longer-wavelength lasers. Raman cross sections increase at shorter wavelengths, even more so as electronic resonances are approached (which tends to occur in the deep UV). Furthermore, background interference from unwanted fluorescence lessens for wavelengths below ~260 nm.[44] The spectrometer-based optical receive systems typically used for Raman detection generally have sufficient rejection of the laser photons required for PD-LIF. Thus, a combined PD-LIF/Raman detection system requires only the development and implementation of the 236.2 nm source laser. Such a system could leverage the advantages of both techniques to expand the variety of detected threats, while maintaining a high sensitivity to military-grade threats.

As already mentioned, PD-LIF has only been demonstrated against materials containing both N and O. It is possible that PD-LIF could be used against other explosive materials by utilizing a different photofragment for detection. Table 6.2 gives an overview of some important explosive materials and their potential detectability via PD-LIF.

In order to detect the peroxide-based materials, one might utilize the OH photofragment, while to detect the chlorate or perchlorate materials, one could use the

TABLE 6.2
Overview of Explosives and their Detectability via PD-LIF

Industrial-Grade Organonitrates			Homemade Explosives (HMEs)	
Nitroaromatic φ-NO$_2$	Nitramines N-NO$_2$	Nitrate Esters O-NO$_2$	Peroxides	Inorganics NO$_3^-$, ClO$_3^-$
2,4-DNT[a]	RDX[a]	PETN[a]	HMTD[b]	Ammonium nitrate/fuel oil[a]
2,6-DNT[a]	HMX[b]	NG[b]	TATP[c]	
DNB[a]		EGDN[b]	DADP[c]	Ammonium nitrate/ nitromethane[a]
TNT[a]		DNDMB[b]	H$_2$O$_2$ mixtures[c]	Urea nitrate[a]
TNB[b]			(i.e., airline liquid threats)	Chlorate/perchlorate variants[c]
Tetryl[b]				Metal (Al, Mg) powders[c]

- Largest common single charge is 155 mm artillery round (~10 kg)
- Suitable for antipersonnel and antivehicle operations
- Access to these materials limited to military
- Larger amount of raw materials obtained from quarrying stocks
- Example: Madrid train bombings

- Small, covert operations (<10 kg)
- Example: Richard Reid shoe bomber
- Example: London 7/7 bombings

- Common choice for large (>1000 kg) terrorist events
- Enables infrastructure attacks
- Example: Khobar Towers
- Example: Oklahoma City
- Example: World Trade Center I
- Example: African embassy bombings

Notes: DNB, dinitrobenzene; TNB, trinitrobenzene; NG, nitroglycerin; EGDN, ethylene glycol dinitrate; DNDMB, dinitro dimethylbutane; HMTD, hexamethylene triperoxide diamine; TATP, triacetone triperoxide; DADP, diacetone diperoxide

[a] Demonstrated PD-LIF detection.
[b] Expected PD-LIF detection (based on photochemistry).
[c] Requires detection channel alternate to PD-LIF: peroxides via OH fragment (280 nm) and chlorates/ perchlorates via ClO fragment (320 nm).

chlorine–oxygen (ClO) photofragment. For these techniques to be effective (as in the case of the NO-based detection technique), these photofragments must be created with additional energy in order to take advantage of a blue-shifted signal channel. Much work remains to be done before it is clear whether PD-LIF can successfully perform this detection. A number of studies discuss the photodissociation products and processes and suggest possible working laser wavelengths.[45–48]

ACKNOWLEDGMENTS

This work was sponsored by the US Army/ECBC under Air Force Contract FA8721-05-C-0002. Opinions, interpretations, conclusions, and recommendations are those of the authors and do not necessarily represent the view of the US government.

REFERENCES

1. P. Mostak, in *Vapour and Trace Detection of Explosives for Anti-Terrorism Purposes*, NATO Science Series II. Mathematics, Physics, and Chemistry, Vol. 167. M. Krausa and A. A. Reznev, eds. pp. 23–30. Kluwer Academic Publishers, Dordrecht, The Netherlands (2004).
2. J. C. Oxley, J. L. Smith, E. Resende, E. Pearce, and T. Chamberlain, Trends in explosive contamination, *J. Forensic Sci.* **48**, 334–342 (2003).
3. T. Tamiri, R. Rozin, N. Lemberger, and J. Almog, Urea nitrate, an exceptionally easy-to-make improvised explosive: Studies toward trace characterization, *Anal. Bioanal. Chem.* **395**, 421–428 (2009).
4. K. Yaeger, Dangerous innovations, in *Trace Chemical Sensing of Explosives*. R. Woodfin, ed., 43–67. Wiley, New York (2007).
5. D. S. Moore, Instrumentation for trace detection of high explosives, *Rev. Sci. Instrum.* **75**, 2499–2512 (2004).
6. A. Mukherjee, S. Von der Porten, C. Kumar, and N. Patel, Standoff detection of explosive substances at distances of up to 150 m, *Appl. Opt.* **49**, 2072–2078 (2010).
7. J. Steinfeld and J. Wormhoudt, Explosives detection: A challenge for physical chemistry, *Annu. Rev. Phys. Chem.* **49**, 230–232 (1998).
8. National Research Council, *Existing and Potential Standoff Explosive Detection Techniques*. The National Academies Press, Washington, DC (2004).
9. A. K. Goyal, M. Spencer, M. Kelly, J. Costa, M. DiLiberto, E. Meyer, and T. Jeys, Active infrared multispectral imaging of chemicals on surfaces, *Proc. SPIE* **8018**, 80180 N (2011).
10. R. Schinke, *Photodissociation Dynamics*. Cambridge University Press, Cambridge (1993).
11. R. N. Zare and D. R. Herschbach, Atomic and molecular fluorescence excited by photodissociation, *Appl. Opt.* **4**, 193–200 (1965).
12. R. N. Zare, My life with LIF: A personal account of developing laser-induced fluorescence, *Annu. Rev. Anal. Chem.* **5**, 1–14 (2012).
13. J. L. Kinsey, Laser-induced fluorescence, *Ann. Rev. Phys. Chem.* **28**, 349–372 (1977).
14. D. Helfinger, T. Arusi-Parpar, Y. Ron, and R. Lavi, Application of a unique scheme for remote detection of explosives, *Opt. Commun.* **204**, 327–331 (2002).
15. C. M. Wynn, S. Palmacci, R. R. Kunz, K. Clow, and M. Rothschild, Detection of condensed-phase explosives via laser-induced vaporization, photodissociation, and resonant excitation, *Appl. Opt.* **47**, 5767–5776 (2008).
16. S. Wallin, A. Pettersson, H. Östmark, and A. Hobro, Laser-based standoff detection of explosives: A critical review, *Anal. Bioanal. Chem.* **395**, 259–274 (2009).

17. S. SenGupta, H. P. Upadhyaya, A. Kumar, S. Dhanya, P. D. Naik, and P. Bajaj, Photodissociation dynamics of nitrotoluene at 193 and 248 nm: Direct observation of OH formation, *Chem. Phys. Lett.* **452**, 239–244 (2008).

18. P. Mostak, in *Vapour and Trace Detection of Explosives for Anti-Terrorism Purposes*, NATO Science Series II. Mathematics, Physics, and Chemistry, Vol. 167. M. Krausa and A. A. Reznev, eds. pp. 23–30. Kluwer Academic Publishers, Dordrecht, The Netherlands (2004).

19. J. C. Mialocq and J. C. Stephenson, Picosecond laser-induced fluorescence study of the collisionless photodissociation of nitrocompounds at 266 nm, *Chem. Phys.* **106**, 281–291 (1986).

20. C. M. Wynn, S. Palmacci, R. R. Kunz, and M. Rothschild, A novel method for remotely detecting trace explosives, *Linc. Lab. J.* **17**, 27–39 (2008).

21. A. Portnov, S. Rosenwaks, and I. Bar, Detection of particles of explosives via backward coherent anti-Stokes Raman spectroscopy, *Appl. Phys. Lett.* **93**, 041115 (2008).

22. J. Shu, I. Bar, and S. Rosenwaks, The use of rovibrationally excited NO photofragments as trace nitrocompound indicators, *Appl. Phys. B* **70**, 621–625 (2000).

23. T. Arusi-Parpar and I. Levy, Remote detection of explosives by enhanced pulsed laser photodissociation/laser-induced fluorescence method, in *Stand-Off Detection of Suicide Bombers and Mobile Subjects*, H. Schubert and A. Rimski-Korsakov, eds. pp. 59–68. Springer, Dordrecht, The Netherlands (2006).

24. D. Wu, J. Singh, F. Y. Yueh, and D. L. Monts, 2,4,6-trinitrotoluene detection by laser-photofragmentation-laser-induced-fluorescence, *Appl. Opt.* **35**, 3998–4003 (1996).

25. G. Boudreaux, T. S. Miller, A. J. Kunefke, J. P. Singh, F. Yueh, and D. L. Monts, Development of a photofragmentation laser-induced-fluorescence laser sensor for detection of 2,4,6-trinitrotoluene in soil and groundwater, *Appl. Opt.* **38**, 1411–1417 (1999).

26. N. Daugey, J. Shu, I. Bar, and S. Rosenwaks, Nitrobenzene detection by one-color laser-photolysis/laser-induced fluorescence of NO (v" = 0–3), *Appl. Spectr.* **53**, 57–64 (1999).

27. T. Arusi-Parpar, D. Helfinger, and R. Lavi, Photodissociation followed by laser-induced fluorescence at atmospheric pressure and 24°C: A unique scheme for remote detection of explosives, *Appl. Opt.* **40**, 6677–6681 (2001).

28. J. Cabalo and R. Sausa, Detection of hexahydro-1,3,5-trinitro-1,3,5-triazine (RDX) by laser surface photofragmentation-fragment detection spectroscopy, *Appl. Spectr.* **57** 1196–1199 (2003).

29. J. Cabalo and R. Sausa, Trace detection of explosives with low vapor pressure emissions by laser surface photofragmentation-fragment detection spectroscopy with an improved ionization probe, *Appl. Opt.* **44**, 1084–1091 (2005).

30. C. M. Wynn, S. Palmacci, R. R. Kunz, J. J. Zayhowski, B. Edwards, and M. Rothschild, Experimental demonstration of remote detection of trace explosives, *Proc. SPIE* **6954**, 695407 (2008).

31. C. M. Wynn, S. Palmacci, R. R. Kunz, and M. Rothschild, Noncontact detection of homemade explosive constituents via photodissociation followed by laser-induced fluorescence, *Opt. Expr.* **18**, 5399–5406 (2010).

32. J. Luque and D. R. Crosley, LIFBASE: Database and spectral simulation program, SRI International Report MP 99-009 (1999).

33. Y. Q. Guo, M. Greenfield, and E. R. Bernstein, Decomposition of nitramine energetic materials in excited electronic states: RDX and HMX, *J. Chem. Phys.* **122**, 244310 (2005).

34. Y. Q. Guo, M. Greenfield, A. Bhattacharya, and E. R. Bernstein, On the excited electronic state dissociation of nitramine energetic materials and model systems, *J. Chem. Phys.* **127**, 154301 (2007).

35. C. M. Wynn, S. Palmacci, R. R. Kunz, and M. Aernecke, Noncontact optical detection of explosive particles via photodissociation followed by laser-induced fluorescence, *Opt. Expr.* **19**, 18671–18677 (2011).

36. M. T. Bremer and M. Dantus, Standoff explosives trace detection and imaging by selective stimulated Raman scattering, *Appl. Phys. Lett.* **103**, 061119 (2013).

37. V. Kumar, R. Osellame, R. Ramponi, G. Cerullo, and M. Marangoni, Background-free broadband CARS spectroscopy from a 1-MHz ytterbium laser, *Opt. Expr.* **19**, 15143–15148 (2011).

38. J. R. Verkouteren, J. L. Coleman, and I. Cho, Automated mapping of explosives particles in composition C-4 fingerprints, *J. Forensic Sci.* **55**, 334–340 (2010).

39. A. Goyal, internal communication, MIT Lincoln Laboratory (2012).

40. A. Tripathi, E. D. Emmons, P. G. Wilcox, J. A. Guicheteau, D. K. Emge, S. D. Christesen, and A. W. Fountain III, Semi-automated detection of trace explosives in fingerprints on strongly interfering surfaces with Raman chemical imaging, *Appl. Spectrosc.* **65**, 611–619 (2011).

41. L. Scott, internal communication, MIT Lincoln Laboratory (2012).

42. Laser Institute of America, American National Standard for Safe Use of Lasers, ANSI Z136.1–2007. Laser Institute of America, Orlando, FL (2007).

43. J. J. Zayhowski, M. Rothschild, C. M. Wynn, R. R. Kunz, Detection of materials via nitrogen oxide, US Patent 8198095 B2 (June 2012).

44. S. A. Asher and C. R. Johnson, Raman spectroscopy of a coal liquid shows that fluorescence interference is minimized with ultraviolet excitation, *Science* **225**, 311–313 (1984).

45. K. Gerick, S. Klee, and F. J. Comes, Dynamics of H_2O_2 photodissociation: OH product state and momentum distribution characterized by sub-Doppler and polarization spectroscopy, *J. Chem. Phys.* **85**, 4463–4479 (1986).

46. F. Dubnikova, R. Kosloff, J. Almog, Y. Zeiri, R. Boese, H. Itzhaky, A. Alt, and E. Keinan, Decomposition of triacetone triperoxide is an entropic explosion, *J. Am. Chem. Soc.* **127**, 1146–1159 (2005).

47. C. Mullen, D. Huestis, M. Coggiola, and H. Oser, Laser photoionization of triacetone triperoxide (TATP) by femtosecond and nanosecond laser pulses, *Int. J. Mass Spectrom.* **252**, 69–72 (2006).

48. O. Johansson, J. Bood, M. Alden, and U. Lindblad, Detection of hydrogen peroxide using photofragmentation laser-induced fluorescence, *Appl. Spectrosc.* **62**, 66–72 (2008).

7 Detection of Explosives Using Laser-Induced Breakdown Spectroscopy (LIBS)

David A. Cremers

CONTENTS

7.1 INTRODUCTION

7.1.1 LIBS FOR EXPLOSIVES DETECTION

In the laser-induced breakdown spectroscopy (LIBS) technique, a powerful laser pulse is projected at a target to induce a set of optical signals characteristic of the target composition.[1,2] As laser light alone is used to induce the signals, the technique has many advantages for the detection of explosives and can be deployed in a variety of concepts of operations (ConOps). In principle, LIBS can be used to detect bulk explosives, explosive residues on surfaces, and explosives in the vapor form. LIBS can also be used to identify materials related to explosives deployment such as land-mine housing materials. The easiest target to identify is the bulk explosive, although the probability of having access to bulk materials is likely to be low. The ability to detect residues on surfaces opens new detection capabilities, as measurements can be carried out at some distance from the target and, by using repetitive laser pulses, large areas can be surveyed. In addition, LIBS can identify small residue masses (down to nanogram quantities) because of its point sampling capability. One example is the detection of residues on car surfaces (e.g., trunk lids, door handles) deposited by hand contamination during deployment of hidden improvised explosive devices (IEDs) in the vehicle. The use of LIBS to interrogate other surfaces such as shipping containers for explosive residues is also a possibility. It should be noted that in the application of LIBS for explosives detection, the main interest is in determining the presence of an explosive (within constraints of the detection limits) rather than quantification of amounts present. The recognition by a LIBS analysis of the presence of explosive residues would most likely serve as a trigger for further measurements, the use of other instrumentation for verification, or both.

7.1.2 ADVANTAGES OF **LIBS** FOR EXPLOSIVES DETECTION

The advantages of LIBS are a result of the unique properties of laser light used to generate the LIBS signal. Some of these advantages are shared by other laser-based methods of explosives detection (e.g., Raman spectroscopy). LIBS advantages include the following:

1. Only optical access to the sample is required.
2. Open path remote interrogation over distances of many tens of meters is possible by projecting the laser pulse onto the target, allowing
   (a) Access to remote targets (e.g., behind a fence, on a wall), and
   (b) A safe operating environment away from local hostile environments (e.g., chemically or radioactively contaminated areas)
3. Analysis results are rapid because the optical signals are induced immediately by the action of the laser light on the target (i.e., lack of sample preparation).
4. Targets can be interrogated that cannot be accessed directly through line of sight but through the use of fiber-optic delivery of laser pulses over many tens of meters.
5. The LIBS spectrum contains information about materials related to explosives, such as casings used in land mines and chemicals used in the manufacture of homemade explosives, thereby providing secondary, sometimes more telltale, optical signatures than the explosive signatures themselves.
6. LIBS can be deployed, in principle, as a universal hazardous materials detector and be programmed to detect nuclear materials (e.g., uranium, plutonium), radiological targets (e.g., americium, cobalt, strontium), and chem-bio agents along with explosives.
7. LIBS instruments can be tailored to fit the ConOps with instrumentation ranging from field-deployable transportable versions to person-portable units.

7.1.3   LIMITATIONS OF **LIBS** FOR EXPLOSIVES DETECTION

As LIBS uses powerful laser pulses to generate the interrogating plasma, safety considerations are paramount in its deployment, as the laser pulses represent skin and ocular hazards.[3] This is especially the case in the use of LIBS for open path standoff detection of remotely located targets. Care must be taken to ensure the operator and personnel in the vicinity of the instrument and the target are properly shielded from direct and reflected laser light. The wavelength predominantly used for LIBS, 1064 nm, is a particular hazard for the eye. Care must also be taken in the use of LIBS around explosive vapors (e.g., gasoline, methane, etc.), as under certain conditions these can be ignited in air. The ignition of explosives by different laser beams has been considered in detail.[4] In our work on a range of bulk explosives using single laser pulses of 25 mJ, we found that ignition was not an issue. Each case must be considered individually, however. The interrogation of microgram and nanogram quantities of explosive residues should not be an issue in terms of safety. In any event, safety concerns must be evaluated for each ConOps in which LIBS is to be deployed.

Besides the safety considerations, some limitations of LIBS for explosives detection include

1. Being a point detector, each pulse interrogates a very small area (focused spot diameter on the order of a few hundreds of microns), which will require the use of a large number of laser pulses to locate explosive residue over a wide area.
2. It is a destructive technique, typically requiring a new area to be interrogated by each laser pulse to record the strongest signals and if signal averaging is desired to provide positive detection.
3. The action of the laser pulse on a target can cause visible damage (e.g., to paint on a car).
4. "Clutter" species (also know as "interferents," "confusants") that may be found along with the explosive that contain the same elements may complicate positive identification (e.g., 2,4,6-trinitrotoluene [TNT] in the presence of diesel fuel, road dust, etc.).

Typically, LIBS can detect very small masses of some elements, down to femtogram quantities in a single shot (e.g., beryllium), whereas concentration detection limits are usually on the order of parts per million (e.g., 1 ppm = 1 μg of chromium/g soil). Because of the low vapor pressure of many common explosives (<0.01 Torr or <1 Pa at 25°C), LIBS detection of explosives in the vapor phase, although attractive for many ConOps, such as the detection of buried IEDs, is not feasible in most cases due to limited concentration detection sensitivity. LIBS analysis of vapors may prove useful for the detection of explosives containing taggants such as ethylene glycol dinitrate (EGDN) or para-mononitrotoluene (p-MNT), however, which have vapor pressures at least 10^4 times greater than those of many explosives.[5,6]

7.2 BACKGROUND

7.2.1 LIBS Technique

LIBS is a form of chemical analysis that utilizes a laser-generated plasma or spark to induce emissions from the target material. Among the suite of analytical chemical analysis methods, it is classified as a form of atomic emission spectroscopy (AES). In any type of AES analysis, there are three basic steps: (1) vaporization of the sample, (2) excitation of the resulting atoms and ions composing the sample, and (3) detection of emission from the excited atomic and ionic species. The detected light can also include emissions from simple molecules that are typically formed as the atoms recombine as the plasma cools (e.g., emission from the molecules cyanide [CN], diatomic carbon [C_2]). As each element and molecular species has a unique emission spectrum, analysis of the spectrum (intensity vs. wavelength) permits emitting species in the plasma to be identified. The very first excitation source (ca. 1860) used a simple Bunsen burner flame for excitation of elements easy to excite (e.g., lithium, calcium, strontium). Conventional laboratory methods of AES analysis use excitation sources that include the inductively coupled plasma (ICP) or the electrode spark (pulsed source) or arc (continuous source). These all utilize some physical device to deliver the excitation energy to the sample. The ICP uses a metal coil that delivers continuous radio-frequency energy to the sample introduced into an argon gas stream flowing through the coil. The spark and arc use metal electrodes between which a hot plasma is formed to excite the material. In contrast, in LIBS, the excitation energy is delivered by focused laser light, thereby accounting for many of the advantages of LIBS compared with conventional forms of AES analysis.

A laser spark formed on a painted surface by a single laser pulse is shown in Figure 7.1a. The plasma at the surface is about 2 mm wide. In addition to atomized material in the plasma, particles are ejected from the surface and can be seen above the plasma. Accompanying the visible bright light from the spark plasma is an audible snap due to the shock wave generated at the surface. The main physical processes involved in LIBS are diagrammed in Figure 7.1b with reference to analysis of a solid target in air. The laser pulse/material interaction begins when the laser pulse strikes the target surface and some fraction of the energy is absorbed. This absorption causes a rapid increase in temperature at the surface. At a sufficiently high temperature, a small amount of the target material is vaporized (nanogram to microgram masses, depending on the laser pulse characteristics and the material properties). As additional laser pulse energy is rapidly delivered to the target, a plasma is formed on the surface (Figure 7.1b). The plasma expands away from the surface in the direction of the incoming laser pulse due to absorption of remaining pulse energy at the air–plasma interface (inverse bremsstrahlung absorption). Typically, nanosecond laser pulses are used to form the LIBS plasma so that heating of the target and plasma formation occur in less than 10 ns. Due to the high initial temperature (e.g., 8000 K) and electron density (e.g., $>10^{16}$ electrons/cm^3) of the laser plasma, the atomized material ablated from the target is efficiently electronically excited, resulting in emissions from the excited species. After the laser pulse, the plasma begins to decay, decreasing in temperature and electron density. A typical useful lifetime of the plasma for

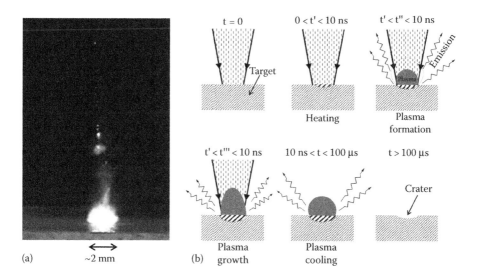

FIGURE 7.1 (a) Laser spark plasma formed on a painted surface by a single laser pulse. The width of the spark on the surface is about 2 mm. (b) Main steps in the laser pulse/material interaction resulting in the generation of LIBS emission signals.

observing spectral emissions is on the order of 20 μs, although some easy-to-excite atomic species can be observed at later times.[1]

A generalized temporal history of the LIBS plasma is shown in Figure 7.2. For a plasma formed on a target in air, emissions observed during the first microsecond are dominated by a spectrally broad continuum radiation due to recombination of free electrons with ionized species (bremsstrahlung radiation). Also observed during the first few microseconds after plasma formation are emissions from once-ionized atomic species (e.g., once-ionized calcium, Ca(II)). As the plasma cools and electrons recombine with ionized species, the background continuum intensity decreases and emissions from neutral atomic species become apparent. In addition, emissions can be observed from simple molecules formed as atoms recombine. A good example is emission from CN, typically observed in the analysis of organic materials such as explosives in nitrogen-rich air. Due to the temporal dependence of emissions from the laser plasma, time-resolved detection is commonly deployed to maximize the signal-to-background ratio. This is typically accomplished using a gated detector such as an intensified charge-coupled device (ICCD). Referring to Figure 7.2, t_d refers to the delay between the laser pulse incident on the target and the start of the detection of emissions, and t_b refers to the detection time or the period over which emissions are recorded.

As noted above, LIBS is a point detection method, as any single pulse samples a small spot on the target, an area on the order of 0.0003 cm². This results in the vaporization of typically less than 1 μg of material with an ablation depth on the order of microns. LIBS is a destructive interrogation method, although vaporization of only a small total mass is required for a complete analysis. The area sampled, mass ablated, and depth probed depend on the laser pulse parameters.[7]

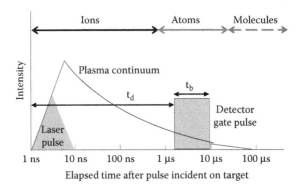

FIGURE 7.2 Temporal evolution of emissions from a laser plasma formed on a solid surface.

7.2.2 LIBS SIGNATURES FROM EXPLOSIVES

The structures and formulas of some common explosives are shown in Figure 7.3. These materials are largely composed of the ubiquitous elements C, N, O, and H (carbon, nitrogen, oxygen, and hydrogen, respectively), common to many organic materials, as well as species normally present in air, the atmosphere in which explosives are typically detected in real-world situations. In addition, the environments in which explosives detection can be expected to be deployed can contain inter-

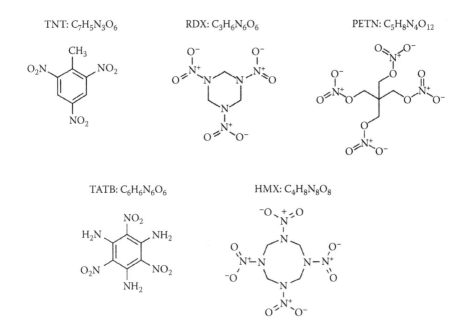

FIGURE 7.3 Formulas and structures of some common explosives. RDX: 1,3,5-trinitro-1,3,5-triazacyclohexane, hexagen; PETN: pentaerythritol tetranitrate; TATB: triaminotrinitrobenzene; HMX: octahydro-1,3,5,7-tetranitro-1,3,5,7-tetrazocine, octogen.

fering or "clutter" organic species having the same elemental constituents, making LIBS-based detection a challenge in these situations.

A LIBS spectrum (single shot) of TNT coated onto silica (SiO_2) is shown in Figure 7.4. Spectra of other explosives show the same spectral features. The presence of silica accounts for the silicon emission lines in the spectrum. Clearly evident are spectral features due to C, H, N, and O elements present not only in the explosive but also in air. Emission of the molecule CN is readily observed and is due to recombination of C and N atoms as the plasma cools. Emissions from minor impurity elements Na, Al, and Ca (sodium, aluminum, and calcium, respectively) are also apparent in the spectrum as these are common elements observed in many LIBS spectra.

As the LIBS spectrum represents emission from elements composing the target, along with emissions from simple molecules formed as the plasma cools, the use of LIBS for the detection of explosives requires an analysis of selected features in the LIBS spectrum. This includes line ratios of species such as C, H, N, and O, which can be attributed to specific explosive targets. It is expected that the measured line intensities for each element will be proportional to the number of atoms in each explosive so that by comparing the relative intensities it should be possible to discriminate between the different explosives. A strong correlation between atomic intensity line ratios and element stoichiometry has been shown for pure explosives and other organic materials.[8] On the other hand, it is clear that the presence of other organic materials, such as fuels and plant matter interrogated along with the explosive, will alter line ratios and complicate the analysis as discussed in Section 7.3.3.

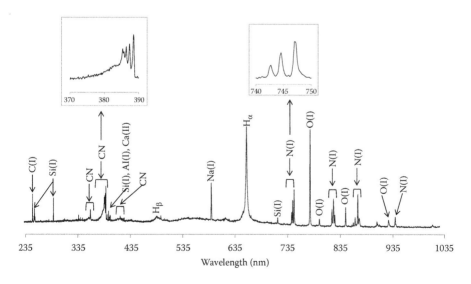

FIGURE 7.4 Single-shot LIBS spectrum recorded from TNT coated on silica. Laser pulse energy was 65 mJ and the detection delay time (t_d) was 1 μs. C, carbon; Si, silicon; CN, cyanide; Al, aluminum; Ca, calcium; H, hydrogen; Na, sodium; N, nitrogen; O, oxygen. The designation (I) refers to the neutral atom and (II) refers to the once-ionized atom.

In addition to analyses attributing specific line ratios to specific explosives, more detailed analyses can be carried out by applying chemometric analysis methods to line ratios or the entire LIBS spectrum.[9] The use of these methods will be discussed below.

As noted in the introduction, LIBS can directly interrogate samples in the form of solids and gases. This is due to the high power densities of the focused laser pulse used for LIBS. In the gas phase, entrained aerosols and vapors will be excited along with the gas. Detection of vapors would be useful for the detection of hidden explosives such as IEDs that may be buried by a roadside and triggered at the passing of a vehicle. Because of the low vapor pressures typical of many explosives, however, detection in the vapor phase, however attractive from a detection standpoint, is in general not feasible. Figure 7.5 lists vapor pressures of some bulk explosives. Also shown are vapor pressures for some *taggants* that may be added to the explosive for identification. These are volatile chemicals that slowly evaporate and are intended for detection in the atmosphere by either detection dogs or vapor-sniffing instruments. Taggants include 2,3-dimethyl-2,3-dinitrobutane (DMDNB) and EGDN. As a benchmark for the sensitivity of LIBS for detecting explosive vapors, the detection limit for carbon in air is on the order of 36 ppm, corresponding to a vapor pressure of 0.027 Torr.[10,11]

7.2.3 METHODS OF DEPLOYING LIBS DETECTION

As the LIBS plasma is formed by focused laser light, there are a number of deployment options. A generalized LIBS apparatus is diagrammed in Figure 7.6. Typically,

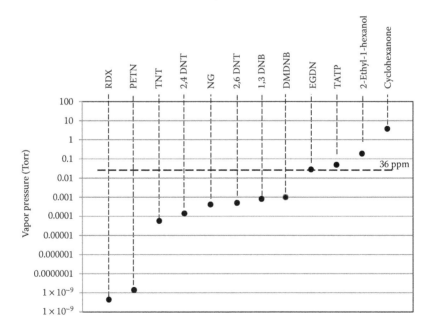

FIGURE 7.5   Vapor pressures of some explosives and related compounds. The dashed line indicates the LIBS limit of detection for carbon in air. DNT, dinitrotoluene; NG, nitroglycerine; DNB, dinitrobenzene; TATP, triacetone triperoxide.

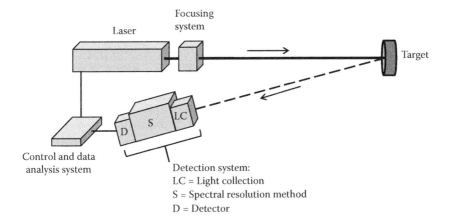

FIGURE 7.6 Generalized LIBS apparatus showing components common to most LIBS systems.

Q-switched nanosecond pulsed neodymium:yttrium aluminum garnet (Nd:YAG) laser systems are used for LIBS. Depending on the LIBS system, energies for a 10 ns laser pulse can range from a few millijoules to a few joules, with focused power densities of a gigawatt (10^9 W) or greater per centimeter squared. Typically, the fundamental wavelength (1064 nm) of the Nd:YAG laser is used for excitation, as the highest power density is available at this wavelength. In certain circumstances, however, harmonic wavelengths of the Nd:YAG laser (532, 355, 266 nm) may be used.

The laser pulse focusing system can be a simple single lens or a more complex, multielement lens system such as a Galilean telescope, which allows the focal position to be adjusted for targets at different distances. The light collection system can be a simple lens, fiber optic, or telescope that collects and directs the plasma light into a spectral resolution method (e.g., line filters, spectrograph), and the detector can be a light-sensitive device (e.g., photodiode, photomultiplier tube) or an array detector (e.g., ICCD, CCD), with the choice dependent on the method of spectral resolution.[1] The LIBS system can be controlled and data processed by a computer system with custom software for more automated analysis configurations. For measurements in air, the spectral range commonly detected extends from 200 to 1000 nm, encompassing the near-ultraviolet (UV), visible, and near-infrared (IR) spectral regions. Over this range, strong spectral features attributable to elements in explosives are observed (see Figure 7.4). Although there is a strong carbon emission at 193 nm, the transmission of light through the atmosphere at wavelengths below 200 nm is greatly reduced.

As noted in the introduction, the method of deploying LIBS can be tailored to a specific ConOps and options are diagrammed in Figure 7.7. Configurations (a) and (b) pertain to instruments that can provide close-up or remote analysis depending on the length of the cables (fiber and electrical) between the probe and the other system components.[12,13] Distances of tens of meters are possible with these configurations, and person-portable instruments for close-up analysis have been developed consisting of a handheld probe connected via umbilical cabling to a backpack, for

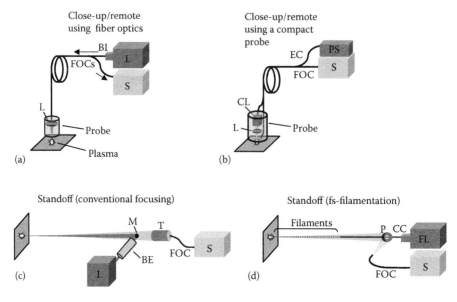

FIGURE 7.7 Deployment configurations for LIBS analysis. L, laser; S, spectrometer; BI, laser beam injection system for optical fiber; FOC, optical fiber; L, lens; M, mirror; T, telescope; BE, beam expander system for laser pulse; CL, compact laser; EC, electrical cabling; PS, laser power supply; P, pierced off-axis parabolic mirror; CC, chirp control; FL, femtosecond laser.

example. Systems (c) and (d) are appropriate for standoff, open-path interrogation of targets to which optical access is possible over an extended distance (r).[12,14] Two configurations are possible depending on the method of transporting the laser pulse to the target to form the laser plasma. In both cases, a large telescope is used to collect the plasma light as the light intensity from the target falls off as the reciprocal of the distance squared (e.g., $1/r^2$). The most widely used method of forming a remote laser plasma is based on conventional focusing (Figure 7.7c) using a lens or mirror optical system to focus the laser pulse on the target. The focusing system is adjustable so that targets at different distances can be interrogated. As an example, a standoff LIBS system using 300 mJ Nd:YAG laser pulses with the beam expanded by a factor of four can generate an analytically useful plasma on solid targets at distances from about 15 to 40 m. LIBS spectra are recorded by this system by collecting the plasma light with a 12 in. (30 cm) telescope coupled to a detection system consisting of a 0.33 m spectrograph/ICCD combination. Using the conventional focusing method (Figure 7.7c), a LIBS system has been described providing analysis of solids at distances out to 120 m.[15] The second method of standoff interrogation uses the filaments formed by a high-power (i.e., terawatt, 10^{12} W) femtosecond laser (Figure 7.7d). The filaments, formed through self-focusing phenomena arising from the high pulse powers, propagate through air in a highly directional mode, forming the laser plasma on the target.[16,17] By controlling the chirp of the laser pulse, the onset of filamentation at a set distance from the laser can be adjusted.

Another variation of LIBS involves the use of double laser pulses to record emission spectra with the focal volumes of the two pulses spatially coincident and the pulses separated in time by a few microseconds. In some cases, these double-pulse methods have been shown to enhance LIBS signals.[18] Double-pulse methods are characterized as collinear or orthogonal configurations. In the collinear arrangement, the two pulses are directed onto the target along the same collinear paths and analytical signals are recorded from the plasma formed by the second pulse. In the orthogonal arrangement, one pulse is typically incident normal to the surface and the second pulse is parallel to the surface and focused to generate an air spark close to the surface and overlapping the volume of the first pulse. For explosives detection in ConOps where LIBS would be deployed, only the collinear arrangement is practical. Collinear double-pulse methods can be deployed on the LIBS configurations shown in Figure 7.7. Often two lasers are used to generate the double laser pulses. In other cases, a single laser can be used to generate the closely temporally spaced pulses. Additionally, through the use of two lasers, the wavelengths of the two pulses can be different, as well as the pulse widths. A number of effects have been reported using double-pulse interrogation depending on the operating parameters. These range from enhancements of plasma temperature, electron density, emission intensities and lifetimes, plasma volume, and mass ablation rates.[12] Characteristics of double-pulse excitation important for explosives detection include a decrease in the entrainment of atmospheric gases in the plasma generated by the second laser pulse and an increase in the signal-to-noise ratio (SNR) of certain explosive emission features.[18,19]

In addition to the use of double Nd:YAG pulses, the fourth harmonic wavelength of Nd:YAG at 266 nm was combined with a pulsed carbon dioxide (CO_2) laser pulse (10.6 μm) to enhance detection capabilities. Enhancements up to 100× were recorded by the action of the CO_2 pulse on the plasma formed by the 266 nm pulse. This was attributed to additional heating of the plasma formed by the 266 nm pulse, heating by the long pulse length, and high energy CO_2 laser pulse via inverse bremsstrahlung absorption (Townsend effect). The use of this dual laser system for the standoff detection of explosives was investigated.[20–23]

7.3   REVIEW OF LIBS DETECTION OF EXPLOSIVES

7.3.1   *In Situ*, Close-Up Detection

One of the first reports of the use of LIBS for explosives detection related to identifying gunshot residue (GSR) on the hands of a shooter.[24] It was shown that by collecting residue samples from the hands of the shooter using sticky tape after multiple firings, or after an uncleaned gun had been fired, it was possible to identify elements related to the gunpowder. Data were collected using a lab-based LIBS apparatus. Emission signals from the clean tape were due to Ca(I), Na(I), and K(I) (potassium), whereas a series of barium lines (e.g., Ba(I), Ba(II)) were observed attributed to the explosive residue. Barium is one of three elements, besides Sb (antimony) and Pb (lead), used in modern methods to identify GSR. By using Monte Carlo simulation techniques, criteria were proposed to indicate positive or negative results from data collected from a shooter firing a single shot from a clean gun.

LIBS for *in situ*, close-up explosives detection has been realized by the development of person-portable prototype instruments.25 These use a compact laser and compact spectrometer (e.g., Avantes, Ocean Optic, Stellarnet, etc.) and typically consist of a backpack coupled to a sampling wand. Two units are shown in Figure 7.8. The units incorporate low-repetition-rate flashlamp-pumped lasers or 10 Hz diode-pumped systems.

Using the second system shown in Figure 7.8b, it was possible to correctly identify 20 types of explosives using only 10 LIBS spectra for each sample. The explosives were four variants of octahydro-1,3,5,7-tetranitro-1,3,5,7-tetrazocine (HMX), hexalites (e.g., hexoctol, cyclotol), three variants of pentaerythritol tetranitrate (PETN), pentalite, 1,3,5-triamino-2,4,6-trinitrobenzene (TATB), 1,3,5-trinitroperhydro-1,3,5-triazine (RDX), C4 explosive (91% RDX + plasticizer and binder), Comp B formulations (typically ~60% RDX, ~40% TNT, and ~1% wax) prepared at different times, and nonhazardous explosives for security training and testing (NESTT) samples of TNT, PETN, and RDX. The identification spectra were acquired in less than 1 min. Additional, separate data were acquired and used to build chemometric models used to identify the explosives. The model used was based on a method of forming recognition algorithms developed specifically for LIBS.[26]

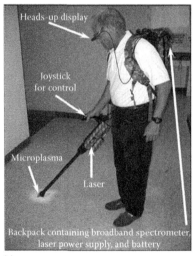
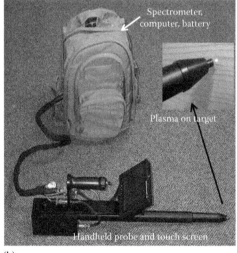

(a)   (b)

FIGURE 7.8 Examples of person-portable LIBS systems developed to detect explosives and other threat materials (e.g., bioagents, nuclear and radiological materials). System (b) can accommodate a flashlamp-pumped low-repetition rate Nd:YAG laser or a 10 Hz diode-pumped Nd:YAG laser. The designs are based on the configuration shown in Figure 7.7b. ((a) Reprinted from *Appl. Geochem.* 21, Harmon, R.S., F.C. DeLucia, C.E. McManus, et al., Laser-induced breakdown spectroscopy—An emerging chemical sensor technology for real-time field-portable, geochemical, mineralogical, and environmental applications, 730–747, Copyright (2006), with permission from Elsevier.)

7.3.2 LAND-MINE DETECTION

In addition to direct detection of explosives (bulk or residue), there have been efforts to use LIBS to identify land-mine casing materials. Such a capability will be useful for demining operations in which handheld or robotic delivered LIBS-based probes are used to interrogate buried targets and determine if they are candidate land mines. Such instruments would most likely be based on the LIBS configurations shown in Figure 7.7a or 7.7b, as these would be amenable to compact instrumentation. One report of the use of LIBS for land-mine detection appears in the work of Harmon et al.[27] In this study, broadband LIBS spectra were recorded in the laboratory for a variety of land-mine casing materials and explosives. The survey spectra revealed observable differences between the different casing types. In a study of a subset of land-mine casings, a library of LIBS spectra was established and then single-shot spectra were recorded from each casing type. The degree of correct identification ranged from 99% to 81%, depending on the material. A concept for a portable LIBS system was described. In a subsequent study by the same group, LIBS spectra were acquired from land-mine materials and non-land-mine plastics.[28] In two blind tests of identification ability, LIBS spectra of the materials were recorded and compared with a material spectral library. Material determinations indicating mine or no mine were correct for over 90% of the samples using linear correlation software.

In another approach to the same problem, a conventional mine prodder system was outfitted with a LIBS system based on a fiber-optic laser (microchip laser + ytterbium-fiber amplifier).[29] Although not person-portable, the system was mobile for field use. Rather than using the entire LIBS spectrum, targeted materials were identified by applying neural network analysis to the temporal behavior of CN emission features near 388 nm. A success rate greater than 80% was achieved in the identification of four types of casing materials. It was noted that identification results could be improved by adding additional classification characteristics such as including observation of the carbon emission signal at 248 nm and a multiexponential decay analysis and by increasing the number of data sets for training the neural network.[30]

Utilizing both laboratory and prototype man-portable LIBS systems, the use of LIBS to discriminate between land-mine casings, nonmine plastics, and other "clutter"-type objects likely to be found in soil during a demining operation was investigated.[31] The land-mine casing materials included a range of antipersonnel and antitank mines from different countries. LIBS spectra of materials in blind tests were collected and compared with a previously assembled spectral library of the materials. Using linear correlation analysis of the spectra, a success rate of greater than 90% was achieved in discriminating between mine-related and non-mine materials using both LIBS systems.

Using the chemometric analysis method partial least-squares discriminant analysis (PLS-DA), it was shown to be possible to discriminate land-mine casings from simulants with a classification success rate of 99.0% and a misclassification rate of only 1.8%.[32] "Unknown" spectra were assigned to a mine type based on a library classification approach. Broadband LIBS spectra were recorded in the laboratory from 41 inert mine casings (antipersonnel and antitank land mines) as well as some land-mine simulants (e.g., wood, black plastic, and nylon). In all, over 12 types of casings from four countries were evaluated.

7.3.3 RESIDUE DETECTION

The simplest ConOps in regard to data interpretation is one that interrogates the bulk explosive. Here, the spectral signature can be attributed solely to the explosive and elements in air. A more difficult scenario, but one perhaps more relevant, is the detection of explosive residue on a surface in air. In this case, almost invariably, the recorded spectrum has contributions from the explosive, air, and the underlying substrate. In addition, the mass of residue interrogated will likely be less than the mass sampled from bulk explosives. Also, coupling of the laser pulse energy into the surface can be strongly affected by the substrate type and it can vary depending on the surface concentration of the residue. Several papers have addressed residue detection, and these are discussed here.

The ability to distinguish between residues of four explosives and four interferent species deposited on aluminum substrates using LIBS line intensity ratios has been demonstrated.[33] Spectra were collected using the person-portable system of Figure 7.8a and a double-pulse standoff LIBS system (20 m) of Figure 7.9. It was found important to exclude N and O due to air to enhance discrimination capabilities. The person-portable system used an argon flow at the interrogation site to remove N and O, whereas emissions from N and O were excluded in the standoff measurements by using double pulses. Standoff data were analyzed using the total recorded LIBS spectrum and specific peak intensity ratios. Chemometric modeling was carried out using principal component analysis (PCA), soft independent modeling of class analogies (SIMCA), and PLS-DA methodologies. It was shown that relative emission line intensities for these measurements were representative of the target stoichiometry and these were useful to differentiate between molecules

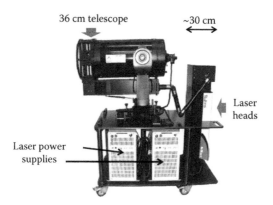

FIGURE 7.9 Army Research Laboratory standoff LIBS detection system used for explosives detection. (Reprinted from *Spectrochim. Acta Part B* 62, Gottfried, J.L., F.C. De Lucia Jr., C.A. Munson et al., Double-pulse standoff laser-induced breakdown spectroscopy for versatile hazardous materials detection, 1405–1411, Copyright (2007), with permission from Elsevier.)

containing the same elements. Also, PLS-DA was shown to differentiate between explosive and other organic materials and to classify samples such as fingerprints, Comp B (RDX and TNT mixture), and *Bacillus subtilis*.

In another study, which used explosives (e.g., TNT, RDX, PETN, HMX, etc.) and nonexplosive interferents (e.g., diesel oil, wax, grease, glue, etc.), LIBS was investigated as a detection/discrimination tool.[34] Small quantities of each material were deposited on aluminum metal substrates. The explosives were deposited in solution form. The interferents were applied as thin residues of unknown thickness. The residues were interrogated by single laser pulses of 250 mJ and a spot size of 0.94 mm. Emission lines used for analysis were C(I) at 247.8 nm, CN at 388.3 nm, C_2 at 516.3 nm, H(I) at 656.2 nm, O(I) at 777.4 nm (triplet), N(I) at 746.8 nm, and Al(I) at 309.3 nm. The spectra were also analyzed using a PCA analysis in two ways. In the first method, line integrals were used for analysis and the best residue separations were obtained for PCA2 plotted as a function of PCA1, although it was not possible to distinguish between the explosives nitroglycerine (NG) and triacetone triperoxide (TATP). By performing a second PCA analysis and introducing parameters taking into account changes in the intensities of C_2, CN, N, and C, it was possible to obtain correct classifications for some TATP and NG data that had been misclassified using PCA analysis alone. For all three types of analyses, 100% success was achieved in the identification of interferents. For the explosives, correct classifications ranged from 0% to 100%, depending on the method used. The estimated threshold for residue detection was in the range of 0.1–1 ng. The results also showed that four of the nine explosives could only be correctly classified when present as a small amount on the aluminum support. In this case, the plasma temperature was higher, leading to more complete atomization of the organic molecules.

The detection of explosive residues and other interferents on painted surfaces was investigated using a double-pulse standoff LIBS system at 30 m.[35] The targets were painted car panels from various vehicles. Explosive residues of RDX, TNT, and Comp B (TNT and RDX mixture) were crushed and then smeared on the panels, providing an estimated coverage of about 10 $\mu g/cm^2$. A variety of atomic emission line intensities and intensity ratios were used as inputs to a PLS-DA model developed to separate explosives from nonexplosives. Using the model, explosives were successfully identified at a level of 98% correct for 908 explosive samples and the false-positive rate for 2876 nonexplosive samples was 2.7%.

In another study, the ability to identify explosive residues on pure metal and alloy substrates was investigated.[36] Residues of RDX were applied to the substrates and plasmas were formed on the surface at short range (10 cm) using double pulses from separate Nd:YAG lasers. The spectra were recorded using an echelle spectrometer and electromagnetic CCD detector. It was demonstrated that although distinct differences in the LIBS spectra of the RDX residue on metallic substrates were observed, it was possible to identify RDX residues independent of the substrate with high fidelity. Approaches used in the chemometric spectral analysis included (1) classifying the RDX residue spectra together in one substrate independent class; (2) using selected emission intensities and ratios to increase the true positive rate (TPR) and decrease the false-positive rate (FPR); and (3) fusing the results from two PLS-DA models generated using the full broadband spectra and selected intensities and ratios related only to RDX. Together, these approaches yielded a TPR of 96.5% and a FPR of 1% for the samples studied.

Although under controlled conditions the identification of explosives residues on a surface has been possible using line ratios, several studies have shown this methodology has limitations. In one study, it was shown that atomic and molecular line intensities as well as their ratios were strongly affected by the extent to which the underlying substrate was ablated.[37] The mechanism is the change in plasma temperature that correlated with the amount of substrate ablated. The extent of substrate ablation was governed, in part, by the surface concentration of explosive residue and it was observed that as the residue decreased, the line intensities of C, H, N, and O increased due to the hotter plasma. In addition, environmental conditions, such as air humidity, were shown to significantly affect the intensity of the hydrogen emission. Atomic emission line intensities are also affected by chemical reactions that consume atomic species. Using a PCA analysis that encompassed six atomic ratios, only three explosives were correctly identified 100%, with NG and TATP identifications failing completely. By reworking the PCA analysis to include chemical reaction effects, these two explosives were correctly identified. Among other findings of this work is that a viable LIBS recognition protocol for organic materials must include consideration of plasma parameters as well as chemical reactions. It was also noted that detection gating parameters (gate delay and width) should also be considered as methods to reduce effects of chemical reactions and recombination processes that can compromise identification fidelity.

In another study, some characteristics of LIBS for the identification of selected organic compounds, including explosives, were investigated.[38] Measurements were carried out in air and helium atmospheres. Emission from carbon was shown to be a poor indicator, whereas hydrogen emission corresponded well to the stoichiometry of the target organic material. In a helium atmosphere, the O emission signal strength was related to the atomic fraction in the compound, but this was not the case for measurements in air. Also, in helium, nitrogen was found to originate from the target species giving rise to emission from CN. Emissions from C_2 and CN were found to be most useful for organic compound identification, with the intensity ratio of CN/C_2 closely correlated to the oxygen balance for five organic species.

7.3.4 STANDOFF DETECTION

The majority of standoff detection studies of explosives have been carried out using the open path configuration based on Figure 7.7c. In one study, the explosives TNT, RDX, C4, Comp B, and PETN were correctly identified in 19 of 21 measurements.[39] The Nd:YAG laser system produced pulses of 350 mJ (10 Hz) and plasmas at distances up to 45 m. The light was collected by a Herschelian telescope (off-axis primary mirror) and then directed to a 125 mm focal length spectrograph coupled to an ICCD. "Confusants" were nonexplosive inorganic and organic materials that included diesel fuel, acetone, beef grease, and paint. LIBS spectral signatures from the C_2, H, O, and N were used to identify the different targets.

The detection of explosives (as well as biomaterials and simulants of nerve agents) was demonstrated at 20 m in the laboratory using a double-pulse LIBS system shown in Figure 7.9.[19] Explosives as bulk materials and residues were identified using PCA analysis of LIBS spectral feature ratios of C, H, O, and N. The use of double-pulse interrogation (3 μs pulse separation) of the target greatly improved the detection sensitivity and selectivity. Double pulses increased the intensity of some emissions (e.g., Al(I)), gave rise to some once-ionized emissions (e.g., Al(II)), and reduced the amount of atmospheric oxygen and nitrogen entrained in the second of the double pulses. Using the same double-pulse LIBS system as described by Gottfried et al.,[19] the detection of explosive residues was demonstrated at an indoor test range at distances out to 50 m.[40] At 50 m, LIBS spectra were significantly reduced in intensity (carbon emission was not observed at this distance), but there was enough remaining spectral information that samples were discriminated using a PLS-DA analysis.

Standoff detection has also been approached using the configuration of Figure 7.7d and IR and UV filaments.[41] The IR filaments were formed using a mode-locked titanium (Ti):sapphire laser coupled to a chirped laser amplifier. The pulses were generated at 10 Hz, had a center wavelength of 795 nm, an energy of 15 mJ, and a pulse width of 100 fs. The UV filaments were generated from a seeded, Q-switched neodymium:yttrium orthovanadate (Nd:YVO$_4$) laser, the amplified pulses of which were frequency doubled and focused into a Brillouin cell. The output of this was frequency doubled to 266 nm. Targets were thin layers of 2,4-dinitrotoluene (DNT) or ammonium perchlorate (NH$_4$ClO$_4$) deposited on metal substrates as solutions of ethanol. The light produced by the filaments on the target was collected by a parabolic mirror located 3 m away and directed into a spectrometer/CCD detection system (spectral range 275–750 nm). Spectra recorded using IR filaments showed mostly spectral features due to the metal substrate with little contribution from DNT or NH$_4$ClO$_4$. In contrast, repeatable spectral features attributable to the two explosives were clearly observed using UV generated filaments regardless of the substrate metal, showing the advantage of these filaments over IR filaments for this application.

7.4   LIBS COMBINED WITH RAMAN SPECTROSCOPY

As we have seen, LIBS is an elemental detection method that produces an emission spectrum consisting of atomic and ionic emission lines plus emissions from simple molecules. By analyzing certain elemental emission features or using chemometric statistical methods, LIBS spectra can be used to identify complex materials such as explosives. The specificity of detection can be increased, however, by combining LIBS detection with other laser-based emission methods that provide what is termed orthogonal data. The goals of combining the two techniques are to increase the probability of detection (PD) and reduce the false-alarm rate (FAR). One such technique is Raman spectroscopy.[42] As noted above, standoff detection of explosives by LIBS has been shown, while Raman-based detection of explosives at a range of 50 m using laser pulses of 140 mJ and a large telescope light collection system has been demonstrated.[43] Because LIBS and Raman use essentially the same detection system components, combining them into a single instrument is feasible.[44]

Briefly, Raman spectroscopy (refer to Chapter 5) uses inelastic scattering of photons between vibrational energy levels and virtual states of a molecular system. There are several variants of Raman spectroscopy, but the simplest and most widely used method is spontaneous Raman spectroscopy. Typically, spontaneous Raman signals are weak, so excitation is usually provided by a laser beam because of its high irradiance. For comparison, Raman differential scattering cross sections are typically on the order of 10^{-30} cm^2/sr, whereas absorption cross sections are on the order of 10^{-18} cm^2. The magnitude of the cross section can be considered a measure of the degree of interaction between light and the molecule. The Raman signal is a shift in the frequency of the incident laser light to either longer (Stokes) or shorter wavelengths (anti-Stokes). The Stokes light is typically the stronger signal as the population of the initial state is greater than that for anti-Stokes emission. Raman spectroscopy yields structural information about a molecular system and has found wide use for a number of applications. One area is forensics, including the identification of controlled substances and narcotics, analysis of inks, explosive materials detection, and investigation of forged documents.[42]

A few studies have investigated the combination of LIBS with Raman for the detection of explosives. In one study, the selectivity and sensitivity of detection using each method was determined.[45] Experiments were carried out using 400 mJ pulses from an Nd:YAG laser (532 nm wavelength) for excitation and light collection provided by a Cassegrain telescope coupled to either a Czerny-Turner spectrograph + ICCD (LIBS detection) or a holographic imaging spectrograph + ICCD (Raman detection). Measurements were carried out at a distance of 20 m on a variety of pure explosive targets and nonenergetic materials sharing elemental compositions with the explosives. LIBS spectra of the targeted samples showed a high degree of similarity between certain materials: RDX was closely related to NH_4NO_3; DNT was similar to nylon; sodium chlorate ($NaClO_3$) was essentially indistinguishable from sodium chloride ($NaCl$) (Figure 7.10). On the other hand, the paired Raman spectra for these materials were distinctly different. Generally speaking, LIBS was superior to Raman spectroscopy at identifying explosives containing inorganic elements (e.g., between sodium and potassium chlorates), whereas Raman is better at identifying organic compounds. The ability of each technique to discriminate between explosives and material such as wood, aluminum, soil, nylon, cotton, and so on was also evaluated. LIBS detection of six explosives and explosive-related materials was evaluated in terms of detectable mass. For all materials, LIBS limits of detection (LOD) were superior. Whereas Raman LOD values were in the range of 9–27 μg (using 25 laser pulses), LIBS LODs ranged from 0.061 to 1.6 ng (single shot). Options for fusing the data from the two techniques were discussed.

In another study, the combined predictive ability of LIBS and Raman was investigated at standoff distances of 20 and 30 m.[46] LIBS and Raman spectra were collected from biowarfare agent simulants, a chem-warfare simulant, RDX, and potential confusant species (e.g., coal, foods, dust, etc.) using separate LIBS and Raman systems. The Raman system used dual excitation wavelengths of 532 and 248 nm. Models were built for both types of data using PLS-DA. Applying the models to a test data set resulted in a score converted to a probability needed to fuse the results from multiple models. Confusion matrices were developed that cataloged the number of correct identifications from 26 different targets. Individual Raman and LIBS classification

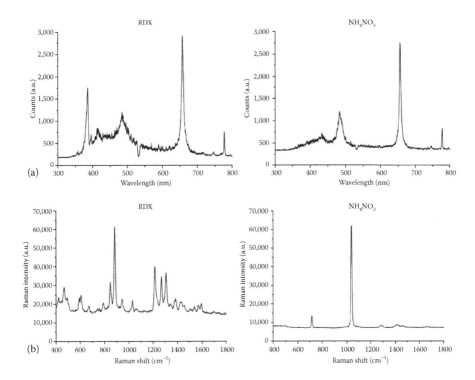

FIGURE 7.10 (a) LIBS spectra of RDX and ammonium nitrate (NH_4NO_3) showing similarities, whereas the (b) Raman spectra of the same materials are distinctly different. (With kind permission from Springer Science+Business Media: *Anal. Bioanal. Chem.* Standoff detection of explosives: Critical comparison for ensuing options on Raman spectroscopy–LIBS sensor fusion. 400, 2011, 3353–3365, Moros, J., J.A. Lorenzo, and J.J. Laserna, figures 3 and 4.)

results based on 10 spectra for each of the samples showed good results for correct classifications using each method separately, with Raman results somewhat better than LIBS results. By fusing the LIBS and Raman results, however, 100% correct classifications were achieved.

The fusion of LIBS and Raman data was studied for a university thesis. Using double-pulse excitation, the LIBS plasma formed by a 266 nm pulse was heated by a CO_2 laser pulse.[20] The signal enhancement afforded by the CO_2 laser pulse permitted the use of lower-energy 266 nm pulses for LIBS while allowing enhancement of the Raman scattering (also using 266 nm excitation) cross section by λ^{-4} compared to longer wavelengths. The multisensor fusion algorithm developed in this study produced more accurate detection results, a higher PD, and lower FAR compared to the results obtained from individual Raman and LIBS measurements.

In a series of published works, the combination of the LIBS plasma enhanced using the Townsend effect coupled with Raman spectroscopy for the detection of explosives in addition to chemical, biological, radiological, and nuclear (CBRN) threats has been discussed.[20,22,23,47]

7.5 SUMMARY

LIBS, typically used for the elemental analysis of materials, can, under certain conditions, be used to identify a range of explosives and discriminate these from non-explosive materials containing similar elements. The identification of these complex materials is accomplished by observing element line ratios and/or by applying one of several types of chemometric analysis methods to the LIBS spectrum. High rates of positive identification of explosives along with low rates of false positives have been achieved. Prototype person-portable LIBS instrumentation has been developed and demonstrated for close-up explosives detection. In addition, larger, transportable, field-deployable LIBS systems have been tested for explosives detection at distances out to 50 m. LIBS explosive recognition capabilities can be increased by combining LIBS with other optical techniques that provide complementary data. One such orthogonal technique investigated is Raman spectroscopy, which also shares many of the same system components. By fusing data from both techniques, preliminary work has demonstrated increased detection fidelity for explosives identification at standoff distances.

REFERENCES

1. Cremers, D.A. and L.J. Radziemski. *Handbook of Laser-Induced Breakdown Spectroscopy*, 2nd edn. Chichester: John Wiley and Sons, 2013.
2. Noll, R.. *Laser-Induced Breakdown Spectroscopy: Fundamentals and Applications*. Berlin: Springer, 2012.
3. Laser Institute of America. American national standard for safe use of lasers. ANSI Z136.1-2007. Washington, DC: American National Standards Institute, 2014.
4. Strakovskiy, L., A. Cohen, R. Fifer, et al. Laser ignition of propellants and explosives. Army Research Laboratory Report ARL-TR-1699, June 1998. http://www.dtic.mil/dtic/tr/fulltext/u2/a348616.pdf. Accessed 30 March 2014.
5. Lai, H., A. Leung, M. Magee, et al. Identification of volatile chemical signatures from plastic explosives by SPME-GC/MS and detection by ion mobility spectrometry. *Anal. Bioanal. Chem.* 396 (8):2997–3007, 2010.
6. Ewing, R.G. and C.J. Miller. Detection of volatile vapors emitted from explosives with a handheld ion mobility spectrometer. *Field Anal. Chem. Technol.* 5 (5):215–221, 2001.
7. Hahn, D.W. and N. Omenetto. Laser-induced breakdown spectroscopy (LIBS), Part I: Review of basic diagnostics and plasma–particle interactions: Still-challenging issues within the analytical plasma community. *Appl. Spectrosc.* 64 (12):335A–366A, 2010.
8. Gottfried, J.L., F.C. De Lucia Jr., C.A. Munson, et al. Progress in LIBS for real-time standoff explosive residue detection. Presented at Explosive Detection Workshop, University of Rhode Island, October 8–9, 2009. http://energetics.chm.uri.edu/?q=system/files/9%20c%20GottfriedLIBS.pdf. Accessed 30 March 2014.
9. Gottfried, J.L. Chemometric analysis in LIBS. In *Handbook of Laser-Induced Breakdown Spectroscopy*, pp. 223–255. Chichester: John Wiley and Sons, 2013.
10. Dudragne, L., Ph. Adam, and J. Amouroux. Time-resolved laser-induced breakdown spectroscopy: Application for qualitative and quantitative detection of fluorine, chlorine, sulfur, and carbon in air. *Appl. Spectrosc.* 52: 1321–1327, 1998.
11. Dikshit V., F.-Y. Yueh, J.P., Singh, et al. Laser-induced breakdown spectroscopy: A potential tool for atmospheric carbon dioxide measurement. *Spectrochim. Acta Part B* 68: 65–70, 2012.

12. Cremers, D.A. and L.J. Radziemski. *Handbook of Laser-Induced Breakdown Spectroscopy.* pp. 94–96. Chichester: John Wiley and Sons, 2006.
13. Fortes, F.J. and J.J. Laserna. The development of fieldable laser-induced breakdown spectrometer: No limits on the horizon. *Spectrochim. Acta B* 65: 975–990, 2010.
14. Cremers, D.A. and L.J. Radziemski. *Handbook of Laser-Induced Breakdown Spectroscopy,* 2nd edn. pp. 115–117, Chichester: John Wiley and Sons, 2013.
15. Laserna, J.J., R. Fernández Reyes, R. González, et al. Study on the effect of beam propagation through atmospheric turbulence on standoff nanosecond laser induced breakdown spectroscopy measurements. *Opt. Express* 17: 10265–10276, 2009.
16. Stelmaszczyk, K., Ph. Rohwetter, G. Méjean, et al. Long-distance remote laser-induced breakdown spectroscopy using filamentation in air. *Appl. Phys. Lett.* 85: 3977–3979, 2004.
17. Rohwetter, Ph., K. Stelmaszczyk, L. Wöste, et al. Filament-induced remote surface ablation for long range laser-induced breakdown spectroscopy operation. *Spectrochim. Acta Part B* 60: 1025–1033, 2005.
18. Babushok, V.I., F.C. De Lucia Jr, J.L. Gottfried, et al. Double pulse laser ablation and plasma: laser induced breakdown spectroscopy signal enhancement. *Spectrochim. Acta Part B* 61: 999–1014, 2006.
19. Gottfried, J.L., F.C. De Lucia Jr., C.A. Munson et al. Double-pulse standoff laser-induced breakdown spectroscopy for versatile hazardous materials detection. *Spectrochim. Acta Part B* 62: 1405–1411, 2007.
20. Singh, A. Standoff Raman Spectroscopy detection of trace explosives. Thesis, Missouri University of Science and Technology. 2010. https://mospace.umsystem.edu/xmlui/bitstream/handle/10355/29468/Singh_2010.pdf?sequence = 1. Accessed 21 March 2014.
21. Waterbury, R.D., A. Pal, D. K. Killinger, et al. Standoff LIBS measurements of energetic materials using a 266 nm excitation laser. In A. W. Fountain and P. J. Gardner (eds), *Proceedings of SPIE vol. 6954 Chemical, Biological, Radiological, Nuclear, and Explosives,* Orlando, FL, 2008.
22. Ford, A., R. D. Waterbury, J. Rose, et al. Results of a UV TEPS-Raman energetic detection system (TREDS-2) for standoff detection. In A. W. Fountain and P. J. Gardner (eds), *Proceedings of SPIE conference on Chemical, Biological, Radiological, Nuclear, and Explosives,* vol. 7304, 2009.
23. Ford, A., R. Waterbury, J. Rose, et al. Extension of a standoff explosive detection system to CBRN threats. In A.W. Fountain and P. J. Gardner (eds), *Proceedings of SPIE conference on Chemical, Biological, Radiological, Nuclear, and Explosives,* vol. 7665, 2010.
24. Dockery, C.R. and S.R. Goode, Laser-induced breakdown spectroscopy for the detection of gunshot residues on the hands of a shooter. *Appl. Opt.* 42(30): 6153–6158, 2003.
25. Harmon, R.S., F.C. DeLucia, C.E. McManus, et al. Laser-induced breakdown spectroscopy—An emerging chemical sensor technology for real-time field-portable, geochemical, mineralogical, and environmental applications. *Appl. Geochem.* 21: 730–747, 2006.
26. Multari, R.A., D.A. Cremers, Methods for forming recognition algorithms for laser-induced breakdown spectroscopy. US Patent 8,655,807 B2.
27. Harmon, R.S., F.C. DeLucia, R.J. Winkel, et al. LIBS: A new versatile, field deployable, real-time detector system with potential for landmine detection. *Proc. SPIE* 5089: 1065–1077, 2003.
28. Harmon, R.S., F.C. DeLucia, A. LaPointe, et al. Discrimination and identification of plastic landmine casings by single-shot broadband LIBS. *Proc. SPIE* 5497: 92–101, 2005.
29. Bohling, C., D. Scheel, K. Hohmann, et al. Fiber-optic laser sensor for mine detection and verification. *Appl. Opt.* 45(16): 3817–3825, 2006.
30. Schade, W., C. Bohling, K. Hohmann, et al. Laser-induced plasma spectroscopy for mine detection and verification. *Laser Part. Beams* 24(2): 241–247, 2006.

31. Harmon, R.S., F.C. DeLucia Jr., A. LaPointe, et al. LIBS for landmine detection and discrimination. *Anal. Bioanal. Chem.* 385: 1140–1148, 2006.
32. Gottfried, J.L., R.S. Harmon, and A. LaPointe. Progress in LIBS for land mine detection, ARL-TR-5127. Aberdeen Proving Ground, MD: US Army Research Laboratory, 2010. http://www.google.com/url?sa=t&rct=j&q=&esrc=s&frm=1&source=web&c d=1&ved=0ccyqfjaa&url=http%3a%2f%2fwww.dtic.mil%2fcgi-bin%2fgettrdoc%3f ad%3dada522281&ei=77c5u-odbbphsatw54gwdg&usg=afqjcnhf8o4dxsrn1uyhc3wq esdpoczyea&bvm=bv.63808443,d.cwc; Accessed 30 March 2014.
33. Gottfried, J.L., F.C. De Lucia, Jr., C.A. Munson, et al. Strategies for residue explosives detection using laser-induced breakdown. *J. Anal. At. Spectrom.* 23: 205–216, 2008.
34. Lazic, V., A. Palucci, S. Jovicevic, et al. Detection of explosives at trace levels by laser-induced breakdown spectroscopy (LIBS). *Proc. SPIE* 7665: 76650V-1–76650V-9, 2010.
35. De Lucia, Jr., F.C. and J.L. Gottfried. Classification of explosive residues on organic substrates using laser induced breakdown spectroscopy. *Appl. Opt.* 51(7): B83–B92, 2012.
36. Gottfried, J.L. Influence of metal substrates on the detection of explosive residues with laser-induced breakdown spectroscopy. *Appl. Opt.* 52(4): B10–B19, 2013.
37. Lazic, V., A. Palucci, S. Jovicevic, et al. Analysis of explosive and other organic residues by laser induced breakdown spectroscopy. *Spectrochim. Acta B* 64: 1028–1039, 2009.
38. P. Lucena, A. Doña, L.M. Tobaria, et al. New challenges and insights in the detection and spectral identification of organic explosives by laser induced breakdown spectroscopy. *Spectrochim. Acta B* 66: 12–20, 2011.
39. Lopez-Moreno, C., S. Palanco, J.J. Laserna, et al. Test of a stand-off laser-induced breakdown spectroscopy sensor for the detection of explosive residues on solid surfaces. *J. Anal. At. Spectrom.* 21: 55–60, 2006.
40. Gottfried, J.L., F.C. De Lucia Jr., C.A. Munson, et al. Detection of energetic materials and explosive residues with laser-induced breakdown spectroscopy: II. Stand-off measurements. US Army Research Laboratory Report ARL-TR-4241, 2007. http://www.dtic.mil/dtic/tr/fulltext/u2/a472708.pdf; Accessed 1 December 2014.
41. Mirell, D., O. Chalus, K. Peterson et al. Remote sensing of explosives using infrared and ultraviolet filaments, *J. Opt. Soc. Am. B* 25(7): B108–B111, 2008.
42. Vandenabeele, P. *Practical Raman Spectroscopy: An Introduction.* Chichester: John Wiley and Sons, 2013.
43. Carter, J.C., S.M. Angel, M. Lawrence-Snyder, et al. Standoff detection of high explosive materials at 50 meters in ambient light conditions using a small Raman instrument. *Appl. Spectrosc.* 59: 769–775, 2005.
44. Hahn, D. W. and N. Omenetto. Laser-induced breakdown spectroscopy (LIBS), Part II: Review of instrumental and methodological approaches to material analysis and applications to different fields. *Appl. Spectrosc.* 66(4): 347–419, 2012.
45. Moros, J., J.A. Lorenzo, and J.J. Laserna. Standoff detection of explosives: critical comparison for ensuing options on Raman spectroscopy–LIBS sensor fusion. *Anal. Bioanal. Chem.* 400: 3353–3365, 2011.
46. Miziolek, A., F. DeLucia, C. Munson, et al. A new standoff CB detection technology based on the fusion of LIBS and Raman. ChemImage. http://www.chemimage.com/resources/publications/threat-detection.aspx. Accessed 21 March 2014.
47. Waterbury, R., J. Rose, D. Vunck et al. Fabrication and testing of a standoff trace explosives detection system. In A.W. Fountain and P. J. Gardner (eds), *Proceedings of SPIE conference on Chemical, Biological, Radiological, Nuclear, and Explosives,* vol 8018, 2011.

8 Active Mid-Infrared Reflectometry and Hyperspectral Imaging[*]

Anish K. Goyal and Travis R. Myers

CONTENTS

* The Lincoln Laboratory portion of this work was sponsored by the Office of the Assistant Secretary of Defense for Research and Engineering under Air Force contract No. FA8721-05-C-0002. The opinions, interpretations, conclusions, and recommendations are those of the authors and are not necessarily endorsed by the US Government.

8.1   INTRODUCTION

Active mid-infrared hyperspectral imaging (active MIR HSI) is discussed as a method for the standoff detection of bulk and trace explosives. This method is sometimes referred to as mid-infrared (MIR) *chemical imaging* because it is an imaging technique that classifies objects in a scene based on chemical composition. Standoff means the measurement is made in a noncontact manner at a distance of at least a few centimeters.

Recent advances in laser technology have made active MIR HSI viable for nonlaboratory field use. This method has many potential benefits that include high sensitivity, high specificity, noncontact and nondestructive detection and identification of chemicals. Furthermore, it is one of the very few methods that can achieve high-speed scanning of surfaces while remaining eye safe. There are, however, challenges to developing and implementing such systems. This chapter investigates both the potential benefits and the challenges of this approach for detecting explosives and related materials.

Specifically considered is measurement of the *spectral reflectance* of a surface when illuminated with a wavelength-tunable MIR laser. This is to be contrasted with photothermal spectroscopy (discussed in Chapter 10), which measures the temperature rise under laser illumination.

8.1.1   Overview

The MIR portion of the optical spectrum roughly spans the wavelength range $\lambda \approx 2.5\text{--}14\ \mu m$ (or in wave numbers, $\tilde{v} = c/\lambda \approx 4000\text{--}700\ cm^{-1}$). The MIR spectrum overlaps the rotational/vibrational frequencies of most molecules. Therefore, MIR spectroscopy is an important class of vibrational spectroscopy and one of the most common laboratory methods for identifying and characterizing chemicals. Optical absorption occurs directly at the vibrational frequencies of the molecule through the optical dipole interaction. The MIR absorption spectrum of a chemical can be rich, and the absorption lines can be extremely strong. Interpretation of MIR spectra is aided by the fact that the absorption lines can often be associated with vibrational modes that are localized to functional groups within a molecule [1]. An example of a functional group is the nitro group ($-NO_2$), which gives rise to a strong absorption band at $\lambda \sim 6.4\ \mu m$ in explosives such as trinitrotoluene (TNT).

Raman spectroscopy is the other major class of vibrational spectroscopy. The Raman effect is an optical scattering process that probes the polarizability of molecular vibrations. It is an inelastic process that shifts the frequency of the scattered light by an amount equal to the vibrational frequency of the molecule (see Chapter 5 for a detailed discussion of Raman spectroscopy). MIR and Raman processes interact with molecular vibrations through different effects (i.e., dipole vs. polarizability), and a particular vibrational mode will have different interaction strengths for the two processes. Qualitatively speaking, the antisymmetric vibrational modes and vibrations due to polar groups are likely to be more *infrared active*, while symmetric vibrational modes are more likely to be *Raman-active* [1]. Therefore, MIR and Raman spectroscopies provide complementary information, and both techniques can be used to detect and identify most chemicals (explosives and their precursors, postblast residues, chemical warfare agents, industrial chemicals, naturally occurring compounds, etc.). However, one great advantage of MIR spectroscopy is that the MIR optical cross sections can be many orders of magnitude larger than the Raman cross sections [2]. Therefore, MIR spectroscopy is suitable for high-sensitivity and high-speed detection with relatively low laser power. Table 8.1 compares MIR and Raman spectroscopy [3].

It is worth mentioning that near-IR spectroscopy probes the overtone bands of the fundamental vibrational modes and also the contributions from electronic transitions. The near-IR spectral features due to vibrational modes consist of broad overlapping

TABLE 8.1
Comparison of MIR Reflectance and Raman Spectroscopies

	Pros	Cons
MIR reflectance	• High chemical specificity • High sensitivity due to large optical cross sections • Ability to scan surfaces rapidly • Standoff trace detection possible • Eye safe	• Complicated reflectance signatures • Need for advanced algorithms • Cannot penetrate many container materials (e.g., glass, plastics) • Strongly absorbed in water • Technologically difficult to access low-wave-number molecular vibrations
Raman scattering	• High chemical specificity • Signatures relatively independent of presentation • Can interrogate through glass and plastic containers • Easily access low-wave-number molecular vibrations • Mature technologies • Compatible with conventional fiber optics	• Very small optical cross sections • Requires high laser intensity • Limited capability for standoff trace detection • Not eye safe in standoff configuration • Potential for burning samples • Not feasible for rapid surface scanning • Fluorescence can obscure signal

Source: J. M. Chalmers, H. G. M. Edwards, and M. D. Hargreaves *Infrared and Raman Spectroscopy in Forensic Science,* Wiley, West Sussex, 2012.

bands that are relatively nonspecific. Also, these are inherently much weaker than the fundamental absorptions and require much larger sampling volumes. The larger sampling volumes can be either an advantage or a disadvantage, depending on the application. Near-IR spectroscopy has been shown to be effective in some applications, such as counterfeit drug detection [3–5]. It is, however, unlikely to be very useful for the high-sensitivity detection and identification of trace chemicals.

Combining MIR reflectance spectroscopy with active HSI enables the high-speed mapping of chemicals on surfaces. To the best of our knowledge, it can provide the highest areal coverage rates (ACRs) of any laser-based method given an eye-safe laser power. The only alternative that is faster is passive HSI. However, passive HSI systems tend to be unreliable, especially in field use, where the measurement geometry and thermal environment cannot be controlled [6].

Resurgence of interest in active MIR HSI to counter the terrorist threat is due in large part to the invention, development, and commercialization of widely tunable quantum cascade lasers (QCLs) [7–10]. QCLs that span almost the entire MIR are now commercially available, and these have enabled MIR spectroscopy in configurations that were not previously possible. Prior to the development of QCLs, hyperspectral microscopy typically utilized Fourier transform infrared (FTIR) spectrometers with globar illumination (or in some cases synchrotron radiation) [11]. With the advent of QCLs and their higher optical power, rapid scanning of surfaces is now possible in both microscopy and standoff configurations. Also important has been continuing development of MIR detector and camera technologies based on both mercury cadmium telluride (MCT) and microbolometer (MB) focal plane arrays (FPAs).

There remain, however, several challenges to implementing practical active MIR HSI systems. Two challenges are related to the detection algorithms. The first is that the reflectance signatures for chemicals on surfaces are more complicated, for example, than gas detection because they depend on a variety of factors that include (1) the presentation of the materials (trace vs. bulk, fine powders vs. pellets, etc.), (2) the surfaces on which the explosives are deposited, (3) the optical fill factor, and (4) the detection geometry. The second is that detection algorithms must be efficient because active MIR HSI can produce a very large number of spectra at high speed. A third challenge is cost. At the time of this writing, MIR lasers and cameras are expensive, which restricts their use to "high-value" applications. For widespread adoption, the cost of these components must be reduced. The potential for lower-cost solutions may exist, for example, by raster scanning the laser beam and using a single-element detector rather than a costly camera.

Despite the challenges, active MIR HSI based on reflectance spectroscopy will likely develop into a valuable and widely adopted method for detecting chemicals such as explosives on surfaces. Furthermore, it is widely applicable in areas such as forensics, biomedical imaging, and process control [3,12–14].

8.1.2 Basic Configuration Using a Wavelength-Tunable Laser

A basic configuration for an active MIR HSI system is shown schematically in Figure 8.1. Alternative configurations are also possible, and these will be discussed

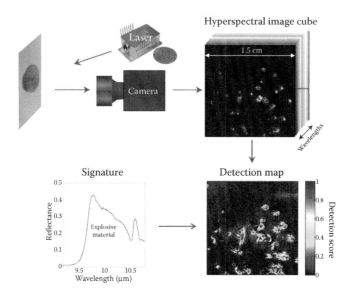

FIGURE 8.1 Basic implementation of active MIR HSI using a wavelength-tunable laser and camera. The surface of interest is illuminated with a wavelength-tunable MIR laser. A frame is captured of the scattered light at each illumination wavelength. The stacking of these frames, one for each illumination wavelength, results in the generation of a three-dimensional hyperspectral image (HSI), sometimes referred to as a *hypercube*. Each pixel in the hypercube is analyzed with respect to a reference signature to generate a two-dimensional detection map that displays the detection score for each pixel in the HSI to indicate the similarity of the measured spectrum with the reference spectrum. In this example of $KClO_3$ particles on a mirror, the detection scores range from about 0 to 1, with 1 corresponding to high similarity with the reference spectrum.

in Section 8.9.2. A wavelength-tunable MIR laser illuminates a surface of interest that may contain, for example, particles of explosives. At each illumination wavelength, the camera captures an image of the reflected light. By sequentially tuning the laser wavelength and capturing images, a hyperspectral image cube (also called a *hypercube*) is generated. Each frame of the hypercube corresponds to the reflectance of the scene at a specific wavelength, and each pixel of the hypercube contains the reflection spectrum at a particular location on the surface.* By analyzing the hypercube on a pixel-by-pixel basis, one can compare the measured reflection spectra to a reference spectral signature from a library of materials. The results are usually presented in the form of a detection map in which a score is assigned to each pixel based

---

* It is worth noting the distinction between the terms *hyperspectral* and *multispectral* imaging. Hyperspectral imaging involves measurement of the spectral response over a continuous spectral range with fine enough resolution to yield a spectrum. This usually involves the measurement of at least tens of wavelengths. Multispectral systems, on the other hand, generally measure only a few bands. But regardless of the number of wavelength bands measured, they do not produce a spectrum of the object over some continuous spectral range. The technique described in this chapter actually encompasses both hyperspectral and multispectral imaging, because in some cases only a few wavelengths may be needed to achieve the required detection capability.

on the similarity of the measured reflection spectrum to the reference spectrum. If the reference spectrum of the chemical is sufficiently specific with respect to the background, a high score (above some threshold) is intended to indicate the presence of the chemical of interest.

Figure 8.1 shows actual experimental data for a hypercube of potassium chlorate ($KClO_3$) particles on a mirror [15]. The hypercube consists of 164 frames that span $\lambda = 9.1$–10.8 μm. Each frame consists of 256×256 pixels at a spatial resolution of 60 μm. The sample was at a distance of 0.1 m from the camera and was illuminated with a QCL at an angle of 30° to the surface normal. The hypercube frame shown in the top right of Figure 8.1 represents the intensity of the diffusely scattered light at $\lambda = 10$ μm. The detection map was calculated using an algorithm called the adaptive cosine estimator (ACE) [16] with the shown reference signature. The particles are clearly identified as $KClO_3$, with detection scores approaching 1 (on a scale of 0–1). In this example, note that the particles appear much larger in the detection map than in the hypercube frame. This is because the light that is forward scattered by the particles and then reflected by the mirror back to the camera also carries the chemical signature of $KClO_3$. This demonstrates that the detection score is related to the shape of the spectral reflectance rather than the scattered intensity.

8.2 SOME CONSIDERATIONS WHEN OPERATING IN THE MID-INFRARED

This section briefly reviews some factors that are important when operating in the MIR: atmospheric transmission, laser eye-safety limits, and the ambient thermal background.

Atmospheric transmission of light is discussed in more detail in Chapter 2. Briefly, the two primary atmospheric transmission windows in the MIR are called the mid-wave infrared (MWIR) and long-wave infrared (LWIR). These extend roughly between $\lambda \approx 3$–5 μm and $\lambda \approx 7.5$–14 μm, respectively. Between these transmission windows, optical radiation is strongly attenuated due to absorption by water vapor, and standoff detection at distances greater than about 10 m becomes difficult. This has direct implications for measuring the absorption band at $\lambda \sim 6.4$ μm in explosives such as TNT, as will be discussed in the following section.

Eye safety is discussed in greater detail in Chapter 2. Briefly, MIR radiation is strongly absorbed by the water in biological tissue such that the maximum permissible exposure (MPE) limits are based on heating of the tissue surface. These limits are much higher than for visible wavelengths, which can cause retinal damage. The MPE for continuous-wave (CW) MIR radiation is 0.1 W/cm^2 [17]. For short-duration exposures, the peak irradiance is even higher. For example, given an exposure of 1 μs, the MPE is 18 kW/cm^2, with a corresponding fluence (energy per unit area) of 18 mJ/cm^2. As will be shown later, the required laser fluence at the target is typically on the order of 1 μJ/cm^2 in active MIR HSI systems when using a high-sensitivity MCT camera as the receiver. The required fluence at the target, therefore, is far below the eye-safety limit. The MPE is most likely to be violated at the transmit aperture of the laser. But even for systems that utilize watt-class lasers, a transmit beam diameter of only a few centimeters is sufficient to remain below the MPE limit.

The ambient thermal background is an important contributor to the MIR signal measured at the camera. For many cases of interest, in fact, the signal received at the camera is dominated by the ambient thermal background and only a small fraction is due to the active return. The total thermal radiation emitted from an object per unit area, per steradian, and per unit bandwidth is given by Planck's expression for the spectral radiance (i.e., radiance per unit wavelength) [18]:

$$\frac{dL}{d\lambda}(\lambda, T) = \frac{2hc^2 \cdot \varepsilon}{\lambda^5 \left[\exp\left(\dfrac{hc/\lambda}{k_B T} \right) - 1 \right]} \tag{8.1}$$

where:

L = radiance (power per unit area per steradian)
ε = emissivity
h = Planck's constant
λ = vacuum wavelength
k_B = Boltzmann's constant
T = temperature

For an object that is optically thick, the object's reflectivity ρ and its emissivity are related by $\varepsilon = 1 - \rho$, because in thermal equilibrium the thermally radiated power must equal the absorbed power. Furthermore, because of the principle of detailed balance, this relationship between emissivity and reflectivity holds for any particular wavelength, angle of incidence, or polarization [19,20]. Integrating the thermal blackbody radiation ($\varepsilon = 1$) over the LWIR atmospheric window between $\lambda = 7.5$ and 14 μm, the radiance $L = 5.9$ mW/cm^2/sr. The total radiated power per unit area (i.e., radiant emittance) is then given by $\pi L = 18.6$ mW/cm^2.

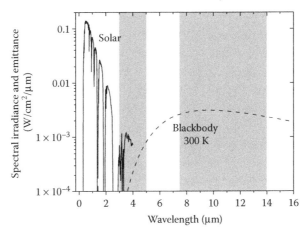

FIGURE 8.2 Comparison of spectral irradiance from direct solar illumination with the spectral emittance from a blackbody at 300 K. The MWIR and LWIR atmospheric transmission windows are indicated as gray shading.

Figure 8.2 compares the spectral irradiance from solar illumination [21] and the blackbody emittance at 300 K. In the LWIR, the thermal background dominates over the solar irradiance.

It bears repeating that in many situations the thermal background will be greater than the active signal of interest. Therefore, great care must be taken to properly subtract the thermal background from the measured signal in order to recover the true spectral reflectance of an object.

8.3   OPTICAL PROPERTIES OF EXPLOSIVES IN THE MID-INFRARED

Several classes of explosives are briefly described in this section with regard to their MIR spectra. For a more complete treatment of explosives, see [22–27] and references therein. Table 8.2 lists examples of some explosives, and components of explosives, along with the approximate location of some of their strongest absorption lines. These are listed in wave numbers (which is customary for MIR spectroscopy) and can be converted to wavelength using $\lambda = 10^4/\tilde{v}$, where $\tilde{v}$ is the photon energy in wave numbers (i.e., cm^{-1}) and the wavelength λ is in microns. The figures, however, are usually plotted versus wavelength for the benefit of those who are not spectroscopy specialists.

The organic explosives that contain the nitro (NO_2) functional group include TNT, tetrahexamine tetranitramine (HMX), cyclotrimethylenetrinitramine (RDX), trinitrophenylmethylnitramine (tetryl), and pentaerythritol tetranitrate (PETN). The asymmetric and symmetric stretching modes of NO_2 give rise to the characteristic absorption frequencies that are common to all of the nitro-based organic explosives

TABLE 8.2

Some Explosives and Components of Explosives Listed by Chemical Class

Class		Material	Absorption Lines (approx. cm^{-1})	References
Organic	Nitro	TNT	1535, 1355	[3,30,31]
		HMX	1550, 1280, 950	[30,31]
		RDX	1580, 1265, 925	[30,31]
		Tetryl	1540, 1300	[30]
		PETN	1625, 1275, 850	[30,31]
	Peroxide	TATP	1200	[31–34]
Inorganic	Nitrate	KNO_3	1380, 835	[30,31,35]
		NH_4NO_3	1390, 825	[30,31,35,36]
		Urea nitrate	1350	[31,37]
	Chlorate	$KClO_3$	965	[30,31,35]
		$NaClO_3$	970	[31,35]
	Perchlorate	$KClO_4$	1100	[30,31,35]
		NH_4ClO_4	1100, 1420	[30,35]

Note: The peak locations for some of the strongest absorptions are also given in wave numbers. Please note that the peak locations are only approximate.

at roughly 1550 and 1350 cm^{-1} ($\lambda \sim 6.4$ and 7.4 μm), respectively [1,12,28]. Triacetone triperoxide (TATP) is an example of a peroxide-based explosive that has a strong absorption peak at about 1200 cm^{-1} (or $\lambda \sim 8$ μm). The inorganic salts are strong oxidizers that are used in combination with chemicals such as sugar, aluminum, and fuel oil to make explosives. These inorganic salts have very strong absorptions associated with their nitrate (NO_3), chlorate (ClO_3), and perchlorate (ClO_4) functional groups that occur at approximately 1350, 975, and 1100 cm^{-1} (or $\lambda \sim 7.4$, 10.3, and 9 μm), respectively.

Most of the explosives have low vapor pressures [29]. For the nitro-based organic explosives that are listed in Table 8.2, the vapor concentration at 25°C varies from about 10 ppb for TNT down to about 10^{-6} ppb for HMX. TATP, however, has a high vapor pressure, with concentrations in air of almost 1 part per thousand (ppt) at 25°C. Therefore, unlike the other explosives listed in Table 8.2, TATP is highly detectable in vapor form.

The optical spectra for several of these materials are shown in Figure 8.3. The complex refractive index, $n = \hat{n} + i\hat{k}$, is plotted for TNT, RDX [38], ammonium nitrate [36], and sodium chlorate [39]. Also shown is the optical absorbance for TATP [32] and nitric acid [40]. The following general observations can be made:

1. TNT has absorption lines due to the NO_2 functional group ($\lambda \sim 6.4$ and 7.4 μm), with peak values for the imaginary index of $\hat{k} \sim 0.2$. This corresponds to an absorption depth, $L_{abs} = \lambda / 4\pi \hat{k}$, of only about 3 μm. When detecting these absorption lines in a standoff configuration, care must be taken to avoid the atmospheric absorption due to water vapor.
2. RDX has a rich MIR spectrum that spans the LWIR.
3. The inorganic nitrates have a strong and broad absorption band at approximately 1350 cm^{-1}. According to published index data, ammonium nitrate (NH_4NO_3) has a strong and broad absorption peak at $\lambda = 7.2$ μm (1385 cm^{-1}) with $\hat{k} \sim 1.7$, which corresponds to an absorption depth in the material of only 0.3 μm. Strong absorption significantly affects the real part of the refractive index such that the optical response must take into account both the real and imaginary parts of the refractive index. As with the nitro-based organic compounds, this peak is close to the atmospheric absorption band of water. There is another peak at $\lambda = 12.1$ μm (825 cm^{-1}). Ammonium nitrate also has absorption at $\lambda \sim 3$ μm, corresponding to the N–H bonds, but these are not specific because of the ubiquitous nature of chemicals with hydrogen bonds.
4. The chlorates and perchlorates are even stronger absorbers than the nitrates. For sodium chlorate ($NaClO_3$), the imaginary index reaches a peak value of $\hat{k} \sim 4$ at $\lambda \sim 10.3$ μm. This gives rise to a Reststrahlen band between about $\lambda \approx 9.7$–10.7 μm in which the reflectance is high [41]. At shorter wavelengths around $\lambda \approx 9.3$ μm, the real part of the refractive index approaches unity and the reflectance approaches zero. Strong absorption in the chlorates and perchlorates makes them relatively easy to detect.
5. TATP has a strong absorption at about 1200 cm^{-1} due to the C–O stretch.

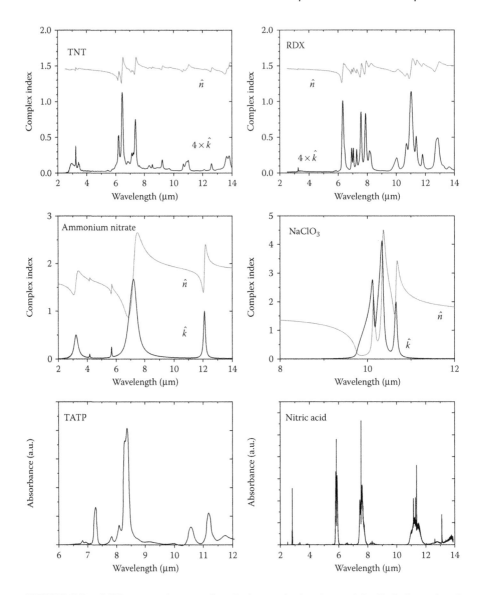

FIGURE 8.3 MIR spectra for several explosives and related materials. Both the real and imaginary parts of the refractive index for TNT, RDX, ammonium nitrate, and sodium chlorate are plotted. Note that the imaginary indexes for TNT and RDX are scaled. The absorbance spectra for TATP and nitric acid are also plotted.

6. The spectrum for nitric acid is included because it is often used to make explosives. Nitric acid has several absorption bands within the LWIR and its spectrum is rich with features.

7. Additional spectra that are not presented in this chapter, but which are critical for understanding the specificity of MIR spectroscopy to detect

explosives in real-world conditions, are the spectra of the common back-grounds, confusers, and interferents.

8.4 SENSOR COMPONENTS, GEOMETRY, AND SIGNAL-TO-NOISE RATIO

The basic elements of an active MIR HSI system are shown in Figure 8.4. This section describes the system geometry, calculation of the signal-to-noise ratio (SNR), and the performance of the component laser and camera technologies. Although the examples are geared toward a detection system with a standoff distance of $R > 1$ m, the discussion is also applicable to close-range imaging.

8.4.1 CAMERA

A camera comprises a two-dimensional array of photodetectors called the FPA, an associated read-out integrated circuit (ROIC), electronics, and optics. The FPA has a number of pixels $N_{pix} = (m_{pix} \times n_{pix})$ that are assumed to be square and contiguous such that the pixel size and spacing are both d_{pix}. The surface of interest is at a distance R from the lens and it is imaged onto the FPA using a lens with effective focal length f and diameter D_{lens} such that its f-number is $f/\# = f/D_{lens}$. Each pixel in the FPA images a region on the ground having a linear size called the ground sampling distance (GSD). The GSD is given by Equation 8.2 [42]:

$$\text{GSD} = \left(\frac{R}{f} - 1\right) d_{pix} \approx \frac{R}{f} d_{pix} \tag{8.2}$$

where the approximation is valid for $R \gg f$. Note that we define the GSD in an object plane that is normal to the viewing direction. The angle between the object plane and the actual surface is θ. The camera's field of view (FOV) is approximately given by $(m_{pix} \times n_{pix})(d_{pix}/f)(180°/\pi)$, while the instantaneous FOV (IFOV) refers to the view of

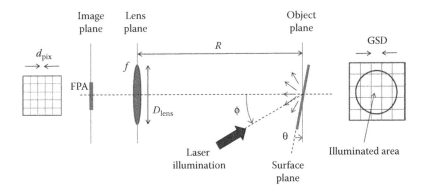

FIGURE 8.4 Geometry of an active MIR HSI system.

single pixel, which is given by $(d_{pix}/f)(180°/\pi)$. As an example, consider a 320×240 FPA with 20 μm pixels and a lens with $f = 100$ mm. The FOV is then $3.7° \times 2.4°$ and the IFOV $= 0.01° = 0.2$ mrad. At a range of 10 m, the imaged footprint on the object plane is 0.64 m $\times$ 0.48 m and the GSD $= 2$ mm.

8.4.2 ILLUMINATION

The active illumination is assumed to come from a laser at an angle ϕ relative to the camera's viewing direction. Following standard nomenclature, the illumination irradiance (power per unit area) is defined in a plane that is normal to the illumination direction.

Illumination configurations can be categorized as being either coaxial ($\phi = 0°$) or oblique ($\phi \neq 0°$). An advantage of coaxial illumination is that the laser beam can be fixed within the camera's FOV for all distances. However, this requires the use of either a beam splitter or an obscuring transmit mirror within the receiver aperture. This may be complicated to implement with commercial off-the-shelf (COTS) camera lenses. Also, if not properly designed, stray light from the transmit beam may interfere with the return signal. Alternatively, oblique illumination may be simpler to implement for a fixed distance to the target and also result in favorable reflection signatures. However, the drawback of oblique illumination is that the laser pointing must be adjusted as the distance to the object is changed in order to center the beam on the camera's FOV. To simplify the discussion that follows, we assume coaxial illumination.

There are also several options for focusing of the laser beam: collimated, diverging, or adaptive focus. For a collimated beam, the laser beam diameter remains constant over a wide range of standoff distances. This, in turn, results in a signal level at each fully illuminated pixel that is independent of distance, as will be shown later. By *fully illuminated pixel*, it is meant that the entire IFOV of a pixel is illuminated by the laser. The distance over which a Gaussian laser beam remains approximately constant is given by the Rayleigh range $z_R = \pi D_0^2/(4M^2\lambda)$ where D_0 is the minimum beam diameter and M^2 is a measure of the beam quality (i.e., a diffraction-limited beam has $M^2 = 1$) [43]. For a diffraction-limited beam at $\lambda = 10$ μm, a 1 cm diameter beam remains collimated over almost 10 m. Increasing the beam diameter to 4 cm increases the Rayleigh range to >100 m. Therefore, it is not difficult to design a system to operate in this mode. Alternatively, the laser beam can be designed to diverge in order to illuminate a fixed FOV as seen by the camera. Since the beam size increases with distance to the object, the signal received at each pixel will also decrease with distance. Therefore, this approach is only suitable for a narrow range of standoff distances. Finally, one can consider an adaptive system that optimally focuses the laser beam depending on the properties of the target. The minimum achievable spot size for a diffraction-limited beam is approximately given by $D_{min} \approx \lambda R/D_T$ where D_T is the beam diameter at the transmit aperture. As an example, for $\lambda = 10$ μm, $D_T = 2$ cm, and a range of $R = 10$ m, the spot size can be as small as 5 mm. Although this approach is more complex than the others, it enables the flexibility to adaptively adjust the signal level from objects depending on their reflectivity.

For standoff detection applications, the beam-pointing stability of the laser source is very important as this will impact the footprint of the illumination on the object and the quality of the generated hypercubes.

8.4.3 RECEIVED SIGNAL POWER

The total power received by the camera, $P_{receiver}$, can be written as in Equation 8.3 [44]:

$$P_{receiver} = P_{laser} \cdot \rho_\Omega \, \Omega \cdot T \qquad (8.3)$$

where:

P_{laser} = incident laser power
ρ_Ω = differential reflectance of the object (i.e., reflectance per unit solid-angle) in the direction of the receiver
Ω = solid angle of the receiver
T = a factor that represents all transmission losses including those due to the receive optics and atmospheric attenuation

The ρ_Ω is equivalent to the bidirectional reflectance distribution function (BRDF) [45]. The solid angle for the receiver is given by Equation 8.4 [18,44]:

$$\Omega = 2\pi \left[1 - \frac{R}{\sqrt{R^2 + \left(D_{lens}/2\right)^2}} \right] \approx \pi \left(\frac{D_{lens}}{2R} \right)^2 \qquad (8.4)$$

where the approximate expression is valid for $R \gg D_{lens}$. For SNR calculations, we are primarily interested in determining the signal power at each pixel, P_{pix}. This is easiest to write in terms of the laser irradiance I_{laser}, as

$$P_{pix} = \left(I_{laser} GSD^2 \right) \rho_\Omega \, \Omega \cdot T \quad \approx I_{laser} \cdot \rho_\Omega \cdot \frac{\pi}{4} \left(\frac{d_{pix}}{f/\#} \right)^2 \cdot T \qquad (8.5)$$

where the approximate expression is valid for $R \gg f$. The first expression makes it clear that the power received at a pixel is proportional to the laser power that is incident within the IFOV of the pixel. The second expression shows that the signal at each pixel in the camera depends upon the laser irradiance, the differential reflectance of the object, and fixed properties of the receiver.

Interestingly, the power received per pixel does not directly depend on the distance to the object for a given laser irradiance (when neglecting effects due to atmospheric attenuation). This result is counterintuitive, because one usually expects that the received signal will drop with increasing distance. This can be understood by noting that the solid angle for collection decreases with range $\Omega \propto R^{-2}$, while the area imaged by a single pixel increases with range as $GSD^2 \propto R^2$—these two effects cancel each other out. That is, the total received power decreases with distance, but the power per pixel remains constant for fixed laser irradiance. As a numerical example, consider the case where $d_{pix} = 20$ μm, $f/\# = 1$, $I_{laser} = 0.1$ W/cm^2, $\rho_\Omega = 0.01$ sr^{-1}, and $T = 1$, then $P_{pixel} = 3$ nW. As discussed below, this received power is suitable for achieving a detector-limited SNR of >100 with an MCT camera. Note that the thermal background at 300 K results in a signal that is 6× greater than the active return. This again underscores the need to carefully subtract the thermal background to accurately measure the active return.

8.4.4 DETECTOR SENSITIVITY

Consider a first-order analysis of the detector noise, neglecting contributions from electronic noise, quantization noise, and so on. The detector-limited SNR is then due to Poisson-distributed statistics and is given by Equation 8.6 [19]:

$$SNR = \frac{S}{\sqrt{S + N_{total}}} = \frac{S}{\sqrt{S + N_{bkgd} + N_{dark}}} \tag{8.6}$$

where:
S = total number of photogenerated electrons due to the active signal of interest
N_{total} = total number of "noise" electrons due to the thermal background, N_{bkgd} and dark current, N_{dark}

In most cases, $N_{total} \gg S$. Since it is reasonable to assume that N_{total} is proportional to both the area of the pixel $A_{pix} = d_{pix}^2$ and the integration time t_{int}, the detector sensitivity is characterized by the specific detectivity [19]

$$D^* = \sqrt{A_{pix}B} \cdot \frac{1}{NEP} = \frac{d_{pix}}{\sqrt{2t_{int}}} \frac{1}{NEP} \tag{8.7}$$

where the electrical bandwidth B is related to the integration time using $B = 1/2t_{int}$ and where the noise-equivalent power (NEP) is related to the detector-limited SNR by

$$SNR = \frac{P_{pix}}{NEP} = P_{pix} \frac{D^*\sqrt{2t_{int}}}{d_{pix}} \tag{8.8}$$

For a Stirling-cycle-cooled LWIR MCT FPA, D^* is typically ~5×10^{10} cm · $\sqrt{Hz}$/W. The NEP is then calculated to be NEP = 9 pW for an integration time of 10 μs and pixel size of $d_{pix} = 20$ μm. MB imagers operate by measuring the temperature rise at each pixel due to the photon flux. The sensitivity of MB cameras is usually given in terms of their noise-equivalent differential temperature (NEdT, or sometimes NETD), which is typically around 50 mK [46]. Because of the nature of MBs, their performance is closely linked to the thermal time constant of the pixels, which is designed to be roughly 10 ms in order to achieve video frame rates. The NEP can be estimated from the change in radiance of a blackbody object when its temperature changes by the NEdT over the spectral band $\lambda = 7.5-14$ µm [46]. Using a differential radiance of $dL/dT = 93$ µW/cm²/sr/K, one calculates NEP = 15 pW. This value is within a factor of ~2 measurements on a commercial MB camera after correcting for pixel size [47].

Most cooled MCT cameras have an NEdT that is only about two to five times better than for MB cameras. It may be surprising that both cooled MCT and uncooled MB imagers have similar NEdT (and NEP), since cooled MCT FPAs have significantly higher cost and complexity. However, an important difference is that a cooled MCT achieves this sensitivity in a measurement time that is about 1000× shorter than an uncooled MB. For active MIR HSI systems, this translates into requiring about 1000× less laser energy to achieve a desired SNR. While an illumination fluence of 1 µJ/cm² (=100 mW × 10 µs) may be sufficient when using a cooled MCT, an uncooled MB requires a fluence on the order of 1 mJ/cm². A related point is that a cooled MCT is capable of much higher frame rates than an uncooled MB camera (>1 kHz vs. 50 Hz). These differences have significant implications for the required laser power and for the ACRs that can be achieved, as will be discussed in Section 8.9.1.

8.4.5 KEY TECHNOLOGIES

The following sections discuss the current state of the art for MIR lasers and cameras.

8.4.5.1 MIR Lasers

Blackbody thermal sources called *globars* are often used in conjunction with FTIR spectrometers for microspectroscopy in the MIR [11]. Thermal sources, however, are generally unsuitable for standoff measurement of surfaces because they have much lower radiance than lasers. A thermal source at 1100 K has a radiance of 0.3 W/cm²/sr over the $\lambda = 8-14$ µm band. In comparison, a 1 mW diffraction-limited laser beam at $\lambda = 10$ µm has a radiance of about 10^3 W/cm²/sr, which is roughly 10^4 greater than for the thermal source. Although thermal sources have been used with FTIR spectrometers for standoff surface detection, this approach becomes unviable at longer ranges [48,49].

As mentioned earlier, the resurgence of interest in active MIR HSI is due in large part to the invention, development, and commercialization of widely tunable QCLs. QCLs are a game-changing MIR laser technology making MIR spectroscopy practical in configurations that were previously not possible [50]. Prior to the invention of QCLs in 1994 [7], wavelength-tunable MIR sources included lead-salt lasers [51], antimonide-based lasers [52], carbon dioxide lasers [53], synchrotron radiation [54], and optical parametric oscillators [55–57]. QCLs will largely supplant these other technologies for MIR spectroscopy applications because they can provide continuous wavelength coverage over much of the MIR (at least $\lambda > 3-14$ µm) with high power and excellent beam quality. Furthermore, they operate at room temperature and are compact, efficient, and reliable [10].

QCLs are capable of watt-class average powers in both the MWIR and LWIR [58,59]. This is good news for active MIR HSI systems because, as will be shown later, the ACR is directly related to the laser power. For HSI systems that utilize MCT cameras, it is desirable to match the operating characteristics of MCT cameras by illuminating the scene with high-pulse energies with a duty factor on the order of 10%. This operating regime is somewhat difficult for QCLs to reach because of their short thermal time constant, small gain volume, and limited maximum current density [10]. To date, a diffraction-limited peak power of 10 W with ~100 ns long pulses has been demonstrated at low duty factor [60].

For spectroscopy, it is desirable that QCLs be widely tunable in wavelength. Much progress has been made to increase the gain bandwidth from a single laser chip [9,61]. Wavelength tuning in QCLs is typically achieved by placing the QCL chip in an external optical cavity that incorporates a dispersive element such as a diffraction grating. In such external-cavity QCLs (ECQCLs), the wavelength is tuned over the gain bandwidth of the laser chip by rotating the grating or mirror within the cavity. Therefore, this approach involves mechanical motion of optical components that will ultimately limit the tuning speed. The widest tuning that has been reported to date from a single laser chip is $\lambda = 7.6$–11.4 μm ($\Delta\tilde{\nu} = 432$ cm^{-1}) [9]. Alternative approaches to wavelength tuning that do not involve an external cavity are areas of active research [62,63].

Multiwavelength arrays of QCLs have also been demonstrated [60,64,65]. These arrays are very suitable for achieving high power because the total power from the array can be the sum of the powers from all devices within the array. Furthermore, extremely high-speed tuning is possible without any moving parts since wavelength tuning is achieved electronically simply by driving the appropriate laser within the array. By applying beam-combining methods to these arrays, the beams from all lasers in the array can be spatially overlapped for standoff detection applications [66–68]. QCL arrays have many attractive features for active MIR HSI applications, but more effort is needed to advance the technology.

Widely tunable ECQCLs are commercially available and some examples are shown in Figure 8.5. Currently, modules are available with peak powers of hundreds of milliwatts in a diffraction-limited beam that can be tuned over spectral ranges of >250 cm^{-1} at scan rates of >100 Hz. Multiwavelength arrays have also recently become commercially available.

8.4.5.2 MIR Cameras

There have also been exciting developments in MIR camera technology based on either thermal or quantum detectors. Thermal-detector FPAs are primarily based on MB technology. Quantum-detector FPAs are based on MCT, indium antimonide (InSb), quantum-well infrared photodetector (QWIP), and the strained-layer superlattice (SLS).

Commercial MB cameras are "uncooled" in that they operate near room temperature. At the time of this writing, the cost of uncooled MB cameras has been dropping significantly and they are becoming widely utilized. As described in the previous section, MB FPAs are characterized by NEdT of ~50 mK, thermal time constants of ~10 ms, and frame rates of ~30 Hz [46]. As an example, Table 8.3 gives

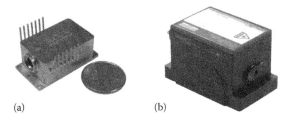

(a)   (b)

FIGURE 8.5   Examples of commercially available widely tunable QCLs from (a) Block Engineering LLC and (b) Daylight Solutions, Inc.

the specifications for a commercially available MB camera having a readout rate of 10 Mpixels/s (FLIR Quark). Many MB FPAs have a short wavelength limit of $\lambda \sim 7.5$ µm, which means that they may not be able to detect some of the primary absorption bands of nitro-based organic explosives.

Imagers based on quantum detectors have the highest performance but these need to be cooled to cryogenic temperatures using Stirling-cycle coolers. The functionality of these imagers is determined to a large extent by their ROICs, which are typically fabricated in silicon CMOS. Conventional ROICs for MIR FPAs are designed such that the photocurrent from each detector in the array charges an associated capacitor. The size of this capacitor determines the *well depth* of the pixel in terms of the maximum number of electrons that can be accumulated. The well depth is typically a few tens of millions of electrons. After integrating the detector current during an exposure, the charge is read out on a row-by-row basis and converted to a digital value using one or more analog-to-digital converters (ADCs). Because of the limited well depth, the NEP cannot be reduced by arbitrarily increasing the integration time. The NEP can be reduced, however, by averaging over multiple frames.

MCT FPAs were first employed for FTIR microspectroscopy in the mid-1990s using the 64×64 element Javelin detector [69]. Table 8.3 gives the performance specifications of a modern, commercially available MCT FPA. This particular FPA

TABLE 8.3

Examples of Commercially Available MB and MCT FPAs

	MB FLIR Quark640	MCT Teledyne NOVA161
Cooling	Uncooled	Stirling-cycle cooled ($T < 80$ K)
Pixels	640×514	128×128
Pixel pitch (µm)	17	40
Spectral band (µm)	7.5–13.5	2–11.5
NEdT (mK)	<50 (*f*/1.0 lens)	<20 (*f*/2.3 lens)
Frame rate (Hz)	30	1610 (max)
Well depth	n/a	51×10^6 electrons
Digital bits	14-bit	

has 128×128 pixels and is capable of frame rates of up to 1610 Hz (Teledyne Nova Sensors, http://www.sbfp.com/fpa_family.html). This corresponds to a readout rate of 26 Mpixels/s. MCT cameras can be configured to be sensitive at shorter MIR wavelengths in order to access some of the important absorption bands for explosives. They also have flexibility in terms of allowing variable exposure times.

An important recent advancement in MCT FPA technology is the digital-pixel FPA. These have digital ROICs to overcome some limitations of analog ROICs [70–72]. Digital-pixel FPAs incorporate an ADC underneath each pixel of the FPA. The ADCs work by counting the number of times a very small capacitor is charged by the detector current. Once the capacitor reaches a preset voltage, the circuitry is designed to discharge the capacitor and trigger a digital counter. The counter is bidirectional (i.e., it can both add and subtract counts), and counting can be dynamically enabled or disabled. In the normal mode of operation, the number of counts accumulated by the counter is proportional to the integrated detector current with a proportionality constant of, for example, a few thousand electrons per count. This architecture enables unique capabilities that are relevant for active MIR HSI systems. First, digital-pixel FPAs can time-gate the optical return from a pulsed laser by enabling the counter to count only while the laser is illuminating the scene. Between pulses, counting is not enabled such that the intrapulse thermal background does not contribute to the signal and, therefore, does not add noise. Second, the digital-pixel FPA has the ability to integrate the signal from multiple laser pulses. The third important capability is *on-chip* subtraction of the thermal background. Typically, subtraction of the thermal background is achieved by taking a first frame with laser illumination and a second frame with no illumination, then differencing the images at the computer [73]. Not only does this increase the number of frames that must be read from the camera, but any motion of the object between frames will result in imperfect registration of the images. The digital-pixel FPA, however, can perform high-speed background subtraction by counting down between laser pulses to result in a counter value that represents only the active signal of interest. A demonstration of these benefits is shown in Figure 8.6. Figure 8.6a shows a conventional thermal image of a hand in which the integration time is 100 μs. During acquisition of the image, a single laser pulse illuminates the surface, but its intensity is too low to be visible. Figure 8.6b shows the integration of 50×2 μs long pulses to increase the active return relative to the passive background. Even with multipulse integration, the laser beam is barely visible because the passive background remains stronger than the active return. Figure 8.6c demonstrates the benefit of combining multipulse integration with on-chip background subtraction. For each laser pulse, the digital-pixel FPA integrates the active return for 2 μs while the laser is on, and it then subtracts thermal background for 2 μs when the laser is off. This process is repeated 50 times to integrate the signal over multiple laser pulses and yields an image of the active return only. This capability is very important because the thermal background is often significantly larger than the active return. Note also that the active return can occupy the full dynamic range of the imager (e.g., 14 bits) without losing headroom to the passive signal. Finally, it is worth mentioning that digital-pixel FPAs have additional advantages such as

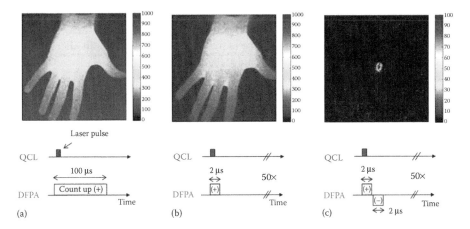

FIGURE 8.6 Demonstration of the benefits of a digital-pixel FPA. (a) A conventional thermal image of a hand that is illuminated by a QCL. The single laser pulse during the 100 μs long integration period is too weak to be visible. (b) Synchronous detection with multipulse integration to increase the active return by a factor of 50 compared to the conventional image. (c) Synchronous imaging with on-chip background subtraction is combined with multipulse integration to extract only the active return. As compared to (b), the ambient thermal background is subtracted on-chip.

very high frame rates (readout rates of >200 Mpixels/s) and rudimentary on-chip processing capabilities. At the time of this writing, these digital-pixel FPAs are not yet commercially available.

The high cost of MCT FPAs has given impetus to the development of FPAs from among the lower-cost III–V semiconductors. InSb FPAs are available with very high sensitivity but their cutoff wavelength is only ~5 μm. QWIP detectors have the limitation of being sensitive only over a relatively narrow band [74]. SLS FPAs are based on InAs/InGaAs superlattice structures and have already been demonstrated in megapixel-class formats with a cutoff wavelength of >9.5 μm [75]. These have good sensitivity over a broad spectral band and hold the promise of being a lower-cost alternative to MCT.

8.5 DEMONSTRATIONS OF ACTIVE MIR HSI FOR EXPLOSIVES DETECTION

There is a growing body of work on the use of active MIR spectroscopy to detect explosives and other chemicals on surfaces [76–79]. In this section, we review the subset of this work that involves imaging.

8.5.1 REVIEW OF REPORTED RESULTS

Microspectroscopy can be considered a class of active MIR HSI. It is conventionally performed using a microscope coupled with a FTIR spectrometer and detector (single-element detector, linear array, or FPA) [69], and it has been successfully used to detect explosive residues in fingerprints [80–83]. More recently, QCLs have been incorporated into microscopes to detect explosives using both transmittance [84,85] and photothermal spectroscopy [86]. One of the earliest uses of QCLs in microscopy was to chemically image biological samples [87].

Standoff active MIR HSI has been demonstrated using both reflectance and photothermal spectroscopy [76,77]. Since photothermal spectroscopy is described in detail in Chapter 10 [77,88,89], this section considers only reflectance spectroscopy. An imaging FTIR has been used in conjunction with a radiant heater as the illumination source to detect liquids on surfaces at a standoff distance of approximately 1 m [48]. For longer-range detection, laser sources are needed. Laser sources such as an optical parametric oscillator (OPO) [90] and QCLs [91] have been used in conjunction with raster scanning to achieve standoff classification of materials based on their reflection spectra. More recently, a system for detecting liquid chemicals on realistic outdoor surfaces such as concrete was demonstrated using QCL illumination and a digital-pixel FPA [92]. The use of QCLs and FPAs for the detection of explosives has been demonstrated by several groups. Oak Ridge National Laboratory, using a MB camera and ECQCL, demonstrated the detection of RDX on polished stainless steel at a distance of 15 cm [93]. Pacific Northwest National Laboratory (PNNL) developed a rapidly tunable ECQCL that was synchronized with a MB camera to capture a 60-frame hypercube in 2 s. Using this system, they demonstrated the ability to extract the absorption spectrum of RDX as deposited on a mirror from the hypercubes [94]. In the following, we discuss in greater detail the work by the Fraunhofer Institute and the MIT Lincoln Laboratory.

8.5.2 FRAUNHOFER INSTITUTE

The Fraunhofer Institute was one of the first organizations to demonstrate standoff detection of explosive residues using reflection-based active MIR HSI with QCLs. In their initial experiments, the QCL wavelength was adjusted via temperature tuning and imaging was achieved using a single-element detector and raster scanning of the sample [95]. Over time, their system evolved to comprise a widely tunable ECQCL, MCT camera, despeckling unit, and adaptive detection algorithms [96]. Their ECQCL spans >300 cm^{-1} to overlap the absorption peaks of several nitro-based organic explosives and was capable of maximum average powers of around 40 mW. The MCT camera had 256×256 pixels and a maximum frame rate of 400 Hz. The maximum integration time of the camera was limited to about 100 μs because of the imager's finite well depth (see discussion in Section 8.4.5.2). Subtraction of the thermal background was achieved by recording alternating frames without laser illumination and subtracting these from the actively illuminated frame after computer acquisition. To manage speckle (which is a topic discussed in Section 8.8), emission from the QCL was coupled to a despeckling unit that incorporated a pair of diamond diffuser plates in which one of the plates rotates [97–99].

Trace residues of PETN, TNT, RDX, and other compounds were measured on a variety of substrates such as metallic plates, painted car panels, and fabric. The samples were typically oriented far from the specular reflection condition in order to capture the diffusely scattered light, since this configuration is most likely to be encountered in practice. It was found that the measured reflection spectra were not readily interpreted with respect to absorption measurements of explosives made using attenuated total reflection (ATR) [100]. Naively, one would expect the reflectivity to be reduced at the absorption lines of the explosive. This situation is observed when the trace explosives (e.g., TNT) are on a highly reflective surface [97]. When the explosives are on a dielectric substrate (e.g., painted car panel or fabric), so-called *contrast reversal* is often observed in which there is an *enhancement* of the reflectance at, or near, the absorption line. This behavior is consistent with the theory, as discussed in Section 8.6. However, it underscores the fact that there is a complex relationship between the reflection signatures from explosive residues and their intrinsic absorption spectra. This is further complicated by the fact that reliable refractive index data for explosives are not readily available. In the end, the group at Fraunhofer relied on an empirical approach to generate reference spectra that were then used in automated detection algorithms.

For smooth substrates such as painted car panels, the underlying substrate has negligible diffuse scattering. Therefore, even with a low fractional surface coverage of the explosive (i.e., low fill factor), most of the light collected by the camera is due to scattering from the explosive particles (see Sections 8.6.1 and 8.6.2.2). This is a favorable condition because the substrate has minimal impact on the measured reflection spectra. Consistent reflection spectra were measured for a variety of explosives (e.g., PETN, TNT, RDX) on painted car panels that were deposited with mass loadings that ranged from about 20 µg to >1 mg within the area of a thumbprint [96]. For substrates such as a polyamide fabric, the substrate itself generated a significant diffuse reflectance signal with spectral features that masked those of the explosive. To deal with this situation, a method was developed to correct for the substrate reflectance that assumes a linear mixing model in which the reflectance of the substrate adds linearly with the reflection spectrum of the explosive [100,101]. Making use of the linear mixing model and combining it with an adaptive detection algorithm, they were able to detect contaminated thumbprints with ~20 µg of PETN on polyamide fabric. Also detected was explosive residue on everyday objects such as keys, a computer mouse, and a smartphone using the adaptive detection algorithm [96]. Finally, it is worth noting that they demonstrated automated detection of PETN on a painted car panel at a range of up to 20 m [96].

8.5.3   MIT LINCOLN LABORATORY

At the MIT Lincoln Laboratory, in collaboration with the Army Research Laboratory (ARL), the authors demonstrated standoff detection of trace levels of chlorates such as $KClO_3$ (which are components of some explosives) on as variety of surfaces [15]. Measurements were made using the configuration shown in Figure 8.1 and described in Section 8.1.2. The QCLs were tuned across the wavelength range $\lambda = 9.1$–10.8 µm in steps of 1 cm^{-1}. At each illumination wavelength, the reflected light was captured by a digital-pixel MCT camera to generate a hypercube with 164 frames. The digital-pixel MCT camera having 256×256, 30-µm pixels was operated synchronously with the QCL to achieve multipulse integration of 50×0.5 µm long pulses with on-chip subtraction of the thermal background to obtain an accurate measurement of the surface reflectance (see details in Section 8.4.5.2) [92]. For the figures shown in this section, the intensity images are plotted relative to the reflectance from a diffuse gold reflection standard. Detection maps were generated using the ACE algorithm [16] with the bulk powder reflectance being used as the reference signature.

For both Figures 8.1 and 8.7, measurements are of $KClO_3$ particles on a metallic mirror at a range of 0.1 m with optics ($f = 50$ mm, $f/\# = 1.4$) to yield GSD = 60 μm. The illumination angle was 30°. In Figure 8.7, the reflection spectrum of an individual particle of $KClO_3$ is compared with the reflectance of bulk powder. These are found to be qualitatively similar.

Figure 8.8 demonstrates the detection of $KClO_3$ particles on cotton fabric at close range with GSD = 60 μm. The top row of images shows the intensity of the reflectance at $\lambda = 9.75$ μm for clean fabric, fabric with coarse $KClO_3$ particles, and fabric with fine $KClO_3$ particles. The particles were applied by wiping a contaminated fingerprint across the surface. The lower row of images shows the ACE detection maps. For the coarse particles larger than about 100 μm, the reflection spectrum of the particles is highly correlated with the reference spectrum, and high-confidence detection is achieved. The fine particles, having a diameter of about 5 μm, can also be detected even though the GSD of 60 μm is much larger than the particle size. This high-sensitivity detection is probably due to the aggregation particles on the cotton fiber and also to the strong absorption lines in $KClO_3$, which yield a high-contrast spectrum.

For quantitative measurements, $KClO_3$ was deposited in a regular array onto a black car panel by ARL as shown in Figure 8.9 using a microdrop generator [102]. The particles are spaced by 1 mm and the mass of each particle varies from left to right in these images from 8 to 80 ng. As deposited, most of the particles have the shape of a spherical cap with a diameter of about 50 μm. However, due to inadvertent flexing of the substrate during testing, many of these particles cracked. The lower panel shows the detection maps at several values of GSD that were obtained by varying the distance to the sample. At close range with a GSD of 60 μm, all of

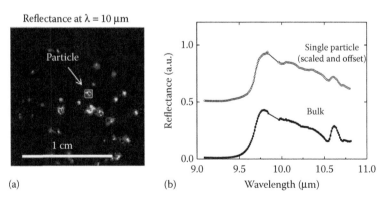

FIGURE 8.7 (a) Image of the reflected intensity for $KClO_3$ particles on a mirror (same image as in Figure 8.1). (b) Comparison of the reflectance from a single particle with the reflectance from $KClO_3$ in powder form.

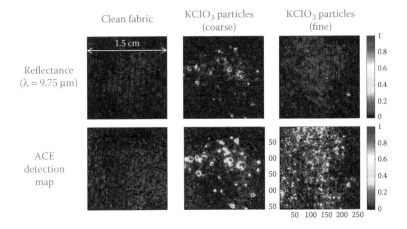

FIGURE 8.8 Demonstration of the detection of $KClO_3$ particles on cotton fabric. The top row shows a frame from the hypercube at $\lambda = 9.75$ μm. The bottom row shows the detection maps. Coarse particles $>\sim100$ μm diameter can easily be detected. The fine particles (~5 μm diameter) can also be detected, but probably only in aggregate when they collect in sufficient concentrations on the cotton fiber. For these measurements, GSD = 60 μm.

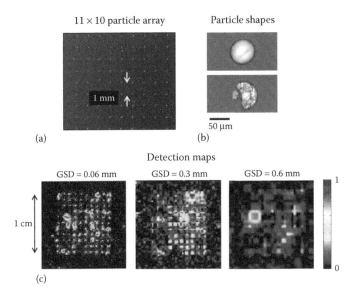

FIGURE 8.9 (a) Photomicrograph of $KClO_3$ particles deposited on a car panel in a regular array using a microdrop generator. (b) Higher magnification images of two particles. The top image is of a particle having the shape of a spherical cap. The lower image is of a damaged particle. (c) Detection maps for the array of particles for several different values of GSD. At close range with GSD = 60 μm, the probability of detecting individual 10 ng-class particles is very high. With increasing GSD, the probability of detection decreases.

the particles can be detected. The observed variability in the strength of the signal is attributed to the fact that some particles are more scattering than others because they have cracked. At GSD = 0.3 mm (range = 0.5 m), where the pixel fill factor is 2%, most of the particles with mass >24 ng are detected. At a GSD = 0.6 mm (range = 1 m), where the pixel fill factor is only 0.6%, almost none of the particles is detected. Therefore, it was possible to detect individual 10 ng class particles at a GSD of 0.3 mm with a fill factor of as little at 2%. This experiment clearly shows that higher sensitivity detection is possible at closer range because of the higher fill factor of the explosive particles. This demonstrates a great benefit of hyperspectral imaging systems as opposed to flood illuminating a surface and measuring the average return.

Figure 8.10 shows detection results for a collection of particles that are deposited onto a car panel via a contaminated fingerprint. Fingerprints were contaminated with either $KClO_3$ or soil. Measurements were made at ranges of 0.1 and 1 m with corresponding GSD of 60 μm and 0.6 mm, respectively. The diameter of the $KClO_3$ particles is only about 5 μm such that detections are due to scattering from multiple particles within the pixel's FOV. The total mass of $KClO_3$ on the surface was determined using liquid chromatography. Robust detection was demonstrated at surface concentrations of ~50 μg/cm². For soil fingerprints, the detection scores are low. By plotting the number of pixels with detection scores above a certain threshold (not shown), one finds that the $KClO_3$ fingerprints can be clearly distinguished from the soil fingerprint. Multipixel detection approaches can significantly reduce the false-alarm rate. This is another important advantage of imaging systems as compared to single-pixel measurements.

Figure 8.11 shows the detection maps for detection of $KClO_3$-contaminated fingerprints at long distances. The $KClO_3$ residue had a concentration of roughly 150 μg/cm², and samples were oriented at approximately 10° from the specular

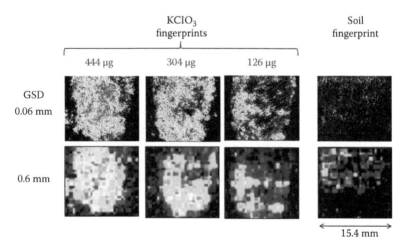

FIGURE 8.10 Detection maps for $KClO_3$-contaminated fingerprints on car panels. The $KClO_3$ surface concentration is varied. Measurements are made at ranges of 0.1 m and 1 m, which correspond to GSD = 0.06 and 0.6 mm, respectively.

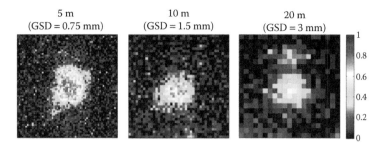

FIGURE 8.11   Detection map for a $KClO_3$-contaminated fingerprint at ranges varying from 5 to 20 m. A lens with $f = 200$ mm was used.

reflection condition. A lens with $f = 200$ mm was used to achieve a GSD of 0.75, 1.5, and 3 mm at ranges of 5, 10, and 20 m, respectively. As the GSD increases, the pixelation becomes more apparent. Nevertheless, robust detection is observed in all cases.

Figure 8.12 gives a demonstration of how advances in laser and camera technology will enable high-speed active HSI. As described in [68], the sample consisted of particles of $KClO_3$ and sand adhered to double-sided tape. The sample was mounted onto a chopper wheel that rotated the sample at linear speeds of up to 10 m/s. Illumination was provided by a multiwavelength QCL array that is capable of extremely high-speed wavelength tuning. The laser illumination was synchronized with the operation of a high-speed digital-pixel FPA to capture the differential spectral reflectance within a measurement time of only 4.1 µs. These are shown in the top row of Figure 8.12 and demonstrate the classification of particles based on chemical

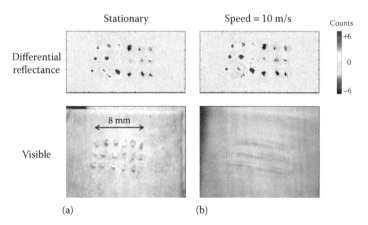

FIGURE 8.12   Differential reflectance images (*top*) and visible images (*bottom*) of particles of $KClO_3$ and sand when (a) stationary and (b) moving at a linear speed of 10 m/s. The differential reflectance image distinguishes the particles based on composition. Since the differential reflectance image is acquired in only 4.1 µs, the image does not significantly blur even when the sample is moving at speeds of 10 m/s. The visible image, however, with the acquisition time of 100 µs is blurred.

composition even when the sample is moving at 10 m/s. The bottom row shows visible images captured with an integration time of 100 μs (i.e., equivalent shutter speed of 1/10,000th of a second). When the sample is moving at 10 m/s, the visible image blurs, while the differential reflectance image remains sharp.

8.5.4 FUTURE WORK

Significant progress has been made in applying active MIR HSI to the standoff detection of explosives. The results to date demonstrate the considerable promise of this method, but much work remains to be done to develop practical systems. Future advances in laser and camera technology will enable high-speed systems that can scan surfaces rapidly. Beyond the development of hardware, an important component of a system is the classification algorithms that are needed to make sense of the large amount of data that are generated. The development of algorithms is especially challenging because the reflection spectra can be complex. Therefore, a significant effort is required to further understand the reflectance spectra and develop efficient and effective classification algorithms. These are discussed in the following sections.

8.6 REFLECTION SIGNATURES OF EXPLOSIVE PARTICLES AND RESIDUES

In order for reflectance-based active MIR HSI to mature as an effective method for identifying explosives, tools must be developed to predict how the shape and magnitude of the reflectance spectrum will change under varying conditions. This is important because the reflection spectra depend on many factors. Modeling of reflection spectra is an area of active research and is discussed in Chapter 9. In this section, we briefly survey the types of samples that may be encountered to highlight the fact that the signatures can be complex.

The MIR reflection spectrum from explosives will depend upon their optical constants, their presentation (e.g., size, shape, density, crystallinity, packing), and the detection geometry. For trace explosives, the signature will also strongly depend on the substrate's physical and optical properties. Given sufficient knowledge about all of these aspects, it is in principle possible to calculate the reflection signatures. But complete knowledge of the sample is almost never available, and this is especially true in real-world nonlaboratory conditions. As the reflection signatures can be complex, modeling is vitally important to provide guidance for estimating the sensitivity and specificity when detecting explosives under varying conditions. It is expected that a particular explosive material will yield a finite number of signature classes that can then be incorporated into a detection algorithm. Supplementary information about the sample (e.g., whether the substrate is fabric or a car panel) will greatly improve detection capability by limiting the number of reference signatures against which to match. Much work remains to be done in this area. A first requirement for making these calculations is knowledge of the optical constants of the materials involved. Unfortunately, even the optical properties of many explosives are still not well established.

8.6.1   BIDIRECTIONAL REFLECTANCE DISTRIBUTION FUNCTION AND FILL FACTOR

Two important concepts for understanding reflection signatures are the bidirectional reflectance distribution function (BRDF) and the fill factor. The scattering from a surface is characterized by the BRDF which is defined in Equation 8.9 [45]:

$$\rho_\Omega = \frac{\text{radiance at detector}}{\text{irradiance of illumination}} \tag{8.9}$$

where the radiance at the detector is the collected power per unit area (in the object plane) divided by the solid angle of the detector. The irradiance of the illumination is the incident power per unit area in a plane that is normal to the illumination direction. This definition is the same as given in Section 8.4.3 for the differential reflectance, ρ_Ω, and has units of inverse steradians, or sr^{-1}. It is a function of illumination angle, viewing angle, polarization, and wavelength.

The fill factor represents the fraction of the surface area that is covered by the chemicals to be detected. In hyperspectral imaging, the fill factor is calculated relative to the GSD. For sparsely distributed explosive particles, imaging can result in a significantly increased fill factor as compared to simply flood illuminating the sample and measuring the average return. In the linear mixing model that was briefly discussed in Section 8.5.2, the reflectance from a surface that is partially covered with particles can be written as the sum

$$\rho_\Omega = F \cdot \rho_{\Omega,\text{particle}} + (1 - F) \cdot \rho_{\Omega,\text{substrate}} \tag{8.10}$$

where:
   F = particle fill factor
   $\rho_{\Omega,\text{particle}}$ and $\rho_{\Omega,\text{substrate}}$ = differential reflectances of the particle and substrate, respectively.

In Section 8.5.3, it was demonstrated that the detection sensitivity is directly related to the fill factor. Detection of 10 ng class particles of $KClO_3$ on car panels was achieved with a fill factor of as little as 2%. Ideally, the measured reflectance will be dominated by the particles such that $F \cdot \rho_{\Omega,\text{particle}} \gg (1 - F) \cdot \rho_{\Omega,\text{substrate}}$, but this is not always the case. The next best situation is that the substrate is spectrally neutral. If this also is not the case, then methods must be implemented to subtract the substrate's spectral contribution.

8.6.2   REFLECTION SPECTRA FROM VARIOUS TYPES OF SAMPLES

This section briefly considers the reflection spectra for a variety of situations: thin film on smooth surface, thin film on rough surface, particles on surfaces, and a collection of closely packed particles (e.g., powder). These are schematically depicted in Figure 8.13.

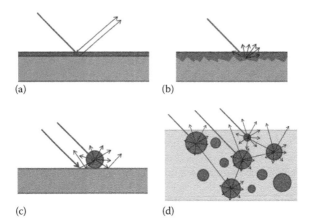

FIGURE 8.13 Four geometries considered with respect to their reflectance signatures are (a) thin film on smooth surface, (b) thin film on rough surface, (c) spherical particles on a smooth surface, and (d) a collection of closely packed particles (e.g., a powder).

8.6.2.1 Scattering from Films

For thin films on surfaces, various factors that affect the reflection spectra include the strength of the absorption features, the thickness of the film, and the refractive index of the substrate [103]. Also important is whether the surface is smooth or rough. The absorption lines in explosives can be classified by their absorption strength. The strong absorption lines result in a significant modification of the real part of the refractive index, and this qualitatively affects the reflection spectrum.

For very thick films, the reflection signature is due to the first-surface Fresnel reflectance, and the substrate can be neglected. For a weak absorption line, the reflection signature has a dispersive shape. For a strong absorption line, the reflection spectrum exhibits a broad Reststrahlen region [41] of high reflectivity that is shifted to higher frequency relative to the absorption line.

For thin films, the effect of the substrate becomes important as summarized below:

- Weak absorption peak: For a thin film on a highly reflective substrate, the signal is due primarily to the double-pass transmittance of the film times the substrate reflectance. This process is termed *transflectance* and the reflectance spectrum has a minimum at the absorption line as would normally be expected. For a very thin film on a weakly reflecting substrate, the first-surface reflectance is not fully developed and a reflection peak occurs at the absorption line. This is an example of *contrast reversal* as discussed in Section 8.5.1. As the film thickness increases, the reflection peak may still be considered to exhibit *contrast reversal* except that the location of the peak shifts to lower frequency.
- Strong absorption peak: For very thin films, the signatures are similar to what is observed with the weakly absorbing material except that the effect is greater and the location of the peak shifts to higher frequencies as the

film thickness increases. For intermediate film thicknesses, the reflection spectra can be complex.

For an underlying substrate that is rough, and when the detector is not oriented to capture the specularly reflected light, the detector will capture only the diffusely scattered signal. This is probably the most straightforward to analyze because the first-surface specular reflectance is not received by the detector. In this case, the observed reflectance is simply the reflectance of the underlying substrate multiplied by the double-pass transmittance of the thin film (i.e., the transflectance spectrum).

8.6.2.2 Scattering from Particles

Many explosives present themselves as solid particles. A good starting point for understanding the reflectance of particles is Mie theory, which describes scattering from isolated spherical particles [104]. For a small nonabsorbing sphere with diameter $d_{particle}$, the scattering cross section $\sigma_{scatter}$ in the Rayleigh limit ($\lambda \gg d_{particle}$) is given by Equation 8.11:

$$\sigma_{scatter} = \frac{32\pi^5 d_{particle}^6}{3\lambda^4}\left(\frac{n^2-1}{n^2+2}\right)^2 \tag{8.11}$$

where n is the refractive index.

The following analysis by the authors was derived from calculations using open-source code to calculate the Mie scattering [105]. In the Rayleigh limit, the angular scattering profile is isotropic in the plane perpendicular to the polarization and dumbbell-shaped in the plane parallel to the polarization. For large particles, the scattering cross section is approximately twice the geometric cross section, and the scattering becomes asymmetric, with a large forward-scattering lobe and a much smaller backward-scattering lobe.

Now consider scattering from particles with an absorption line. There are two competing phenomena that affect the scattered intensity. An increasing imaginary index causes more light to be absorbed by the particle and decreases the scattering; but the larger magnitude of the complex refractive index increases the scattering. The shape of the spectra depends primarily on three factors: the size of the particle, the strength of the absorption feature, and whether the light is forward or backward scattered. In all cases, the general shape of the reflectance spectrum can be described as follows: there is a baseline reflectance that represents the reflectance from a nonabsorbing particle. Relative to this baseline, there is a reflectance valley due to absorption at the absorption line. Near the bottom of this valley, there is a reflectance peak. To a first approximation, differences in the shape of the reflectance spectrum for weakly absorbing and strongly absorbing particles can be understood as differences in the magnitude of the reflectance valley and the reflectance peak relative to the baseline:

- Weak absorption peak
  - For small particles, the reflectance valley is much less pronounced than the reflectance peak. The reflectance spectrum is very similar to that

            observed from the first-surface reflectance of a material for both for-
            ward- and backward-scattered light.

- For large particles, the reflectance valley dominates for both forward-
  and backward-scattered light. Thus, the reflectance spectrum is simply
  a broad valley centered at the absorption line with a small peak at the
  bottom of the valley. For forward-scattered light, the reflectance valley
  is slightly less pronounced than for backward scattering because the primary contribution of the forward-scattered light is light that has traversed
  the particle once, whereas the primary contribution of the backward-scattered light is light that has traversed the diameter of the particle twice.

- Strong absorption peak
  - For small strongly absorbing particles, the reflectance valley is
    so strong and broad that it essentially brings the reflectance to zero in
    the vicinity of the absorption line. Thus, the reflectance peak is the
    noticeable feature for both forward- and backward-scattered light.
    The peak occurs at the Frohlich frequency, which is the frequency
    at which the imaginary part of the refractive index is closest to $\sqrt{2}$
    [104].
  - For large particles, the reflectance valley for backward scattering once
    again brings the reflectance nearly to zero so that the reflectance peak is
    the dominant feature. The shape of the backward-scattered reflectance
    peak is similar to that for a thick film in that it exhibits a Reststrahlen
    band. The forward scattering is much larger than the backward scatter-
    ing, but the forward-scattering lobe has a narrow angular distribution.
    The shape of the spectrum is more complicated than the other cases, but
    there is a reflectance peak that is shifted to a higher frequency as com-
    pared with the backscattered reflectance peak. Also, there is a sharp dip
    in the spectrum at the Christiansen frequency, which occurs when the
    refractive index is nearly equal to 1 [45].

To a first approximation, the reflectance of particles on a smooth surface can be visu-
alized as the sum of light that is directly backscattered from the particle and light
that is scattered from the particle and then reflected from the smooth surface back
toward the receiver. When the sample is nearly specularly oriented and the substrate
is highly reflective, then forward scattering will dominate the reflectance spectra for
larger particles. Since the forward-scattering lobe is relatively narrow for large particles, as the sample is rotated away from normal, the magnitude of the reflectance
will rapidly decrease and the shape of the spectrum will approach that of backward
scattering. This change is more dramatic for p-polarized light as compared to s-polarization. The changes with angle are less dramatic if the surface is weakly reflective
and also if the particles scatter isotropically (as is the case for small or highly irregular particles). Figure 8.14 compares measurement and calculation of the reflection
spectrum of fine $NaClO_3$ particles on a gold mirror versus illumination angle. $NaClO_3$
was used because published values for the refractive index are available [39]. The
solid and dashed curves represent the measured and calculated spectra, respectively.
Calculations were based on a simple model using Mie scattering, which adds the

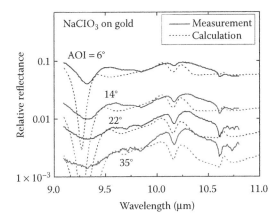

FIGURE 8.14 Reflectance spectrum of $NaClO_3$-contaminated fingerprint on a car panel for several illumination angles. The solid curves are measurements made using the active MIR HSI system described in Section 8.5.3. The dashed line is a calculation based on Mie theory.

forward- and backward-scattering contributions as described above. The main point to note is that the intensity of the return is a strong function of the illumination angle and that this simple model does a good job of predicting this dependence. The model is also reasonably good at predicting the shape of the spectra, although it is unclear what gives rise to the deviation at $\lambda = 9.3$ μm. In [106], the backscattered reflectance was measured as a function of sample angle for TNT, RDX, and tetryl on a car panel. It was found that the magnitude of the reflectance decreased by many orders of magnitude as the sample was rotated away from normal incidence.

8.6.2.3 Reflectance from Bulk Powders

The reflectance from bulk powders can be calculated using a parameter called the single-scatter albedo and the angular scattering profile [45,107]. The single-scatter albedo is the ratio of the scattered light to the sum of the scattered *and* absorbed light for a single representative particle within the powder. In the general case where the single-scatter albedo varies throughout the material, a Monte Carlo method can be used to solve for the BRDF. If the single-scatter albedo varies only with depth, then the radiative transfer equation gives the exact solution for the BRDF; however, numerically solving the equation can be time-consuming. Therefore, various analytic approximations to the solution have been proposed, including the Kubelka-Munk and Hapke models [45]. The Hapke model gives the exact solution to the radiative transfer equation in the limit of isotropic scattering. For nonisotropic scattering, the asymmetry of the scattering profile can be approximated using the Henyey-Greenstein function. Other models are also available. One thing that should be noted when using these models is that the particle size distribution that gives the best approximation to the actual reflectance is often much smaller than the size of the actual particles due to the fact that irregularly shaped particles are approximately equivalent to a collection of closely spaced smaller particles.

8.7 DETECTION ALGORITHMS

The detection algorithms are a critical component of an HSI system [42,82,94,108, 109]. Given the complexity of the reflectance signatures expected from real-world environments, further advancement in detection algorithms is likely necessary. It is beyond the scope of this chapter to discuss algorithms in detail, but a few points are worth making.

The Fraunhofer Institute has applied well-developed hyperspectral analysis algorithms to detect trace explosives [96]. They utilized the adaptive target generation process (ATGP) [110] to account for the background and the adaptive matched subspace detector (AMSD) [111] as the detection algorithm. For reference spectra, they utilize measurements of explosive residues on, for example, painted car panels. At the MIT Lincoln Laboratory, the detection algorithm used was based on the ACE algorithm [16]. These algorithms determine the similarity of the spectral response at a given pixel relative to the reference spectrum and the clutter background. These algorithms, therefore, do not take into account any spatial correlation between pixels. Future detection algorithms can be expected to make use of the rich data set provided by HSI to utilize multipixel detection to improve detection capability.

8.8 LASER SPECKLE

Laser speckle is a concern for almost all active HSI systems and, therefore, we discuss methods for its suppression in some depth. This section is based on analysis and numerical modeling of speckle by the authors under various conditions of interest for hyperspectral imaging using the theory found in [112,113]. Speckle is a high-contrast, fine-scale granular pattern that results from constructive and destructive optical interference observed when coherent light reflects from a randomly rough surface. After reflecting from a rough surface, a speckle pattern is formed at the pupil of the camera lens. The speckle pattern at the lens pupil is then focused by the lens and results in a speckle pattern at the FPA. The speckle pattern at the FPA arises because any practical lens is not capable of capturing all of the scattered light. For fully developed speckle, the standard deviation of the intensity σ_s at the lens pupil is equal to the average optical intensity $\langle S \rangle$! Thankfully, the speckle noise can be significantly reduced through the process of *speckle averaging*. A measure of the speckle noise at the FPA is given by the contrast

$$C = \frac{\sigma_s}{\langle S \rangle} = \frac{1}{\sqrt{M}} \qquad (8.12)$$

which is the inverse of the speckle-limited SNR, and where M is a parameter that represents the number of uncorrelated speckle patterns that are added on an intensity basis to yield speckle averaging. For fully developed speckle pattern with no speckle averaging, $M = 1$. We consider the following averaging factors which contribute to M in an imaging configuration: (i) spatial, (ii) polarization, (iii) angular, (iv) spectral, and (v) temporal. Each of these terms is examined in the following sections.

8.8.1 SPATIAL AVERAGING

Speckle has a characteristic size at the imaging plane of a camera that is equal to the point spread function due to diffraction by the circular aperture of the lens. Speckle averaging occurs when multiple speckles occupy the detector area $A_{pix} = d_{pix}^2$. The diffraction-limited spot size at the FPA can be written as

$$d_{diffraction} = \frac{\lambda}{D_{lens}} \frac{1}{1-(R/f)} \approx \frac{\lambda f}{D_{lens}} \tag{8.13}$$

where the second expression is valid for $R \gg f$.

For a circular aperture, the diffraction-limited spot is an Airy function and the effective area of a speckle is [112]

$$A_{speckle} = \frac{4}{\pi} d_{diffraction}^2 \tag{8.14}$$

The following expression is then found to be a very good approximation for the effect of spatial averaging at the pixel

$$M_{spatial} \cong 1 + \frac{A_{pix}}{A_{speckle}} \tag{8.15}$$

As expected, larger pixel sizes yield greater speckle averaging. Figure 8.15 shows a simulated speckle pattern in the image plane of a FPA for $\lambda = 10$ μm and $f/\# = 1$. The rectangles represent 10- and 100- μm pixels. The plot shows the simulated reflection spectrum at a single pixel from a rough surface having unity reflectance. The smallest pixel exhibits a great deal of speckle noise such that it is a poor representation of the actual reflection spectrum. As the pixel size increases to 1 mm, the speckle noise is considerably reduced due to spatial averaging such that the speckle-limited SNR = 89.

8.8.2 POLARIZATION AVERAGING

When light is depolarized after reflecting from an object, the speckle pattern is the sum of two partially independent speckle patterns and speckle averaging occurs. The degree of polarization is characterized by P, which varies from 0 for unpolarized light to 1 for polarized light. In this case, M due to polarization diversity is given by Equation 8.16 [113]:

$$M_{polarization} = \frac{2}{1+P^2} \tag{8.16}$$

which can vary from 1 to 2. Greater depolarization is expected for volume scattering in bulk powders because of the multiple scattering events involved, as compared to surface scattering. Polarization averaging is, however, a relatively minor effect.

8.8.3 Spectral Averaging

As the wavelength of illumination is changed, the speckle pattern observed in the image plane will also change. The magnitude of this change is related to the round-trip optical pathlength difference (OPD) of the light within an optical resolution element of the lens. The size of a resolution element at the object is approximately given by $R\lambda/D_{lens}$. The greater the OPD within a resolution element, the more the speckle pattern will change as the laser wavelength is changed. We consider two sources of OPD: (1) random height variations of the object and (2) viewing the surface at a nonnormal angle. For both cases, it is useful to define a characteristic spectral width $\Delta\tilde{v}_o$ (in terms of wave numbers), which results in significant decorrelation of the speckle pattern. The speckle averaging due to spectral diversity can then be written as

$$M_{spectral} = \frac{\Delta\tilde{v}_{laser}}{\Delta\tilde{v}_o} \quad \text{for } \Delta\tilde{v} \gg \Delta\tilde{v}_o \qquad (8.17)$$

where it is assumed that the laser spectrum has a rectangular profile with spectral width of $\Delta\tilde{v}_{laser}$. In the following, for simplicity, it is assumed that the illumination and viewing angles are the same.

For the case of random Gaussian-distributed height variations, the characteristic spectral width is inversely proportional to the standard deviation of the roughness, σ_h, and is given by Equation 8.18 [112]:

$$\Delta\tilde{v}_o = \frac{1}{4\sqrt{\pi}\,\sigma_h} \quad \text{(Gaussian surface, normal incidence)} \qquad (8.18)$$

Interestingly, the average separation of the peaks in the observed spectrum that arises due to speckle is approximately given by $0.7\Delta\tilde{v}_o$ [114]. For the example spectrum in Figure 8.15, the calculated separation between peaks of 10 cm^{-1} is in agreement with the simulation results.

For the case of a large observation angle relative to the surface normal, the OPD changes linearly in the direction of sample tilt with laser wavelength. When the speckle pattern at the lens pupil has shifted by one pupil diameter, the speckle pattern in the image plane will be uncorrelated with the original speckle pattern. The characteristic spectral width for this case is given by Equation 8.19:

$$\Delta\tilde{v}_o = \frac{D_{lens}}{3\lambda R \cdot \tan\theta} \quad \text{(nonnormal viewing)} \qquad (8.19)$$

where θ is the observation angle. Significant spectral averaging can occur for modest values of $\Delta\tilde{v}_{laser}$ when viewing objects at long range.

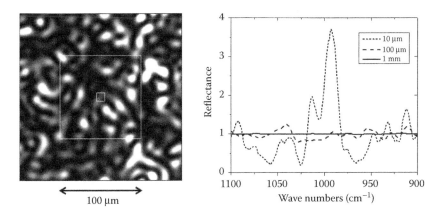

FIGURE 8.15 Simulation of a speckle pattern in the image plane of a camera at a wavelength of 10 μm using a focusing lens with $f/\# = 1$. The rectangles represent 10 and-100-μm pixels. The simulated speckle pattern is formed by normal-incidence illumination of a surface having Gaussian roughness with standard deviation of $\sigma_h = 100$ μm. The plot shows the simulated reflectance spectra from a material with a uniform reflectance of unity. Each curve represents the spectrum at a single pixel for three different pixel sizes: 10 μm, 100 μm, and 1 mm. The speckle noise decreases with increasing pixel size due to speckle averaging.

There are at least two approaches for overcoming speckle when considering spectral diversity. One approach is to operate in the regime where the spectral resolution, which is given by $\Delta\tilde{v}_{\text{laser}}$, is much larger than $\Delta\tilde{v}_o$ to obtain spectral averaging. Another approach is to measure a spectral region that is narrow compared with $\Delta\tilde{v}_o$ such that the speckle pattern will not change significantly. In this case, the shape of the reflectance spectra within a single pixel will not change much over this small spectral range, but the pixel-to-pixel variation in the received power can be large.

8.8.4 ANGULAR AVERAGING

Reduction in speckle contrast through angular averaging is very similar to reduction in speckle contrast through spectral averaging at a large observation angle. Both cases involve shifting the speckle pattern at the lens pupil. In the case of spectral diversity, the speckle pattern shifts as the wave number changes; in the case of angular averaging, the lens pupil itself shifts as the observation angle changes. Similar to the characteristic spectral width, we can define a characteristic *tangent width*, which is the change in the tangent of the observation angle that will move the speckle pattern by one pupil diameter. This characteristic tangent width is given by $\Delta u_o = D_{\text{lens}}/2R$. Averaging over all observation angles between θ_1 and θ_2, one obtains Equation 8.20:

$$M_{\text{angular}} = \frac{3}{2}\frac{\Delta u}{\Delta u_o} \tag{8.20}$$

where $\Delta u = \left|\tan(\theta_2) - \tan(\theta_1)\right|$.

8.8.5 TEMPORAL AVERAGING

Temporal averaging can be achieved by varying in time the phase of the light incident onto each point of the surface so that different speckle patterns are realized at different moments in time. A common method to achieve temporal averaging is to place a moving diffuser and a stationary diffuser in front of the illumination source. In [98], for example, two polycrystalline diamond plates were used. The stationary diffuser ensures that the illumination has a uniform distribution of phases and the moving diffuser scrambles those phases with time. If the moving diffuser moves with a speed v and the diffuser is characterized by a phase correlation radius r_o, then [112]

$$M_{\text{temporal}} \propto \frac{vT}{r_o} \tag{8.21}$$

where T is the integration time.

8.8.6 TOTAL AVERAGING

These speckle averaging factors are multiplicative such that $M_{\text{total}} = \Pi_n M_n$. To achieve maximum speckle noise reduction, M_{total} should be made as large as possible using all the available methods discussed in this section. Since the laser emission is highly coherent, and it is this coherence that gives rise to speckle, methods to destroy the spatial coherence of the source may be effective for reducing the speckle noise. This approach is most relevant for close-range detection where spatial coherence (i.e., good beam quality) of the laser is not needed.

8.9 SYSTEM DESIGN

In the following, the areal coverage rate is calculated for the system configuration shown in Figure 8.1, which utilizes a wavelength-tunable laser and camera. Other system configurations that can be used for active MIR HSI are also mentioned.

8.9.1 AREAL COVERAGE RATE

The ACR is an important parameter characterizing a system for surface detection. The ACR will depend on the properties of the object, range to the object, and the characteristics of the laser and camera. Here, a simple comparison is made between a system that uses a wavelength-tunable laser in combination with either an MB or MCT camera. These examples in no way represent a system optimization. Their purpose is simply to explore some of the trade-offs that are possible.

Building on the example in Section 8.4.4, consider an FPA with $d_{\text{pix}} = 20$ μm, camera lens with $f/\# = 1$ and ideal transmission for the optical system $T = 1$. Furthermore, assume that the differential reflectance of the object is $\rho_\Omega = 0.01$ sr^{-1} and that the illumination intensity at the object is maintained below the eye-safety limit of

$I_{laser} = 0.1$ W/cm^2. Then the power collected at each fully illuminated pixel is $P_{pixel} = 3$ nW. Since both MCT and MB cameras have a noise-equivalent power of roughly NEP ~ 10 pW, the detector-limited SNR is given by P_{pixel}/NEP ~ 300. For definiteness, also assume that the illumination laser is capable of a peak power of 1 W in 1 μs long pulses at a duty factor of 10% such that the maximum average power is 100 mW.

Many factors influence the ACR, but it can be written in a simple way as the measurement area divided by the dwell time to acquire the full hypercube:

$$ACR = \frac{\text{measurement area}}{\text{dwell time}} \approx \frac{\left(P_{laser}/I_{laser}\right)F_{beam}}{\left(t_{image} \cdot N_{avg}N_\lambda N_{ref}\right) + t_{pointing}} \tag{8.22}$$

The measurement area is given by the ratio of the laser power to laser intensity at the object times a factor $F_{beam} \leq 1$ that accounts for the nonrectangular beam shape and nonuniform intensity profile of the laser beam. The dwell time consists of the time to acquire a hypercube (i.e., the term given in parentheses in the denominator of Equation 8.22) plus the time to point the system to a new position on the ground, $t_{pointing}$. The time to acquire a hypercube is the product of the time to capture a single camera frame at a given wavelength t_{image}, the number of camera frames that are averaged to create a single frame of the hypercube N_{avg}, the number of wavelengths in the hypercube N_λ, and a factor N_{ref}, representing the additional blank reference frames that are required for subtraction of the passive thermal background. Furthermore, the time to capture a single image, t_{image}, can be decomposed into several contributions, including the camera's readout time t_{read}, time to tune the laser wavelength t_{laser}, and the measurement time $t_{measure}$. Since some of these time periods can be overlapping, it can at least be stated that t_{image} is greater than the maximum of any one of these three terms, that is, $t_{image} \geq \max(t_{read}, t_{laser}, t_{meas})$, but that it is less than or equal to their sum.

In the following, it is assumed that the hypercube consists of 10 frames, $N_\lambda = 10$, and that a single camera frame is per wavelength sufficient to achieve the desired SNR such that $N_{avg} = 1$. To simplify the problem, it is further assumed that the time to capture a frame is equal to the read time $t_{image} \approx t_{read}$. Both the pointing time and the laser's nonuniform intensity distribution are neglected such that $t_{pointing} = 0$ and $F_{beam} = 1$.

First, consider an MB camera operating at a frame rate of 50 Hz ($t_{read} = 20$ ms) in which every other frame corresponds to an image of the actively illuminated scene, while the intervening frames are for subtraction of the passive background such that $N_{ref} = 2$. Since the MB camera has a slow response, the camera senses only the laser's average power. For an average power of 100 mW and operating at the eye-safety limit at the target of 100 mW/cm^2, the measurement area is 1 cm^2. In this case, one calculates ACR = 2.5 cm^2/s.

Now consider a high-end MCT camera with a digital-pixel FPA as described in Section 8.4.5.2 that is capable of a frame rate of up to 2 kHz ($t_{read} = 0.5$ ms). Further assume that the digital-pixel FPA is capable of multipulse integration and on-chip background subtraction. Since the laser operates under pulsed conditions with 10% duty factor, and since the digital-pixel FPA can operate synchronously with the laser's pulsing, the FPA effectively samples the peak laser power of 1 W. The measurement area can therefore be increased by 10× to 10 cm². To integrate the signal over 10 μs using 1 μs long laser pulses at 10% duty factor, the measurement time is 100 μs. This measurement time is less than the camera's read time by 5× such that utilization of the FPA is only 20%. The combined effect of the laser and camera duty factors is to reduce the average laser irradiance at the object to 50× below the eye-safety limit. Since background subtraction is performed on-chip without the need for a reference frame, $N_{ref} = 1$. The resulting ACR = 2000 cm²/s. This is 800× higher than with the MB camera. The enhancement is due to a factor of 10× from the ratio of the laser's peak-to-average power, 2× from on-chip background subtraction, and 40× from the camera's higher frame rate.

These calculations are very simplistic. Nevertheless, the goal of this comparison was simply two-fold: (1) to give an appreciation for how high the ACR can be using active MIR HSI and (2) to illustrate the factors that influence the achievable ACR. One additional factor that was not considered, but which can be important, is that the readout time for the FPA can be reduced by reading out only a portion of the FPA in a process called windowing. Windowing can enable even higher ACR when using a MCT camera.

Finally, it is worth noting that although an active MIR HSI system can achieve very high ACRs, it does so by generating a very large amount of data at high data rates. The computational resources and the algorithms to analyze the data must be fast enough to keep up.

8.9.2 ALTERNATIVE APPROACHES

The focus of this chapter has been on a particular implementation of active MIR HSI that utilizes a wavelength-tunable laser and camera as shown in Figure 8.1. It is worth noting that alternative configurations are also possible. One alternative approach is to focus the laser beam to a small spot and to raster scan the beam across the scene while the reflected light is detected at a single-element detector. For raster scanning, either the wavelength can be tuned at each position prior moving to the next position, or the entire scene can be scanned at a single wavelength and then rescanned at the next wavelength. The choice of approach will depend on many factors, including the relative speeds of wavelength tuning and raster scanning.

Instead of using a wavelength-tunable laser, it is also possible to use a broadband illumination source and a spectrometer at the receiver. This approach is taken with FTIR microspectroscopy in which the receiver can be a FPA, a linear array, or a single-element detector [69]. This approach has also been applied for standoff detection using an imaging FTIR spectrometer [48]. For longer-range standoff applications, the thermal source can be replaced with a broadband laser source. Also, rather than using an FTIR spectrometer, the receiver could incorporate a dispersive spectrometer. It is beyond the scope of this chapter to discuss the pros and cons of these various approaches.

8.10 SUMMARY

Active MIR HSI based on reflectance spectroscopy has been explored as a method to detect bulk and trace explosives. For trace detection, active MIR HSI has been demonstrated to detect 10 ng class explosive particles on painted car panels at close range and to have limits of detection on the order of 10–100 $\mu g/cm^2$. A great advantage of this technique is that a system can be eye safe while achieving high-speed scanning of surfaces. Areal coverage rates of >1 m^2/s appear feasible with watt-class laser transmitters and high-performance MCT cameras, but further work is required to develop the component technology. Since the reflection spectra of trace explosives on surfaces can be relatively complex, advances in the understanding of the reflection spectra, coupled with the development of algorithms, are also needed. Speckle can be problematic in laser-illuminated imaging systems because it adds speckle noise that can obscure the reflection spectrum of interest. Further understanding of speckle and the development of methods for managing it are needed. Despite these challenges, active MIR HSI has many desirable attributes and it will likely develop into a valuable and widely adopted method for detecting chemicals such as explosives on surfaces.

REFERENCES

1. D. Lin-Vien, N. B. Colthup, W. G. Fateley, and J. G. Grasselli, *The Handbook of Infrared and Raman Characteristic Frequencies of Organic Molecules* (Academic Press, San Diego, 1991).
2. I. M. Craig, M. C. Phillips, M. S. Taubman, E. E. Josberger, and M. B. Raschke, Infrared scattering scanning near-field optical microscopy using an external cavity quantum cascade laser for nanoscale chemical imaging and spectroscopy of explosive residues, *Proceedings of SPIE* **8631**, 863110 (2013).
3. J. M. Chalmers, H. G. M. Edwards, and M. D. Hargreaves (eds), *Infrared and Raman Spectroscopy in Forensic Science* (Wiley, West Sussex, 2012).
4. N. Eisenreich, H. Kull, and J. Hertz, Rapid NIR measurement of explosives, in *25th International Annual Conference of ICT* (Karlsruhe, 1994), pp. 86/81–86/87.
5. T. Rohe, T. Grunblatt, and N. Eisenreich, Near infrared transmission spectroscopy on propellants and explosives, in *27th International Annual Conference of ICT* (Karlsruhe, 1996), pp. 85/81–85/10.
6. T. Blake, J. Kelly, N. Gallagher, P. Gassman, and T. Johnson, Passive standoff detection of RDX residues on metal surfaces via infrared hyperspectral imaging, *Analytical and Bioanalytical Chemistry* **395**, 337–348 (2009).
7. J. Faist, F. Capasso, D. Sivco, C. Sirtori, A. L. Hutchinson, and A. Cho, Quantum cascade laser, *Science* **264**, 553–556 (1994).
8. F. Capasso, C. Gmachl, D. L. Sivco, and A. Y. Cho, Quantum cascade lasers, *Physics Today* **55**, 34–40 (2002).
9. A. Hugi, R. Terazzi, Y. Bonetti, A. Wittmann, M. Fischer, M. Beck, J. Faist, and E. Gini, External cavity quantum cascade laser tunable from 7.6 to 11.4 μm, *Applied Physics Letter* **95**, 061103-061103 (2009).
10. J. Faist, *Quantum Cascade Laser* (Oxford University Press, New York, 2013).
11. R. Bhargava, Infrared spectroscopic imaging: The next generation, *Applied Spectroscopy* **66**, 1091–1120 (2012).
12. K. L. McNesby and R. A. Pesce-Rodriguez, Application of vibrational spectroscopy in the study of explosives, in J. M. Chalmers and P. R. Griffiths (eds), *Handbook of Vibrational Spectroscopy* (John Wiley & Sons, Chichester, 2002).
13. M. Diem, P. R. Griffiths, and J. M. Chalmers (eds), *Vibrational Spectroscopy for Medical Diagnosis* (Wiley, Chichester, 2008).
14. S. Sasic and Y. Ozaki, eds., *Raman, Infrared, and Near-Infrared Chemical Imaging* (John Wiley& Sons, Hoboken, NJ, 2010).

15. A. K. Goyal, T. Myers, M. Spencer, R. R. Kunz, S. Park, B. Tipton, M. Kelly, E. L. Holthoff, M. E. Farrell, and P. M. Pellegrino, Active infrared hyperspectral imaging of solid particles on surfaces, *Proceedings of SPIE* (2013).

16. D. Manolakis, M. Pieper, E. Truslow, T. Cooley, M. Brueggeman, and S. Lipson, The remarkable success of adaptive cosine estimator in hyperspectral target recognition, *Proceedings of SPIE* **8743**, 874302 (2012).

17. American National Standards Institute, American National Standard for Safe Use of Lasers, ANSI Z136.1-2007 (Laser Institute of America, 2007).

18. R. W. Boyd, *Radiometry and the Detection of Optical Radiation* (John Wiley and Sons, New York, 1983).

19. R. H. Kingston, *Detection of Optical and Infrared Radiation*, 2nd edn. (Springer-Verlag, Berlin, 1979).

20. C. Kittel and H. Kroemer, *Thermal Physics*, 2nd edn. (W. H. Freeman, New York, 1980).

21. ASTM International, Standard tables for reference solar spectral irradiances, ASTM G173-03(2012) (ASTM International, West Conshohocken, PA, 2012).

22. A. Beveridge, (ed.), *Forensic Investigation of Explosives*, 2nd edn., International Forensic Science and Investigation Series (CRC Press, Boca Raton, 2012).

23. N. R. Council, *Existing and Potential Standoff Explosives Detection Techniques* (The National Academies Press, Washington, DC, 2004).

24. J. I. Steinfeld and J. Wormhoudt, Explosives detection: A challenge for physical chemistry, *Annual Review Physical Chemistry* **49**, 203–232 (1998).

25. E. C. Bender and A. Beveridge, Investigation of pipe bombs, in A. Beveridge (ed.), *Forensic Investigation of Explosions*, 2nd edn. (CRC Press, Boca Raton, 2011).

26. R. B. Hopler, The history, development, and characteristics of explosives and propellants, in A. Beveridge (ed.), *Forensic Investigation of Explosives*, 2nd edn. (CRC Press, Boca Raton, 2011).

27. R. A. Isbell and M. Q. Brewster, Optical properties of energetic materials: RDX, HMX, AP, NC/NG, and HTPB, *Propellants, Explosives, Pyrotechnics* **23**, 218–224 (1998).

28. J. Janni, B. D. Gilbert, R. W. Field, and J. I. Steinfeld, Infrared absorption of explosive molecule vapors, *Spectrochimica Acta—Part A: Molecular and Biomolecular Spectroscopy* **53**, 1371–1381 (1996).

29. D. S. Moore, Recent advances in trace explosives detection instrumentation, *Sensing and Imaging* **8**, 9–38 (2007).

30. D. E. Chesan and G. Norwitz, Qualitative analysis of primers tracers igniters incendiaries boosters and delay compositions on a micro scale by use of infrared spectroscopy (Dept. Army Frankford Arsenal, Philadelphia, 1971).

31. S. Zitrin and T. Tamiri, Analysis of explosives by infrared spectrometry, in A. Beveridge (ed.), *Forensic Investigation of Explosions*, 2nd edn. (CRC Press, Boca Raton, 2012).

32. C. Bauer, U. Willer, R. Lewicki, A. Pohlkotter, A. A. Kosterev, D. Kosynkin, F. K. Tittel, and W. Schade, A mid-infrared QEPAS sensor device for TATP detection, *Journal of Physics* **157**, 012002 (2002).

33. J. Oxley, J. Smith, J. Brady, F. Dubnikova, R. Kosloff, L. Zeiri, and Y. Zeiri, Raman and infrared fingerprint spectroscopy of peroxide-based explosives, *Applied Spectroscopy* **62**, 906–915 (2008).

34. I. Dunayevskiy, A. Tsekoun, M. Prasanna, R. Go, and C. K. N. Patel, High-sensitivity detection of triacetone triperoxide (TATP) and its precursor acetone, *Applied Optics* **46**, 6397–6404 (2007).

35. F. A. Miller and C. H. WIlkins, Infrared spectra and characteristic frequencies of inorganic ions, *Analytical Chemistry* **24**, 1253–1294 (1952).

36. M. A. Jarzembski, M. L. Norman, K. A. Fuller, V. Srivastava, and D. R. Cutten, Complex refractive index of ammonium nitrate in the 2–20 μm spectral range, *Applied Optics* **42**, 922–930 (2003).

37. J. Oxley, J. Smith, S. Vadlamannati, A. C. Brown, G. Zhang, D. S. Swanson, and J. Canino, Synthesis and characterization of urea nitrate and nitrourea, *Propellants, Explosives, Pyrotechnics* **38**, 335–344 (2013).

38. Index data for TNT and RDX were derived from attenuated total reflection (ATR) FTIR spectra. Courtesy of Block MEMS, LLC.

39. G. Andermann and D. A. Dows, Infrared reflectance spectrum and optical constants of sodium chlorate, *Journal of Physics and Chemistry of Solids* **28**, 1307–1315 (1967).

40. Pacific Northwest National Laboratory, Northwest-infrared vapor-phase infrared spectral library (Pacific Northwest National Laboratory, 2014). Available at: http://nwir.pnl.gov/.

41. E. H. Korte and A. Roseler, Infrared reststrahlen revisited: commonly disregarded optical details related to n<1, *Analytical and Bioanalytical Chemistry* **382**, 1987–1992 (2005).

42. M. T. Eismann, *Hyperspectral Remote Sensing* (SPIE Press, Bellingham, 2012).

43. A. E. Siegman, *Lasers* (University Science Books, Philadelphia, PA, 1986).

44. R. M. Measures, *Laser Remote Sensing: Fundamentals and Applications* (Krieger Publishing, Malabar, FL, 1992).

45. B. Hapke, *Theory of Reflectance and Emittance Spectroscopy*, 2nd edn. (Cambridge University Press, Cambridge, 2012).

46. M. Vollmer and K.-P. Mollmann, *Infrared Thermal Imaging* (Wiley-VCH, Weinheim, 2011).

47. M. C. Phillips and N. Ho, Infrared hyperspectral imaging using a broadly tunable external cavity quantum cascade laser and microbolometer focal plane array, *Optics Express* **16**, 1836–1845 (2008).

48. R. Harig, R. Braun, C. Dyer, C. Howle, and B. Truscott, Short-range remote detection of liquid surface contamination by active imaging Fourier transform spectrometry, *Optics Express* **16**, 5708–5714 (2008).

49. L. Pacheco-Londoño, W. Ortiz-Rivera, O. Primera-Pedrozo, and S. Hernández-Rivera, Vibrational spectroscopy standoff detection of explosives, *Analytical and Bioanalytical Chemistry* **395**, 323–335 (2009).

50. C. K. N. Patel, A. Lyakh, R. Maulini, A. Tsekoun, and B. Tadjikov, QCL as a game changer in MWIR and LWIR military and homeland security applications, *Proceedings of SPIE* **8373**, 83732E (2012).

51. M. Tacke, Lead-salt lasers, *Philosophical Transactions of the Royal Society A* **359**, 547–566 (2001).

52. L. J. Olafsen, Antimonide mid-IR lasers, in H. K. Choi (ed.), *Long-Wavelength Infrared Semiconductor Lasers* (John Wiley and Sons, New York, 2004).

53. A. Bell, C. Dyer, A. W. Jones, and K. Kinnear, Standoff liquid CW detection, *Proceedings of SPIE* **5268**, 302–309 (2004).

54. C. R. Liao, M. Rak, J. Lund, M. Unger, E. Platt, B. C. Albensi, C. J. Hirschmugl, and K. M. Gough, Synchrotron FTIR reveals lipid around and within amyloid plaques in transgenic mice and Alzheimer's disease brain, *Analyst* **138**, 3991–3997 (2013).

55. C. R. Howle, D. J. M. Stothard, C. F. Rae, M. Ross, B. S. Truscott, C. D. Dyer, and M. H. Dunn, Active hyperspectral imaging system for the detection of liquids, *Proceedings of SPIE* **6954**, 69540L (2008).

56. E. Lippert, H. Fonnum, G. Arisholm, and K. Stenersen, A 22-watt mid-infrared optical parametric oscillator with V-shaped 3-mirror ring resonator, *Optics Express* **18**, 26475–26483 (2010).

57. C. R. Swim, Review of active chem-bio sensing, *Proceedings of SPIE* **5416**, 178–185 (2004).

58. Y. Bai, N. Bandyopadhyay, S. Tsao, S. Slivken, and M. Razeghi, Room temperature quantum cascade lasers with 27% wall plug efficiency, *Applied Physics Letters* **98**, 181102–181103 (2011).

59. A. Lyakh, R. Maulini, A. Tsekoun, R. Go, and C. K. N. Patel, Multiwatt long wavelength quantum cascade lasers based on high strain composition with 70% injection efficiency, *Optics Express* **20**, 24272–24279 (2012).

60. P. Rauter, S. Menzel, A. K. Goyal, C. A. Wang, A. Sanchez, G. Turner, and F. Capasso, High-power arrays of quantum cascade laser master-oscillator power-amplifiers, *Optics Express* **21**, 4518–4530 (2013).

61. T. Dougakiuchi, M. Fujita, N. Akikusa, A. Sugiyama, T. Edamura, and M. Yamanishi, Broadband tuning of external-cavity dual-upper-state quantum-cascade lasers in continuous wave operation, *Applied Physics Express* **4**, 102101 (2011).

62. T. S. Mansuripur, S. Menzel, R. Blanchard, L. Diehl, C. Pflugl, Y. Huang, J.-H. Ryou, R. D. Dupuis, M. Loncar, and F. Capasso, Widely tunable mid-infrared quantum cascade lasers using sampled grating reflectors, *Optics Express* **20**, 23339–23348 (2012).

63. P. Fuchs, J. Seufert, J. Koeth, J. Semmel, S. Hofling, L. Worschech, and A. Forchel, Widely tunable quantum cascade lasers with coupled cavities for gas detection, *Applied Physics Letters* **97**, 181111 (2010).

64. B. G. Lee, H. A. Zhang, C. Pflugl, L. Diehl, M. A. Belkin, M. Fischer, A. Wittmann, J. Faist, and F. Capasso, Broadband distributed-feedback quantum cascade laser array operating from 8.0 to 9.8 mm, *IEEE Photonics Technology Letters* **21**, 914–916 (2009).

65. P. Rauter, S. Menzel, B. Gokden, A. K. Goyal, C. A. Wang, A. Sanchez, G. Turner, and F. Capasso, Single-mode tapered quantum cascade lasers, *Applied Physics Letters* **102**, 181102 (2013).

66. B. G. Lee, J. Kansky, A. K. Goyal, C. Pflügl, L. Diehl, M. A. Belkin, A. Sanchez, and F. A. Capasso, Beam combining of quantum cascade laser arrays, *Optics Express* **17**, 16216–16224 (2009).

67. A. K. Goyal, M. Spencer, O. Shatrovoy, B. G. Lee, L. Diehl, C. Pfluegl, A. Sanchez, and F. Capasso, Dispersion-compensated wavelength beam combining of quantum-cascade-laser arrays, *Optics Express* **19**, 26725–26732 (2011).

68. A. K. Goyal, T. Myers, C. A. Wang, M. Kelly, B. Tyrrell, B. Gokden, A. Sanchez, G. Turner, and F. Capasso, Active hyperspectral imaging using a quantum cascade laser (QCL) array and a digital-pixel focal plane array (DFPA) camera, *Optics Express* **22**, 14392 (2014).

69. J. Sellors, R. A. Hoult, R. A. Crocombe, and N. A. Wright, FT-IR imaging hardware, in S. Sasic and Y. Ozaki (eds), *Raman, Infrared, and Near-Infrared Chemical Imaging* (John Wiley, Hoboken, NJ, 2010).

70. M. W. Kelly, R. Berger, C. Colonero, M. Gregg, J. Model, D. Mooney, and E. Ringdahl, Design and testing of an all-digital readout integrated circuit for infrared focal plane arrays, *Proceedings of SPIE* **5902**, 59020J (2005).

71. B. Tyrrell, R. Berger, C. Colonero, J. Costa, M. Kellly, E. Ringdahl, K. Schultz, and J. Wey, Design approaches for digitally dominated active pixel sensors: leveraging Moore's law scaling in focal plane readout design, *Proceedings of SPIE* **6900**, 69000W (2008).

72. B. Tyrrell, K. Anderson, J. Baker, R. Berger, M. Brown, C. Colonero, J. Costa, et al., Time delay integration an in-pixel spatiotemporal filtering using a nanoscale digital CMOS focal plane readout, *IEEE Transactions Electron Devices* **56**, 2516–2523 (2009).

73. F. Fuchs, S. Hugger, J. Jarvis, V. Blattmann, M. Kinzer, Q. K. Yang, R. Ostendorf, et al., Infrared hyperspectral standoff detection of explosives, *Proceedings of SPIE* **8710**, 87100I (2013).

74. S. D. Gunapala, S. V. Bandara, J. K. Liu, J. M. Mumolo, C. J. Hill, S. B. Rafol, D. Salazar, J. Woolaway, P. D. LeVan, and M. Z. Tidrow, Toward dualband megapixel QWIP focal plane arrays, *Infrared Physics & Technology* **50**, 217–226 (2007).

75. M. Sundaram, A. Reisinger, R. Dennis, K. Patnaude, D. Burrows, J. Bundas, K. Beech, and R. Faska, Infrared imaging with quantum wells and strained layer superlattices, *Proceedings of SPIE* **8268**, 82682L (2012).

76. L. A. Skvortsov, Active spectral imaging for standoff detection of explosives, *Quantum Electronics* **41**, 1051–1060 (2011).

77. L. A. Skvortsov and E. M. Maksimov, Applications of laser photothermal spectroscopy for standoff detection of trace explosive residues on surfaces, *Quantum Electronics* **40**, 565–578 (2010).

78. S. Wallin, A. Pettersson, H. Östmark, and A. Hobro, Laser-based standoff detection of explosives: a critical review, *Analytical and Bioanalytical Chemistry* **395**, 259–274 (2009).

79. C. Bauer, A. K. Sharma, U. Willer, J. Burgmeier, B. Braunschweig, W. Schade, S. Blaser, L. Hvozdara, A. Muller, and G. Holl, Potentials and limits of mid-infrared laser spectroscopy for the detection of explosives, *Applied Physics B* **92**, 327–333 (2008).

80. R. Bhargava, R. Schwartz Perlman, D. Fernandez, I. Levin, and E. Bartick, Non-invasive detection of superimposed latent fingerprints and inter-ridge trace evidence by infrared spectroscopic imaging, *Analytical and Bioanalytical Chemistry* **394**, 2069–2075 (2009).

81. T. Chen, Z. D. Schultz, and I. W. Levin, Infrared spectroscopic imaging of latent fingerprints and associated forensic evidence, *Analyst* **134**, 1902–1904 (2009).

82. P. Ng, S. Walker, M. Tahtouh, and B. Reedy, Detection of illicit substances in fingerprints by infrared spectral imaging, *Analytical and Bioanalytical Chemistry* **394**, 2039–2048 (2009).

83. Y. Mou and J. W. Rabalais, Detection and identification of explosive particles in fingerprints using attenuated total reflection Fourier transform infrared spectromicroscopy, *Journal of Forensic Sciences* **54**, 846–850 (2009).

84. M. C. Phillips and B. E. Bernacki, Hyperspectral microscopy of explosives particles using an external cavity quantum cascade laser, *Optical Engineering* **52**, 061302 (2013).

85. M. C. Phillips, J. D. Suter, and B. E. Bernacki, Hyperspectral microscopy using an external cavity quantum cascade laser and its applications for explosives detection, *Proceedings of SPIE* **8268**, 82681R (2012).

86. R. Furstenberg, C. Kendziora, M. Papantonakis, V. Nguyen, and A. McGill, Advances in photo-thermal infrared imaging microspectroscopy, *Proceedings of SPIE* **8729**, 87290H (2013).

87. B. Guo, Y. Wang, C. Peng, H. Zhang, G. Luo, H. Le, C. Gmachl, D. Sivco, M. Peabody, and A. Cho, Laser-based mid-infrared reflectance imaging of biological tissues, *Optics Express* **12**, 208–219 (2004).

88. R. Furstenberg, C. A. Kendziora, J. Stepnowski, S. V. Stepnowski, M. Rake, M. R. Papantonakis, V. Nguyen, G. K. Hubler, and R. A. McGill, Stand-off detection of trace explosives via resonant infrared photothermal imaging, *Applied Physics Letters* **93**, 224103 (2008).

89. C. A. Kendziora, R. M. Jones, R. Furstenberg, M. Papantonakis, V. Nguyen, and R. A. McGill, Infrared photothermal imaging for standoff detection applications, *Proceedings of SPIE* **8373**, 83732H (2012).

90. G. P. A. Malcolm, G. T. Maker, G. Robertson, M. H. Dunn, and D. J. M. Stothard, Active infrared hyperspectral imaging system using a broadly tunable optical parametric oscillator, *Proceedings of SPIE* **7486**, 74860H (2009).

91. Y. Wang, Y. Wang, and H. Le, Multi-spectral mid-infrared laser stand-off imaging, *Optics Express* **13**, 6572–6586 (2005).

92. A. K. Goyal, M. Spencer, M. Kelly, J. Costa, M. DiLiberto, E. Meyer, and T. Jeys, Active infrared multispectral imaging of chemicals on surfaces, *Proceedings of SPIE* **8018**, 80180N (2011).

93. M. E. Morales-Rodriguez, L. R. Senesac, T. Thundat, M. K. Rafailov, and P. G. Datskos, Standoff imaging of chemicals using IR spectroscopy, *Proceedings of SPIE* **8031**, 80312D (2011).

94. B. E. Bernacki and M. C. Phillips, Standoff hyperspectral imaging of explosives residues using broadly tunable external cavity quantum cascade laser illumination, *Proceedings of SPIE* **7665**, 76650I (2010).

95. F. Fuchs, C. Wild, Y. Rahmouni, W. Bronner, B. Raynor, K. Kohler, and J. Wagner, Remote sensing of explosives using mid-infrared quantum cascade lasers, *Proceedings of SPIE* **6739**, 673904 (2007).

96. S. Hugger, F. Fuchs, J. Jarvis, M. Kinzer, Q. K. Yang, R. Driad, R. Aidam, and J. Wagner, Broadband-tunable external-cavity quantum cascade lasers for the spectroscopic detection of hazardous substances, *Proceedings of SPIE* **8631**, 86312I (2013).

97. B. Hinkov, F. Fuchs, J. M. Kaster, Q. Yang, W. Bronner, R. Aidam, and K. Kohler, Broad band tunable quantum cascade lasers for stand-off detection of explosives, *Proceedings of SPIE* **7484**, 748406 (2009).

98. F. Fuchs, S. Hugger, M. Kinzer, B. Hinkov, R. Aidam, W. Bronner, R. Losch, et al., Imaging stand-off detection of explosives by quantum cascade laser based backscattering spectroscopy, *Proceedings of SPIE* **7808**, 780810 (2010).

99. F. Fuchs, S. Hugger, M. Kinzer, R. Aidam, W. Bronner, R. Losch, Q. Yang, K. Degreif, and F. Schnurer, Imaging standoff detection of explosives using widely tunable midinfrared quantum cascade lasers, *Optical Engineering* **49**, 111127 (2010).

100. S. Hugger, F. Fuchs, J. Jarvis, M. Kinzer, Q. K. Yang, W. Bronner, R. Driad, R. Aidam, K. Degreif, and F. Schnurer, Broadband tunable external cavity quantum cascade lasers for standoff detection of explosives, *Proceeding of SPIE* **8373**, 83732G (2012).

101. F. Fuchs, S. Hugger, M. Kinzer, Q. K. Yang, W. Bronner, R. Aidam, K. Degreif, S. Rademacher, F. Schnurer, and W. Schweikert, Standoff detection of explosives with broad band tunable external cavity quantum cascade lasers, *Proceedings of SPIE* **8268**, 82681N (2012).

102. E. L. Holthoff, M. E. Farrell, and P. M. Pellegrino, Standardized sample preparation using a drop-on-demand printing platform, *Sensors* **13**, 5814–5825 (2013).

103. M. C. Phillips, J. D. Suter, B. E. Bernacki, and T. J. Johnson, Challenges of infrared reflective spectroscopy of solid-phase explosives and chemicals on surfaces, *Proceedings of SPIE* **8358**, 83580T (2012).

104. C. F. Bohren and D. R. Huffman, *Absorption and Scattering of Light by Small Particles* (John Wiley and Sons, New York, 1983).

105. C. Matzler, MATLAB functions for Mie scattering and absorption, Version 2, IAP Research Report, No. 2002-11 (University of Bern, Bern, 2002). Available at: http://omlc.org/software/mie/maetzlermie/Maetzler2002.pdf.

106. J. D. Suter, B. E. Bernacki, and M. C. Phillips, Spectral and angular dependence of mid-infrared diffuse scattering from explosives residues for standoff detection using external cavity quantum cascade lasers, *Applied Physics B* **108**, 965–974 (2012).

107. B. Philips-Invernizzi, C. Caze, and D. Dupont, Bibliographical review for reflectance of diffusing material, *Optical Engineering* **40**, 108–1092 (2001).

108. C.-I. Chang, *Hyperspectral Imaging: Techniques for Spectral Detection and Classification* (Kluwer Academic, Boston, 2003).

109. C.-I. Chang, *Hyperspectral Data Processing: Algorithm Design and Analysis* (Wiley-Interscience, New York, 2013).

110. H. Ren and C. Chang, Automatic spectral target recognition in hyperspectral imagery, *IEEE Transactions on Aerospace and Electronic Systems* **39**, 1232–1249 (2003).

111. D. Manolakis, C. Siracusa, and G. Shaw, Hyperspectral subpixel target detection using the linear mixing model, *IEEE Transactions on Geoscience and Remote Sensing* **39**, 1392–1409 (2001).

112. J. W. Goodman, *Speckle Phenomena in Optics: Theory and Applications* (Ben Roberts, Greenwood Village, CO, 2007).
113. J. C. Dainty, ed., *Laser Speckle and Related Phenomena* (Springer-Verlag, Berlin, 1984).
114. A. A. Maradudin and T. Michel, The transverse correlation length for randomly rough surfaces, *Journal of Statistical Physics* **58**, 485–501 (1990).

9 Infrared Spectroscopy of Explosives Residues
Measurement Techniques and Spectral Analysis

Mark C. Phillips and Bruce E. Bernacki

CONTENTS

9.1 INTRODUCTION

Standoff or noncontact techniques are needed for detection and identification of explosives and other materials when physical sampling is impossible, undesired, or not permitted. Reasons to avoid contact-based sampling include a need to remain a safe distance from a potential explosive threat, limited physical access to a surface, a desire for rapid *in situ* detection of explosives on a surface, or the need to conserve the analyte for additional laboratory methods. Optical techniques are inherently non-contact and nondestructive. A light source may be propagated over large distances before interacting with a sample and then may be propagated over a distance to a detector. The response of the material to optical radiation is very fast, as is the propagation time of the light to and from the sample. Material properties including transmission, reflection, and emission may be probed using optical techniques. By varying the wavelength of incident or detected light to record a spectrum, it is possible to define and measure a characteristic response of many materials for detection, identification, and sometimes quantification.

Infrared (IR) spectroscopy in particular provides a powerful technique for explosives detection. The mid-infrared (MIR) spectral region, defined here as spanning wavelengths from 3 to 25 μm, contains spectral features arising from vibrations and rotations of molecules. The spectral response of molecules in the MIR is highly dependent on their molecular structure and therefore provides a spectral "finger-print" of a chemical species. Explosives such as N-methyl-N,2,4,6-tetranitroaniline (tetryl), 1,3,5-trinitroperhydro-1,3,5-triazine (RDX), trinitrotoluene (TNT), dinitro-toluene (DNT), pentaerythritol tetranitrate (PETN), and 1,3,5,7-tetranitro-1,3,5,7-tetrazacyclooctane (HMX) all have different and distinctive spectral absorption and reflectance. Complex mixtures of materials can be analyzed by either finding regions of minimal spectral overlap of the constituent spectra, or by multivariate spectral analysis algorithms to determine the constituents.

The high brightness, high spectral radiance, and coherence properties of laser sources enable sensitive detection of materials including explosives and per-mit propagation over large distances [1,2]. Quantum cascade lasers (QCLs) are a semiconductor-based laser source in the MIR [3], and continued advances in per-formance have provided expanded wavelength coverage, high output power, and room-temperature operation [4]. These features make QCLs highly desirable for standoff explosives detection applications, especially when a compact instrument is needed. The external-cavity QCL (ECQCL) provides the narrow-linewidth, broadly tunable MIR source needed for spectroscopic measurements of solid materials [5]. In this chapter, we are concerned with reflectance-based spectroscopic detection of explosives; however, it is worth noting that photothermal and photo-acoustic tech-niques (refer to Chapter 10) have also been demonstrated using ECQCLs and other laser sources [6–13].

Optical reflectance spectroscopy is well suited for noncontact applications, and comparing an observed material's spectrum with a reference or library spectrum allows high-confidence identification in many cases. The spectroscopy of homoge-neous gases, liquids, and solids is well established and provides the basis for these identification techniques. However, direct application of these techniques to more complicated systems such as residues on surfaces often does not provide satisfactory results, except in highly controlled and contrived situations. For standoff optical detection, the goal is often to detect small quantities of explosive particles or residues situated on top of a second material, or substrate. One challenge with creating a model for this optical system is that it is highly dependent on the geometry of the observation and the morphology of the sample, including the sample thickness, particle size, and crystallinity. For example, a thin, uniform residue of the target material will behave differently from a collection of particles, which will also behave differently from the bulk material.

In laboratory measurements, it is often possible to prepare materials and design experiments in such a way as to measure isolated absorption or reflection spectra of solid materials. However, for practical standoff explosives detection applications, there is usually little or no control over sample preparation or measurement geometry. Measured spectra will typically not consist of pure absorption or reflection spectra, but will be determined by both effects. To make matters worse, the resulting spectrum is usually not a linear combination of absorption and reflection spectra. Characterizing and understanding the variations in spectral signatures with differences in material preparation and observation conditions is essential for designing spectral identification algorithms with a high probability of detection and low false-alarm rates. For targets consisting of residues on substrates, the fundamental spectroscopy has not been thoroughly studied, and additional research is required to develop high-confidence automatic threat-detection algorithms.

In this chapter, we describe our ongoing efforts at Pacific Northwest National Laboratory (PNNL) to develop solutions for standoff detection of explosives using ECQCL spectroscopy. Section 9.2 provides an introduction to theoretical considerations for reflectance spectroscopy of explosives residues and films. Section 9.3 gives a summary of many prior measurements of MIR reflectance spectroscopy of explosives, using both ECQCL and Fourier transform infrared (FTIR) spectroscopy. Section 9.4 summarizes the experimental work performed at PNNL for measurements of MIR reflectance spectroscopy of explosives using ECQCLs. Early work focused on demonstrating standoff hyperspectral imaging of explosives in both specular and diffuse reflectance geometries using ECQCL sources. Quantitative measurements of diffuse reflectance were performed to allow prediction of radiometric performance of standoff detection systems. Later work has focused on developing spectral analysis algorithms to account for the observed variability in measured reflectance spectra. This goal has also motivated a number of microspectroscopy studies aiming to isolate reflection or absorption spectra from individual explosives particles. Results from these experiments have demonstrated the high sensitivity of ECQCL-based explosives detection, with detection of sub-nanogram particles of explosives in hyperspectral transmission microscopy. The microscopy studies also reveal a number of spectral effects due to particle morphology and crystallinity. These include differences in spectra between α- and β-crystalline forms of RDX and also effects due to relative orientation of the PETN crystal axis and polarization axis of the probe light. We also describe nanospectroscopy experiments that demonstrate sub-femtogram detection limits for particles of explosives. Section 9.5 describes the application of various spectral analysis techniques, originally developed for hyper-spectral imaging analysis, to the problem of explosives detection.

9.2 THEORETICAL CONSIDERATIONS FOR REFLECTANCE SPECTROSCOPY

Figure 9.1 shows a simplified picture of a hypothetical standoff reflectance spectroscopy system. For the purposes of this chapter, we assume the target material to be detected is a film, residue, or collection of particles of explosive material that sits on a substrate material of different composition (as opposed to large quantities of bulk explosive material). A tunable laser illumination source is directed at the sample at a given incident angle with respect to the surface normal, and these two lines further define the plane of incidence (note that the laser could also be replaced by broadband collimated illumination for active FTIR spectroscopy). In the absence of the target material and for a perfectly smooth substrate, a specular reflection is directed away from the substrate. This specular reflection lies in the plane of incidence and has an angle equal in magnitude to the angle of incidence but opposite in sign with respect to the surface normal. For rough substrates and target materials, a portion of the incident laser illumination may still be directed in the specular reflection direction; however, there is also a diffuse reflectance component directed into the hemisphere away from the surface. The quantitative distribution of this diffuse reflectance is often described by a bidirectional reflectance distribution function (BRDF), details of which may be found in [14] and other references therein. The light reflected from the surface may be detected with a point detector or imaging detector, and

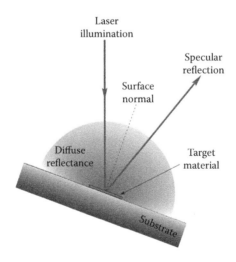

FIGURE 9.1 Geometry for standoff reflectance spectroscopy of explosives. A target explosive material is present as a film, residue, or particles deposited on a different substrate material. Laser illumination is directed at the target material at an angle with respect to the surface normal. Both specular and diffuse reflectance may be detected.

considerations such as spatial resolution and field of view for the detection optical system are important for optimization and evaluation of performance.

The goal of standoff spectroscopic detection is to measure spectral features characteristic of the target explosive material in the reflected light from a system so that the explosive may be detected, identified, and possibly quantified. These spectral features may take the form of attenuation or absorption at wavelengths characteristic of the vibrational modes (resonances) of the explosive compound. Alternatively, these spectral features may take the form of dispersive features characteristic of surface reflections near these same absorption resonances (discussed further below). In general, the light reflected from an overall system such as the one shown in Figure 9.1 may interact with the explosive material in various ways and thus influence the measured spectrum. For example, the incident light may reflect from the first surface interface between the air and explosive residue. The incident light may also penetrate into the explosive, be absorbed at some wavelengths, and may then reach the substrate material. Depending on the reflectivity and roughness of the substrate, this light may be reflected or scattered back to the detector. For particulate explosives, the situation is even more complicated, and the light may experience multiple reflections at air/explosive/substrate interfaces, as well as propagation through the particles. The term transflection is sometimes used to denote the light that has interacted with the structure and returned to the detector [15]. The transflection can contain contributions from the first surface reflection, subsequent reflections at material boundaries, and absorption within the bulk of the target material. It may also include effects of multiple-beam interference, depending on the geometry.

In classical electrodynamics, the interaction between light and materials is described by the complex refractive index $\tilde{n}$, or equivalently the complex dielectric function ε, which are related by $\varepsilon = \tilde{n}^2$. These quantities are functions of wavelength and describe the modification of electromagnetic waves propagating through a uniform medium. The real part of the complex refractive index $n = \text{Re}[\tilde{n}]$ describes the phase of the propagating wave, while the imaginary part $k = \text{Im}[\tilde{n}]$ describes the attenuation of the wave, leading also to the notation of (n, k) to express the complex refractive index. It is often convenient to model the optical response of a material using dispersion theory [16–18], in which the complex dielectric constant of a material is expressed as a sum of Lorentzian oscillators, where in IR spectroscopy each oscillator represents a vibrational mode, or *absorption resonance*, of the material:

$$\varepsilon(\omega) = n_e^2 + \sum_j \left[\frac{\omega_{pj}^2}{\omega_{0j}^2 - \omega^2 - i\omega\gamma_j} \right] \tag{9.1}$$

where:

ω_{0j} = center frequency
γ_j = width
ω_{pj} = strength of the jth oscillator in the summation.

The term n_e represents the residual refractive index due to all contributions not included in the sum. A particular advantage of using an oscillator model is that the

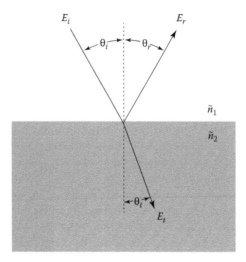

FIGURE 9.2   Reflection and transmission of light at an interface.

complex refractive index is represented by a small set of oscillator parameters, rather than as tabulated spectral data, and also satisfies Kramers–Kronig causality relationships by construction [18].

The Fresnel formulas provide the reflection and transmission coefficients between two isotropic materials with complex refractive indices, and their derivation is discussed in many textbooks [18,19]. Figure 9.2 shows the geometry of reflection and transmission at the boundary of two materials with complex refractive indices $\tilde{n}_1$ and $\tilde{n}_2$. An electromagnetic plane wave with electric field amplitude E_i and propagation direction θ_i with respect to the surface normal is incident on this boundary, and outgoing reflected and transmitted waves are denoted by E_r and E_t, with angles θ_r and θ_t, respectively. The three rays shown in Figure 9.2 lie in a plane defined as the plane of incidence. The reflection and transmission amplitude coefficients $r_{12}=E_r/E_i$ and $t_{12}=E_t/E_i$ depend on the angle and polarization of the light incident on the boundary.

$$r_{12}^s = \frac{\tilde{n}_1 \cos\theta_i - \tilde{n}_2 \cos\theta_t}{\tilde{n}_1 \cos\theta_i + \tilde{n}_2 \cos\theta_t}$$

$$t_{12}^s = \frac{2\tilde{n}_1 \cos\theta_i}{\tilde{n}_1 \cos\theta_i + \tilde{n}_2 \cos\theta_t}$$

$$r_{12}^p = \frac{\tilde{n}_2 \cos\theta_i - \tilde{n}_1 \cos\theta_t}{\tilde{n}_1 \cos\theta_t + \tilde{n}_2 \cos\theta_i}$$

$$t_{12}^p = \frac{2\tilde{n}_1 \cos\theta_i}{\tilde{n}_1 \cos\theta_t + \tilde{n}_2 \cos\theta_i} \tag{9.2}$$

Here, s and p denote electric fields polarized perpendicular and parallel to the plane of incidence, respectively. The corresponding intensity reflection and transmission coefficients, referred to as the reflectance and transmittance, are related to the complex magnitude squared of the above quantities, for example, $R_{12}=|r_{12}|^2$.

The Fresnel formulas capture the fundamental challenge of reflectance spectroscopy: the reflection of light from an interface is a nonlinear function of the complex refractive indices of both materials, as well as the angle of incidence, polarization state of the illumination light, and angle of detection. The complex refractive index spectrum is the characteristic quantity of the explosive material itself. Ideally, we would like to use the complex refractive index for identification; however, we typically cannot measure it directly, especially in a standoff detection scenario. Instead, we measure the reflectance, transmittance, or transflectance, which contain additional information about the system geometry and substrate material. For a single smooth interface, or for smooth homogeneous films, it is possible to analytically determine reflectance and transmittance spectra, and thus the measured quantities are related to the complex refractive indices in a straightforward manner. However, for rough surfaces or particles, the relationship of measured reflectance to complex refractive index cannot be solved analytically except for highly simplified geometries unlikely to be encountered for real-world explosives detection scenarios.

At wavelengths far away from absorption resonances where $k=0$, the Fresnel reflectance depends only on n, the real part of the complex refractive index, and there is little wavelength dependence of the reflectivity. Similar approximations can be made for $k \ll 1$, which is often the case for very weak absorption resonances. However, for spectroscopic detection of materials such as explosives, the distinguishing spectral features are the strongest absorption resonances. In this case, the k-dependence of the reflectance cannot be ignored. Furthermore, the behavior of the reflectance becomes highly nonlinear for $k > n$, which often occurs near strong absorption resonances in solids and leads to strong modifications of the reflectance spectrum relative to the reflectance spectrum near weaker absorption resonances. Such behavior is related to a range of observed effects, such as IR Reststrahlen bands [20], phonon modes in solid-state materials [20], and Fano resonances [21].

As a simple example, Figure 9.3 shows plots of calculated reflectance spectra for two cases of complex refractive index near absorption resonances. The first case (left column) describes a weak absorption resonance ($n_e = 1.4$, $\omega_{0j} = 245.044 \times 10^{12}$, $\gamma_j = 1.50796 \times 10^{12}$, $\omega_{pj} = 1 \times 10^{13}$), whereas the second case (right column) describes a strong absorption resonance ($n_e = 1.4$, $\omega_{0j} = 245.044 \times 10^{12}$, $\gamma_j = 1.50796 \times 10^{12}$, $\omega_{pj} = 5 \times 10^{13}$). These values for oscillator parameters are reasonable for solid explosives with published (n, k) data [17]. The first row of Figure 9.3 plots the real (n) and imaginary (k) parts of the complex refractive index calculated using the oscillator parameters along with Equation 9.1. The second row shows the reflectance spectrum for normal incidence reflection from a single interface between air and the materials. Figure 9.3c shows the typical "dispersive" spectrum characteristic of reflectance near weak absorption resonances. However, the reflectance near the strong absorption resonance shown in Figure 9.3d exhibits a strongly peaked shape more representative of a Reststrahlen feature. Therefore, we cannot generalize a single spectral peak shape which would be observed for reflectance even from a single interface, because the peak shape changes depending on the magnitudes of n and k. Accounting for the dependence of peak *shape* with variations in complex refractive index is crucial for interpreting reflectance spectra.

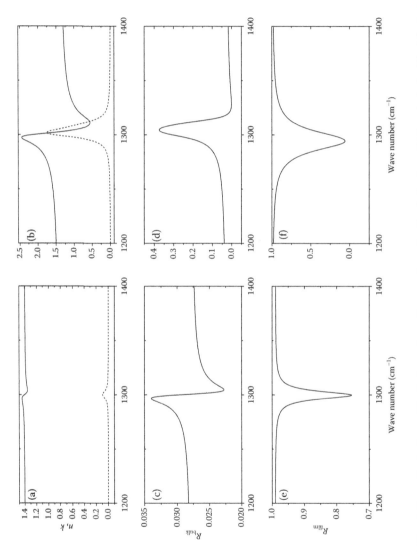

FIGURE 9.3 (a–f) Simulations of reflectance. The first row shows the complex refractive index (n, solid line; k, dashed line) for two model Lorentzian oscillators of different magnitudes. The second row shows the calculated normal incidence reflectance from an interface between air and the materials. The third row shows the calculated normal incidence transflectance from a 1 μm thick film of material in optical contact with a gold substrate.

The third row of Figure 9.3 shows calculated normal incidence transflectance from a 1 μm film of the material on top of a highly reflective gold substrate. In this case, the transflectance from the entire system was calculated using standard matrix-based methods for layered materials (see e.g., [22]). In this case, the trans-flectance spectra for both materials exhibit absorption of light near the resonance. This simulation corresponds to a double-pass transmission measurement, in which the incident light passes through the material, is reflected from the substrate, and then passes through the material again before returning to the detector. Because the film is relatively thin and the substrate highly reflective, in the simulations shown in Figure 9.3e and f the dominant spectral feature results from the double-pass transmission and not the first surface reflection. It is important to note that the calculated film transflectance includes the effect of multiple reflections within the layered structure, of which the first surface reflection is one component; furthermore, the calculated results are highly dependent on the film thickness. In some cases, for example, thick films or highly absorbing substrates, the first surface reflection spectrum can become the dominant spectral feature that is observed [23].

The spectra shown in Figure 9.3c and d can be regarded as arising from surface reflections at an air–material boundary. In contrast, the spectra shown in Figure 9.3e and f can be regarded as more characteristic of absorption within the volume of a material. Predicting or modeling the observed spectrum of an inhomogeneous material requires not only the complex refractive index, but also consideration of the geometry of the measurement. Diffuse reflectance or scattering arises from multiple reflections at interface surfaces combined with absorption by the volume of the material. The relative contribution of these two effects depends not only on the fundamental optical constants, but also on sample morphology effects including particle size, sample concentration, and sample thickness. It is often observed that the spectrum of specular reflection from a clean crystal face can differ dramatically from the spectrum of diffuse scatter from a powdered sample of the same material, which also varies with particle size [24]. In a diffuse scattering geometry, and for an inhomogeneous distribution of particles, the observed spectrum will include contributions from both surface reflections and volume absorption, typically in non-linear combinations, which can lead to peak shifts and even complete peak inversions. In addition, reflection and absorption by the substrate on which the residues exist can also dramatically influence the diffuse reflectance spectrum. A number of models have been developed for diffuse reflectance from particulate materials ([25] provides an overview); however, they are often limited to particular systems and may not be directly applicable to the problem of predicting the spectra of explosives residues on surfaces across a broad spectral range. Development of physics-based models for predicting the diffuse reflectance spectra of explosives particles and films remains an active area of research [15,23], and will enable improvements in detection algorithms.

The variability in observed transmittance and reflectance spectra of explosives makes defining a single reference (or library) spectrum for each material difficult, if not impossible. In many cases neither the reference transmission spectra of thin films nor the integrated reflectance spectra of unmixed solids provide reference spectra that adequately resemble the instrumental signals obtained from real-world explosives and surfaces. One potential approach for explosives detection is to measure multiple reference spectra for each explosive compound, attempting to capture the expected variability in detected spectra. The library spectra might contain reference spectra of explosives with variable particle sizes, variable substrates, and/or variable measurement geometries. However, this approach requires a large spectral library to account for the large number of configurations of samples on background substrates likely to be encountered. In some cases, the spectral variation may be small or consist of simply a linear scaling factor, such as for changes in particle size or film thickness over small ranges. However, because the physical basis for observed reflectance spectra arises from multiple reflection and transmission events, it is unlikely that the library size can be reduced greatly by using linear combinations of a few characteristic library spectra. An additional risk with using large library sets is the increased probability of false positives; it becomes more and more likely an algorithm will fit an observed spectrum with an incorrect linear combination of library spectra. Nevertheless, the approach of using multiple reference spectra to capture spectral variability serves as an important first step in developing more advanced detection algorithms.

This section has provided a brief and simplified description of the theory of reflectance spectroscopy to introduce the basic concepts, with an emphasis on phenomenology. While necessarily incomplete (an entire book could easily be written on the topic), the contents of this section highlight many of the challenges associated with interpretation of reflectance spectra. Despite the complexity of reflectance spectroscopy, it provides a very sensitive detection technique for pure materials such as explosives with strong absorption resonances. As shown in the next sections, standoff detection of trace residues of explosives using reflectance spectroscopy has been demonstrated by multiple research groups at meter-scale distances. Although the measured spectra are often complicated, they nevertheless provide distinctive signatures of different explosives and substrates. Rather than ignoring the richness and variability inherent to reflectance spectra, we believe that additional research to develop both better phenomenological understanding and better quantitative modeling of diffuse reflectance will enable better detection algorithms and further improve performance of standoff detection systems.

9.3   REVIEW OF IR REFLECTANCE SPECTROSCOPY OF EXPLOSIVES

Standoff reflectance spectroscopy of explosives has generally fallen into two categories, depending on whether the specular or diffuse reflectance was detected. As discussed below in Sections 9.4.3 and 9.4.4, the specular reflection component typically provides power that is orders of magnitude higher than the diffuse reflectance component, making detection at meter-scale distances possible without using cryogenically cooled detectors. In addition to higher returned power, the spectra measured in this configuration are often easier to interpret than those measured in diffuse reflectance. As discussed in Section 9.2, measurements of thin films of explosives on reflective substrates result in a double-pass transmission configuration, in which the incident light passes through the explosives, reflects from the substrate, and passes through the explosives a second time before returning to the detector. However, many of the references listed below show high variability in measured spectra and results often match reference spectra only in general location of peaks and not in spectral shapes.

ECQCL-based standoff detection of explosives in the specular reflectance geometry has been demonstrated in both nonimaging (point detector) and hyperspectral imaging (array detector) measurements. Quartz tuning forks have been used as low-cost photodetectors to detect ECQCL light reflected from surfaces coated with explosives: reflectance spectra of RDX, PETN, and TNT at 20 m in the specular reflection geometry have been measured using ECQCLs combined with a quartz tuning fork as a nonimaging photodetector [26,27]. More recently, standoff reflectance spectroscopy at 25 m using an ECQCL and a quartz tuning fork as a photodetector was used for detection of DNT and PETN on roughened aluminum plate and wood surfaces [28]. Even though the surfaces were rough, the received power and spectral shape (transmittance-like) indicate that the results are for measurement in a specular reflection geometry. In [29], standoff reflectance spectroscopy of RDX, TNT, and PETN on stainless steel plates was reported using ECQCLs and a thermoelectrically cooled nonimaging photodetector. An ECQCL and a microbolometer camera were used for hyperspectral imaging of RDX on an aluminum plate at distances up to 4 m in [30]. An ECQCL and a cooled mercury cadmium telluride (MCT) array were used for hyperspectral reflectance imaging of TNT on reflective and absorbing substrates in [31]; particularly notable were the large variations in reflectance spectrum from TNT observed depending on the sample substrate.

Standoff specular reflectance spectroscopy of explosives on surfaces has also been demonstrated using FTIR-based systems, with both active and passive illumination. Nonimaging reflectance spectroscopy at distances up to 15 m of PETN, RDX, TNT, and DNT deposited on aluminum plates was demonstrated in [32]. Standoff reflectance hyperspectral imaging of RDX on steel plates was performed in [33] at distances up to 50 m using an IR imaging spectrometer with passive illumination. FTIR-based reflectance spectroscopy of RDX and TNT on stainless steel surfaces at distances up to 3.7 m was studied in [34]. Far-IR (50–300 cm^{-1}) FTIR-based reflectance spectroscopy and spectroscopic ellipsometry of pelletized HMX was studied in [35], which is noted here because it included measurements of the complex refractive index.

Compared to measurements of specular reflectance, there are fewer studies of diffuse reflectance from explosives on surfaces. Angular dependence of diffuse reflectance spectra from TNT, RDX, and tetryl on painted automotive surfaces was studied using ECQCL illumination with both single-element [36] and imaging [37] detectors. The angular dependence of reflectance spectra from PETN deposited on aluminum plates was studied using both ECQCL and FTIR spectroscopy in [38,39] and showed large variations in reflectance spectra depending on substrate roughness and sample aerial density. Hyperspectral imaging (refer to Chapter 8) of diffuse reflectance from RDX on stainless steel was performed using an ECQCL and microbolometer camera at a 1.5 m standoff distance in [40]. Measurements of diffuse reflectance of TNT, DNT, RDX, and PETN particles on silicon carbide substrates at <2 m standoff distance were performed using ECQCL illumination and a cooled MCT detector, combined with a Risley prism scanner to provide steering of the illumination beam [41]. An ECQCL and a cooled MCT array detector were used for hyperspectral imaging at 2 m of diffuse reflectance from TNT and PETN on aluminum and painted automotive panel surfaces [42,43]. In [42–48], diffuse reflectance of TNT, DNT, PETN, RDX, and various nonexplosive interferents on painted automotive, plastic, and denim surfaces were measured using an ECQCL and cooled MCT array and included measurements of diffuse reflectance up to a 20 m standoff distance. Recently, a high-performance digital pixel focal plane array has been used in combination with an array of distributed feedback QCLs to perform differential reflectance imaging of diffuse reflectance from $KClO_3$ particles [49].

9.4 MEASUREMENTS OF EXPLOSIVES USING ECQCL SPECTROSCOPY

9.4.1 EXTERNAL CAVITY QUANTUM CASCADE LASERS

Grating-tuned ECQCLs [50–57] typically provide a tuning range greater than 10% of the center wavelength, enabling detection of molecules with broad absorption features, and detection of multiple chemicals with absorption features within the ECQCL tuning range. PNNL has developed a custom broadly tunable ECQCL platform to provide wide wavelength coverage and enable simultaneous detection of multiple chemicals. The PNNL-ECQCL was originally developed for trace gas spectroscopy and sensing and has been integrated in sensors for in-field detection in demanding environments [58,59]. The same features that make the PNNL-ECQCL useful for trace gas sensing of large molecules are also attractive for spectroscopy and sensing of solid materials. Despite broader MIR absorption bands of molecules in the solid phase, the tuning range of the ECQCL is large enough to measure multiple features of many solid chemical compounds, including explosives. The PNNL-developed ECQCL systems typically provide 100–200 cm^{-1} of tuning in the MIR spectral region. Recent developments in QCLs exhibiting wide gain bandwidths have enabled ECQCL tuning ranges >400 cm^{-1}, providing access to much larger portions of the MIR spectrum in single ECQCLs [60,61]. It is also possible to combine several QCL gain media into a single package to provide broader wavelength coverage in a compact instrument [62].

The PNNL-ECQCL has a few distinctive features that make it well suited for spectroscopy of explosives. The ECQCL is based on a Littman–Metcalf cavity design [63], which provides high wavelength selection with no beam steering as the wavelength is tuned. The current to the QCL is modulated with a large amplitude sinusoidal waveform typically at 100 kHz. This current modulation serves to mitigate the effect of mode-hops as the wavelength is tuned, resulting in a smooth output of power versus wavelength without discontinuities, which is highly desirable for spectroscopy. In addition, the amplitude modulation of the ECQCL output power permits lock-in detection with single-element detectors, or the detected power can be time-averaged using slower detectors such as focal plane arrays (FPAs), in which case it behaves like a continuous-wave (CW) source [64]. The high scan stability and reproducibility demonstrated by the ECQCL provides numerous benefits for spectroscopy [59].

The ECQCL systems were designed to provide fast wavelength sweeps; the full tuning range is scanned at frequencies up to 20 Hz. The ECQCL systems provide spectral resolution of 0.1–0.3 cm^{-1}, which is sufficient for resolving the pressure-broadened absorption lines of even small molecules at atmospheric pressure with minimal instrumental broadening. Although this high spectral resolution is not required for resolving spectral features of solid explosives, it is beneficial for separating out effects of atmospheric absorption. For example, we have demonstrated gas-phase spectroscopy in the 7–8 μm spectral region at pathlengths up to 50 m [65]. This spectral region is often considered opaque due to atmospheric H_2O absorption. However, the high spectral resolution of the ECQCL allows many of the H_2O absorption lines to be resolved, and the ECQCL is attenuated far less in the windows between these lines.

9.4.2 SAMPLES

In our work, we have focused on the problem of detection of residues and particles of explosives deposited on surfaces. Experiments have been performed on the explosive compounds tetryl, RDX, TNT, and PETN. Explosives dissolved in solvents of acetone, methanol, and/or acetonitrile were purchased from AccuStandard [66]. The liquids used contained calibrated amounts of explosives, typically in 100–1000 μg/mL concentrations. Measured quantities of liquids were deposited on substrates, and the solvent was allowed to evaporate. After the solvent evaporated, the explosives remained on the surface as crystals, residues, or films, depending on the solvent and substrate. As is typical in such drop-cast samples, the resulting films were highly nonuniform, often exhibiting the so-called "coffee-ring" structure. Thus prepared, the explosives were usually bound tightly to the substrate, but discrete particles could be readily scraped from the surface for deposition on a different substrate. For additional sample preparation and characterization, please refer to Chapter 3.

9.4.3 STANDOFF HYPERSPECTRAL REFLECTANCE IMAGING

The most favorable configuration for performing standoff detection of explosives occurs when a specular reflection from a target can be captured by the detector. This measurement geometry has two main advantages. The first is that large standoff distances are easily achieved, because the typical R^{-2} dependence of diffusely scattered light is avoided and nearly all of the illumination light incident on the target can be detected. In the MIR spectral region, many targets have strong specular reflection components, even though they may appear rough at visible wavelengths. The second advantage is that the spectroscopic interpretation of results is often straightforward. As discussed above, in the specular reflection geometry, detection of thin films or residues of explosives often approximates a double-pass transmission, as diagrammed in the inset to Figure 9.4. For optically thin layers of explosives on highly reflective (metallic) substrates, the measured spectrum contains reduced intensities corresponding to regions of high absorption by the explosives. While the spectra can be complicated by thick explosives layers, leading to first-surface reflections from the explosives, or by dielectric substrates that absorb light instead of reflecting it,

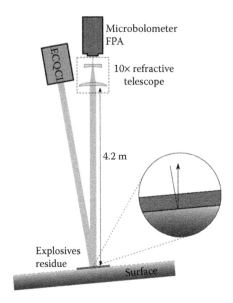

FIGURE 9.4 Experimental configuration for standoff hyperspectral imaging of specular reflection (not to scale). The ECQCL and sample were aligned to direct the specular reflection back to the microbolometer FPA detector. A 10× refractive telescope was used to provide magnification of the image. The inset shows the beam path for the double-pass transmission measured in this configuration.

in general the spectra are easier to both interpret and model than when considering diffuse reflectance geometries. Unfortunately, the specular reflection geometry is almost always impractical for realistic measurement scenarios, as it requires precise alignment of source, target orientation, and detector. Nevertheless, the specular reflection geometry offers a useful "first step" for testing standoff explosives detection instrumentation.

PNNL has used ECQCLs to investigate hyperspectral imaging-based detection of explosives residues [30]; the experimental geometry is shown in Figure 9.4. An ECQCL operating in the 9–10 μm range was used as the tunable source for these studies. The imager was an FLIR Systems Thermovision A40 uncooled microbolometer FPA. For this work, the laser was tuned in 2 cm^{-1} steps from 984 to 1103 cm^{-1}, providing 60 bands of spectral data for the image hypercube. Due to the limit of the image capture rate of our experimental apparatus of 30 frames/second, 2 s were required to record the full 60 bands of image data. A portion of the ECQCL output was sampled by a monitor detector to provide power normalization to subsequently acquired images. The average output power of the laser at the peak of the tuning curve was 5 mW. The output of the ECQCL was directed to the sample located at various standoff distances from 0.1 to 4.2 m, limited by the size of the laboratory, and the specular reflection from the target was directed to the detector by carefully aligning the angle of the aluminum plate. The reflected light was collected using a 10× telescope constructed from zinc selenide (ZnSe) lenses in combination with the germanium (Ge) lens attached to the camera (50 mm diameter, 38 mm focal length),

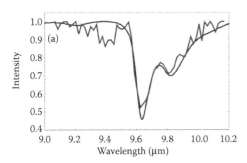
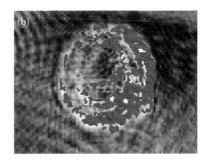

FIGURE 9.5 Standoff reflectance hyperspectral imaging of RDX at 4.2 m. (a) Experimental spectrum from 3×3 pixel average in image (*red*), along with library transmission spectrum of RDX. (b) Image of RDX residue after a minimum noise fraction (MNF) transform. The MNF 4 component is shown, revealing the RDX stain (grayscale image). Also shown is the result of a MTMF transform, using 20 of the 60 possible eigenvectors for increased signal to noise. The red pixels have a high matched filter score, and low infeasibility values, indicating a positive detection of RDX. The RDX residue in (b) is approximately 25 cm in diameter with an average aerial density of 100 µg/cm^2.

to adjust the magnification and field of view. These initial investigations used refractive optics with no attempts to mitigate the coherence effects occurring due to laser illumination.

Figure 9.5 shows an example of specular reflectance hyperspectral imaging of RDX. Samples of RDX were prepared by solvent drop-casting onto uncoated aluminum sheets with 75×100 mm dimensions, resulting in an approximately circular residue with a diameter of 25 mm and an approximate areal density of 100 µg/cm^2; the residue was not uniformly distributed but clustered in somewhat concentric rings. Figure 9.5a shows the measured spectrum from a 3×3 pixel average in a region of the RDX residue (red trace). The spectrum is consistent with a double-pass transmission measurement, and also shown in the figure is a reference spectrum obtained from FTIR transmission through a thin film of RDX (blue trace). Despite the presence of interference fringes due to the coherent laser illumination, the experimental and reference spectra show good agreement. Figure 9.5b shows an image of the RDX residue after processing via a minimum noise fraction (MNF) transform. The MNF 4 component is shown, revealing the RDX stain (grayscale image). Also shown is the result of a mixture-tuned matched filter (MTMF) transform, using 20 of the 60 possible eigenvectors for increased signal to noise. The red pixels have a high matched filter score, and low infeasibility values, indicating a positive detection of RDX. A detailed discussion of these spectral analysis techniques will be presented in Section 9.5, as will additional experimental results from hyperspectral reflectance imaging.

9.4.4 Diffuse Reflectance Spectroscopy

As discussed above, standoff detection of explosives using the specular reflection geometry is impractical for most detection scenarios. We have also investigated

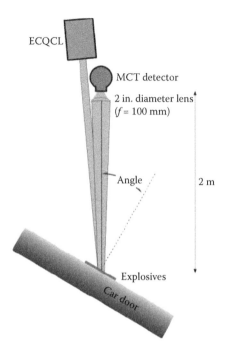

FIGURE 9.6 Experimental configuration for standoff diffuse reflectance measurements of explosives. The ECQCL illumination beam and scattered light collected by the MCT detector were aligned nearly collinear and at an angle to the normal of the surface.

effects of diffuse reflectance on standoff explosives detection [36,37]. The purpose of the research was twofold. The first purpose was to evaluate radiometric detection efficiency based on scattering targets consisting of explosives residues on real-world targets. The second was to investigate differences in recorded spectra of explosives residues between specular and diffuse reflectance geometries.

The experimental configuration for these studies is shown in Figure 9.6. The ECQCL for these experiments operated in the 7–8 μm spectral region with an average power of 10.5 mW at the center of the tuning range. The ECQCL output was collimated to a beam with a diameter of 10 mm and directed to the target surface. The target surface was a white-painted automotive door, on which residues of RDX, TNT, and tetryl were deposited. The RDX, tetryl, and TNT had estimated surface loadings of 81, 37, and 42 μg/cm², respectively, and were nonuniform.

For angle-dependent reflectance spectroscopy, the ECQCL light scattered from the target surface was collected using a 50 mm diameter, 100 mm focal length ZnSe lens and focused onto a 1 mm² cryogenically cooled MCT single-element detector. The ECQCL beam and the collection lens were mounted nearly collinear and held stationary while the sample was rotated at angles between 0° (specular) and 50° from the normal. A 2 m standoff distance was used for these studies. Modulation of the ECQCL current at 100 kHz enabled lock-in detection of the scattered light, to reject both thermal background radiation and laser-induced photothermal signals. The ECQCL wavelength was repeatedly swept across its full

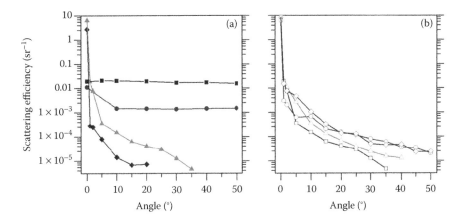

FIGURE 9.7 Measurements of quantitative diffuse scattering efficiency. Plots show the scattering efficiency relative to a silver mirror, in units of steradian⁻¹, as a function of angle relative to surface normal. (a) Bare surfaces of closed-cell polyurethane foam (*squares*), sand (*circles*), dirty portion of car door (*triangles*), and clean portion of car door (*diamonds*). (b) Different areas of car door containing no explosives (*squares*), RDX (*circles*), tetryl (*diamonds*), and TNT (*triangles*).

tuning range with a scan duration of 100 ms, which provided an effective spectral resolution of 0.4 cm⁻¹, determined primarily from the lock-in amplifier time constant of 300 μs. For small scattering angles, a single 100 ms laser scan produced spectra with high signal-to-noise ratios (SNRs). At large scattering angles, up to 20 spectra were averaged over 2 s. The fast acquisition speeds highlight the advantages of using a laser source for data collection when compared with thermal broadband illumination.

Figure 9.7 shows examples of quantitative angle-resolved scattering efficiency recorded with this setup. The results were normalized to the specular reflection from a silver mirror and spectrally integrated over the full ECQCL tuning range. It was observed that the scattering efficiency decreased by 4–5 orders of magnitude between specular and diffuse reflection angles, highlighting the radiometric challenge of detection of scattered MIR light. Nevertheless, for many targets, signals could be detected at angles up to 50°. Figure 9.7a shows scattering efficiency measured for different target surfaces, without explosives residues. Differences in scattering efficiency were observed based on surface roughness, with the smooth car door substrate providing the lower scattering efficiency, as expected. Dirt particles present on the car door were found to increase the scattering efficiency by up to an order of magnitude and allowed detection of reflectance at higher angles. In contrast, sand and closed-cell polyurethane foam approached Lambertian scattering targets [14], with nearly constant scattering efficiency out to large angles. The results for angle-resolved reflectance from the explosives residues are shown in Figure 9.7b. While the presence of the explosives residues was found to increase scattering efficiency, the large difference in specular versus diffuse scattering efficiency was still observed. In the MIR spectral range, the high specular reflection component contains a large

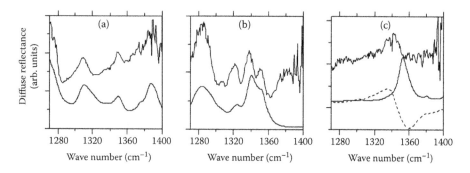

FIGURE 9.8 Diffuse reflectance spectra measured in backscattering geometry (angle of 5° relative to surface normal). The explosives residues were deposited on an automotive door. Experimental spectra are shown as the top traces, and library absorption spectra are shown as the lower traces for (a) RDX, (b) tetryl, and (c) TNT. The *dashed line* in (c) is a library reflectance spectrum for TNT.

fraction of the total reflected light, making detection of diffuse scatter a significant challenge at large standoff distances. Quantitative data on scattering efficiency are important, because it allows prediction of system performance for different operating configurations.

In addition to the quantitative scattering efficiency measurements, spectral measurements of the diffuse reflectance were also made at various angles. Figure 9.8 shows diffuse reflectance spectra obtained in the backscattering geometry for RDX, tetryl, and TNT at an angle of 5° from the surface normal. Qualitatively similar spectra were observed for larger angles up to 40°. For these measurements, the specular reflection was not measured by the detector, as indicated by the low measured values of reflectance (normalized to the specular reflectance from a silver mirror). Also shown in Figure 9.8 are library absorption spectra for RDX, tetryl, and TNT obtained from FTIR transmission measurements of explosives films deposited on ZnSe windows. For RDX and tetryl, the diffuse reflectance is higher in spectral regions corresponding to high absorption, which is the opposite behavior to that observed for the specular reflection geometry. In addition, many of the spectral features appear narrower in the diffuse reflectance versus the library absorbance spectra. The diffuse reflectance spectrum for TNT does not resemble the library absorption spectrum, but instead appears similar to a library reflectance spectrum obtained by specular reflectance from a thin TNT film on a galvanized steel surface. Similar effects have been noted in the literature previously and were found to be strongly dependent on sample and substrate surface properties [31,43]. These results indicate that surface roughness has an important effect on the relative contributions from reflection, absorption, and scattering.

The complex nature of scattering makes interpretation of the diffuse scattered spectra a challenging problem. In the results presented here, enhanced scattering was observed near absorption peaks for tetryl and RDX. However, the TNT residue did not exhibit this behavior. As discussed in Section 9.2, reflection and transmission from a material near an absorption resonance is determined by the complex refractive index and the relative magnitudes of its real and imaginary components. For strong absorption resonances in the MIR spectral region, the reflectivity can have a strongly peaked shape leading to Reststrahlen-like effects [20,24,67]. Strong reflectivity peaks in the MIR have been calculated for many explosives compounds based on measurements of the full complex refractive index [16,17,68]. Different spectral features can also be observed based on the relative absorption and reflectivity of the explosive sample and substrate. The same material can exhibit either peaks or dips in the specular reflection spectrum. This effect has been previously observed for TNT films on surfaces, in which it was referred to as "contrast reversal" [31,43]. We also observe differences in the sign of RDX peaks between the diffuse reflectance and the specular reflection shown in Section 9.4.3. The stronger absorption strength in the 7–8 μm spectral region investigated here may contribute to the increased reflectance observed near the RDX resonances.

The results collected using the MCT single-point detector should be comparable to the performance achievable with an MCT FPA and indicate that it is possible to capture scattered spectra at large angles beyond the surface normal. In addition, increasing the power of the ECQCL illumination and increasing the aperture of the collection optics would provide additional increases in signal, allowing operation with larger field of view or larger standoff distance. ECQCL sources have been demonstrated that provide hundreds of milliwatts of power [69], and future developments are expected to increase the power further. It is worth noting that even for high laser powers, typical hyperspectral imaging applications keep the laser intensity at eye-safe levels since the maximum permissible exposure (MPE) in this wavelength range is 100 mW/cm^2 for exposure durations >10 s [70].

Quantitative results such as those obtained from the angle-resolved scattering efficiency can be used to predict the radiometric performance of systems with variations in parameters such as collection aperture, standoff distance, illumination power, and detector sensitivity. The radiometric performance of a standoff explosives detection system can be expressed by the amount of light detected by each detector element (pixel) divided by the noise-equivalent power (NEP) of the detection system. A general equation for the detected power can be written based on a simplified form of the light detection and ranging (lidar) equation [71].

$$P_D = \frac{A}{4\pi r^2} \cdot S \cdot P_S \tag{9.3}$$

where:
P_D = detected power
A = area of the collection optics
r = standoff distance from target to collection optics
S = scattering efficiency of the target in steradian^{-1}
P_S = power of the source on the target

Note that P_S can be considered an effective power on the target that is imaged onto the detector, which depends on the illumination area on the target, the system magnification, and the detector area (pixel size in the case of an FPA). This equation ignores atmospheric absorption and scattering effects, but is a useful starting point for calculations of detected power for a given standoff detection system.

9.4.5 REFLECTION–ABSORPTION INFRARED SPECTROSCOPY (RAIRS) MEASUREMENTS OF EXPLOSIVES RESIDUES

The large differences observed in the spectra measured for specular versus diffuse reflectance geometries of explosives residues highlight the complex nature of reflectance spectroscopy for nonuniform films. To support efforts to measure and interpret the spectra of explosives particles and residues, we have applied various spectroscopic and imaging techniques using ECQCLs. In this section, we describe experiments using reflection–absorption infrared spectroscopy (RAIRS) to measure absorption spectra of explosives particles with greatly reduced reflectance contributions [72]. RAIRS is commonly used to measure thin films or monolayers of materials adsorbed on highly reflective substrates [73–75], often in combination with FTIR spectroscopy [73,76]. The ECQCL source provides similar data, but at a much higher acquisition speed. Nonuniform thin films or residues of materials such as explosives are not traditionally measured using RAIRS, but we find that the measured spectra resemble RAIRS spectra from uniform films and provide useful information to characterize the explosives. However, additional analysis will be required to account fully for effects of scattering and reflections on the measured RAIRS spectra.

The RAIRS technique uses a beam incident on a surface at large angles of incidence, which significantly enhances the absorption of p-polarized light, while suppressing the absorption of s-polarized light [74]. Our measurements of RAIRS spectra were performed using an ECQCL tuned over the spectral range of 1270–1400 cm^{-1} (7.14–7.87 μm), with average power of 6 mW at the center of its tuning range. Figure 9.9 shows a diagram of the experimental setup for these measurements.

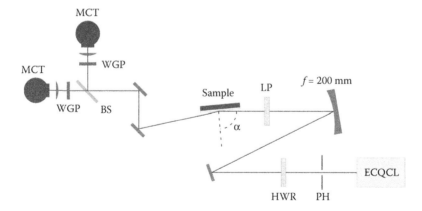

FIGURE 9.9 Schematic of experimental setup for performing RAIRS spectroscopy at angle of incidence α. The optical elements are labeled as follows: pinhole (PH), half-wave rhomb (HWR), linear polarizer (LP), beam splitter (BS), wire-grid polarizer (WGP), and MCT detector. Two MCT detectors were used to simultaneously monitor the s- and p- polarized reflections from the surface.

For these experiments, the ECQCL power was attenuated by placing a pinhole in the beam path, to avoid saturating the MCT detectors. The ECQCL output was focused onto the sample using a spherical mirror with 200 mm focal length placed 300 mm from the sample, providing a 160 µm diameter beam at the sample location. The RAIRS measurements were performed using an incident angle of 86°, and a 1 s acquisition time was used to measure the spectra.

The axis of the linear polarization incident on the sample was rotated using the half-wave rhomb (HWR) to include both s- and p-components. The reflected beam from the sample was split using a nonpolarizing ZnSe beam splitter (BS), and the transmitted and reflected beams were focused onto two MCT detectors using 50 mm focal length lenses. In each beam path before the detector, a wire grid polarizer was placed and adjusted so that one detector monitored the s-polarization and the other monitored the p-polarization reflected from the sample. The HWR and linear polarizer were adjusted to balance the final detector signals between the s- and p-reflections.

An absorbance spectrum was calculated via Equation 9.4:

$$A = -\left\{ \ln\left[\frac{I_p}{I_s}\right] - \ln\left[\frac{I_p^0}{I_s^0}\right] \right\} \tag{9.4}$$

where:

I_p = reflected power from the measured sample for p-polarized light
I_s = the reflected power from the measured sample for s-polarized light
I_p^0 = the reflected power from a bare gold surface for p-polarized light
I_s^0 = the reflected power from a bare gold surface for s-polarized light

Measuring the s- and p-polarized reflection signals simultaneously accounted for drifts in the ECQCL output power, and subtracting the absorbance spectrum from a reference bare gold surface accounted for any differences caused by the optics between the s- and p-signals, such as a polarization dependence in the beam splitter reflectivity.

Samples were prepared by drop-casting solutions of RDX, PETN, and tetryl onto a gold-coated silicon wafer substrate and allowing the solvent to evaporate. Measured volumes of 2.5 or 10 µL were deposited on a nominally 1 cm² piece of gold-coated silicon wafer using stock solutions from AccuStandard with concentrations of 1000 µg/mL. After the solvent evaporated, this provided samples with average surface concentration of explosives of 2.5 and 10 µg/cm². However, as noted in Section 9.4.2, this method of sample preparation results in highly non-uniform films, consisting of residues of small particles or crystals. In an effort to produce more uniform films, we prepared additional samples by spin-coating solutions onto the substrates. In this case, the surface concentration of the final films is only an estimate, due to unknown amounts of solution lost from the surface during the spin-coating procedure. The films produced were visibly more uniform than those prepared by drop-casting, but still possessed significant spatial nonuniformity.

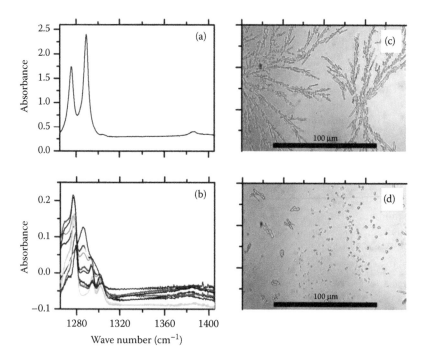

FIGURE 9.10 Variation in RAIRS spectra and crystal morphology of PETN residues. (a) RAIRS spectrum from 10 µg/cm² residue on gold surface. (b) RAIRS spectra from various locations on 2.5 µg/cm² residue. (c) Reflection-mode optical microscope image of one location on 10 µg/cm² residue. (d) Reflection-mode optical microscope image of one location on 2.5 µg/cm² residue.

Figure 9.10 shows example RAIRS spectra from PETN samples at two different average surface concentrations. The spectra possess excellent SNRs at the 1 s acquisition time used and without any additional spectral smoothing or filtering. In this case, for the higher concentration the spectrum has two strong peaks with a few additional weaker peaks. At the lower concentration, large variations between the spectra were observed, with many additional peaks present or absent depending on the position probed on the sample. Inspection under an optical microscope showed highly nonuniform PETN deposits with multiple crystalline forms. Even though the RAIRS technique averages over large areas of the surface, significant variation in spectrum is observed for nonuniform samples.

Figure 9.11 shows RAIRS spectra obtained for RDX residues of varying surface concentration. The spectra have not been offset, and the increasing baseline indicates an increase in nonresonant scattering with sample concentration. A number of absorption features are easily identified in the spectra. At the lowest concentration, only two absorption peaks are present within the ECQCL tuning range. As the concentration increases, additional peaks become dominant. It is likely that the spectra represent variations in the crystalline form of RDX. Previous studies have shown that for drop-cast samples on smooth surfaces, the β-form of RDX can predominate

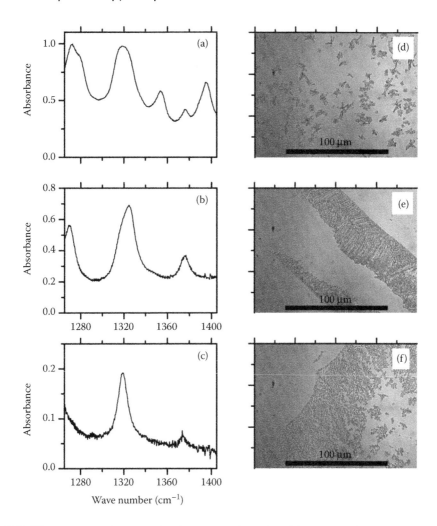

FIGURE 9.11 Variation in RAIRS spectra of RDX residues expressing both α- and β-crystalline forms. (a) RAIRS spectrum from 10 μg/cm² RDX residue on gold surface. (b) RAIRS spectrum from 2.5 μg/cm² RDX residue on gold surface. (c) RAIRS spectrum from spin-coated RDX residue on gold surface. Reflection-mode optical microscope image of one location on (d) 10 μg/cm² residue and (e) 2.5 μg/cm² residue. (f) Reflection-mode optical microscope image of a different location on 2.5 μg/cm² residue.

for lower surface concentrations, while the α-form becomes dominant at higher concentrations [77]. Comparison with the IR absorption band assignments for α- and β-RDX is in good agreement within experimental uncertainties [78].

The obtained spectra of explosives residues demonstrate that the ECQCL-RAIRS technique can measure high-SNR absorption spectra of low-concentration samples. From the size of the ECQCL beam at the sample, the measurement probes an area <0.03 cm² on the surface, representing a probed mass of <300 ng for 10 μg/cm² surface concentrations. The high SNR of the measured spectra, even without additional spectral smoothing or filtering, indicates sub-nanogram detection limits are possible. While obviously not suitable for standoff detection of explosives, the RAIRS spectra can provide valuable information for sample characterization and measurement of absorption spectra.

9.4.6 MICROSPECTROSCOPY OF EXPLOSIVES RESIDUES

The visible-light micrographs shown in Figure 9.11 indicate the high degree of spatial inhomogeneity of solvent-cast explosives. Similar inhomogeneity is expected for the case of detecting small particles of explosives on surfaces. Therefore, it is useful to also consider microspectroscopy techniques to characterize and understand the effects of sample inhomogeneity on IR spectra. While not truly a standoff detection technique due to the close proximity of samples to the measurement apparatus, we note that the techniques are still noncontact and nondestructive. Using an ECQCL provides a high-brightness source for these microspectroscopy and imaging experiments, which permits rapid measurements with high SNR.

In [66], we presented results from hyperspectral microscopy of explosives particles and residues in a transmission geometry. These experiments used an ECQCL operating in the 9–10 μm spectral region, coupled with a microbolometer FPA for detection of light transmitted through samples placed on an IR-transparent ZnSe window, as illustrated in Figure 9.12. Hypercubes were obtained in 4 s with 100

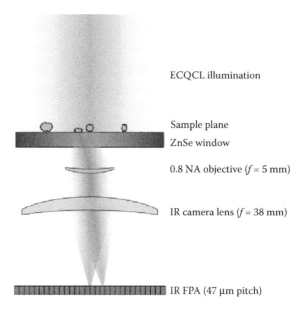

ECQCL illumination

Sample plane
ZnSe window

0.8 NA objective (f = 5 mm)

IR camera lens (f = 38 mm)

IR FPA (47 μm pitch)

FIGURE 9.12 Experimental configuration for hyperspectral transmission microscopy. The ECQCL beam was directed through a transparent ZnSe window on which particles of explosives were placed. The transmitted light was imaged using magnifying optics and a microbolometer FPA. Images were recorded as the ECQCL wavelength was changed, resulting in image hypercubes with 100 spectral bands.

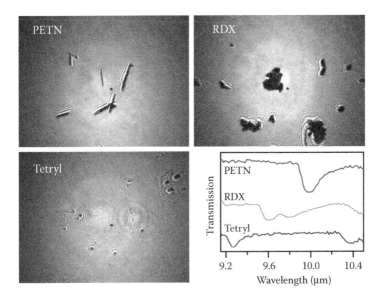

FIGURE 9.13 Transmission microspectroscopy of explosives particles. The three images (2×1.5 mm) show the mean IR transmission for particles of PETN, RDX, and tetryl. Transmission spectra from image points inside particles of PETN, RDX, and tetryl show the clear differences in spectral features.

wavelength bands and a 2×1.5 mm field of view. The spatial resolution was 13 μm, demonstrating near-diffraction limited imaging. A range of explosives particles and residues were measured using the technique. Both solution-cast residues and discrete particles of explosives such as RDX, tetryl, and PETN were measured, and example images and spectra are shown in Figure 9.13. Imaging and spectroscopic identification of particles with few- to sub-nanogram masses were demonstrated, and a minimum detectable thickness of 125 nm was calculated. Additional analysis of hyperspectral microscopy images of PETN crystals will be presented in Section 9.5.

In addition to transmission microscopy, reflection-mode microspectroscopy was performed to characterize particles and residues of explosives deposited on gold substrates. In this case, hyperspectral imaging was not performed but instead a single-element detector was used to measure spatially resolved reflection spectra of the samples under high magnification. Figure 9.14 shows a diagram of the experimental configuration. A 0.7 numerical aperture (NA) ZnSe doublet was used to focus and collect the ECQCL beam. A barium fluoride (BaF_2) window coupled a portion of the ECQCL beam into the microscope objective. The back-reflected beam was sampled using the reflection from a metallic neutral density (ND) filter and detected using a cryogenic MCT detector, as with the RAIRS experiments. The focal position was adjusted to maximize the reflected signal and minimize the focused spot size, which was measured to be 40 μm ($1/e^2$ diameter).

Registration of the ECQCL beam to the sample was performed by coaligning a visible-light microscope with the IR microscope. An incandescent light source was coupled into the ZnSe lens using the reflective ND filter. The back-reflected light was

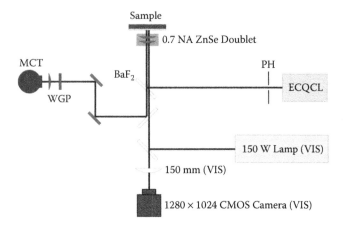

FIGURE 9.14 Experimental configuration for reflectance microspectroscopy of explosives residues. The ECQCL beam was directed through a pinhole (PH) for attenuation, reflected from a BaF_2 window, and focused onto the sample surface using a 0.7 NA doublet lens. The reflected light was collected using the same lens and directed to a MCT detector. A wire-grid polarizer (WGP) was placed in front of the detector, aligned along the ECQCL polarization axis. Coaligned with the MIR ECQCL beam was a white-light microscope consisting of a 150 W quartz tungsten halogen lamp and a CMOS camera, in addition to imaging optics.

directed to a CMOS camera. A RG780 long-pass filter was used to limit the detection bandwidth and reduce chromatic aberrations. A 50 mm diameter 150 mm focal length achromatic doublet was used in the visible beam path to reimage the sample onto the camera. It should be noted that the optical system was not diffraction limited at either the visible or MIR wavelengths. It was not possible to both minimize the spot size of the IR beam and have the visible image in proper focus simultaneously using our refractive optics setup. Nevertheless, despite the visible images' poor quality, they were valuable for locating particular regions on the samples and for registration of the ECQCL spot for microspectroscopy. An optimized optical system could be designed to perform much better at both the visible and the MIR wavelengths.

Spectra $I(\lambda)$ from sample regions with explosives particles were obtained by scanning the ECQCL over its tuning range in 1 s with a 1 ms lock-in time constant. A region of the sample surface without explosives particles was selected for a background spectrum, $I_0(\lambda)$. The data were normalized and plotted as absorbance: $A(\lambda) = -\log_{10}[I(\lambda)/I_0(\lambda)]$. In this case, due to the thin samples and metallic substrate, the measured reflectance approximated the double-pass transmission configuration, in which the reflection was attenuated at wavelengths of high absorption. Contributions to the spectrum from first-surface reflections of the explosives were not typically observed, but may have been present in some cases.

Figure 9.15 shows microabsorption spectra obtained from 2.5 $\mu g/cm^2$ PETN samples prepared by drop-casting and solvent evaporation. As previously observed, the PETN exhibited an elongated crystalline form. The figure shows a comparison between the spectra of crystals oriented perpendicular and parallel to the IR beam polarization. Two sets of resonances are clearly visible for the two polarization

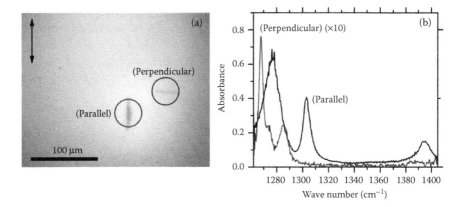

FIGURE 9.15 Reflectance microspectroscopy of PETN crystals. (a) Reflection-mode visible microscope image of PETN microcrystals. The ECQCL polarization direction is shown by the *arrow*. The two *circles* show the approximate areas probed the ECQCL beam. In one case, the PETN crystal axis is parallel with the ECQCL polarization, while in the other case the PETN crystal axis is perpendicular. (b) Absorbance spectra measured for the two locations shown in (a). The perpendicular spectrum has been scaled by a factor of 10. Note the difference in peak locations and overall magnitude of absorbance.

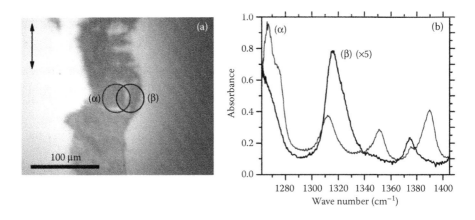

FIGURE 9.16 Reflectance microspectroscopy of RDX crystals. (a) Reflection-mode visible microscope image of RDX microcrystals. The ECQCL polarization direction is shown by the *arrow*. The two *circles* show the approximate areas probed the ECQCL beam. (b) Absorbance spectra measured for the two locations shown in (a). The β spectrum has been scaled by a factor of 5. Note the difference in peak locations.

states. There is also a difference in the absorption strength. The scale bar indicates the size of the particles and the circles show the nominal regions probed by the 40 µm diameter beam.

Figure 9.16 shows microabsorption spectra for an RDX sample also deposited on the gold substrate. In this case, it was also possible to observe two distinct sets of absorption resonances depending on the sample location probed. Figure 9.16 shows a particularly striking example, in which the two absorption spectra are observed from locations in close spatial proximity. Inspection of the visible image shows a small gap between the two regions, consistent with a difference in crystalline form. Based on the differences in spectra, we conclude that the two regions correspond to the α- and β-crystalline forms of RDX.

9.4.7 NANOSPECTROSCOPY OF EXPLOSIVES

The results on microspectroscopy and imaging of explosives demonstrate that detection of trace particles of explosives with nanogram to sub-nanogram mass is possible with ECQCL-based techniques. However, the techniques are fundamentally limited to spatial resolutions on the order of the wavelength of the light, which is on micrometer scales in the MIR. It is clear that many of the explosives particles contained spatial structure at shorter length scales than were probed using the far-field microscopy techniques. In addition, many of the particles themselves were not fully resolved by the imaging systems, which affected the detection limits.

Recently, we have also demonstrated spectroscopy and imaging of explosives residues at the nano scale [79]. In these experiments, a PNNL-ECQCL operating in the 7–8 μm wavelength region was integrated into a scattering-type scanning near-field optical microscopy (s-SNOM) instrument. The s-SNOM experiments were conducted by focusing the ECQCL beam onto the metallic tip of an atomic force microscope (AFM), as diagrammed in Figure 9.17a. The strong near-field localization near the AFM tip provided spatial resolution on the order of the tip radius (~25 nm), allowing spectroscopy below far-field classical diffraction limits. The detected light scattered from the tip-sample region was demodulated at a harmonic of the tip oscillation frequency to isolate the scattered light originating from the region below the tip. By scanning the wavelength of the ECQCL, the near-field spectral response of the explosives was measured.

Using the ECQCL with the s-SNOM, we demonstrated spectroscopy and imaging of residues of tetryl, RDX, and PETN deposited on gold surfaces. Figure 9.17 shows example results obtained from measuring small particles of RDX with ~0.5 μm diameter, 50 nm height, and an estimated mass of 20 fg. Figure 9.17b shows an s-SNOM spectrum obtained near the center of the RDX particle, normalized to the s-SNOM signal from gold. Figure 9.17c shows the AFM topographic image of the particle, with spatial scales indicated in the figure. Figure 9.17d and e show s-SNOM images obtained at fixed wavelengths on and off the strongest peak of RDX observed in the s-SNOM spectrum. The high SNR of both spectra and s-SNOM images indicate that detection limits were well below the femtogram level, and the technique would be able to detect a single monolayer of explosives. In addition, using the ECQCL allowed the s-SNOM measurements to be made with simultaneous high-speed, high spectral resolution, and high SNR.

Due to the measurement configuration, the spectra obtained from the explosives contained contributions from both n and k, making interpretation of the spectra somewhat challenging. However, good correlation was found between the resonances observed in s-SNOM and those measured in RAIRS spectra. In some

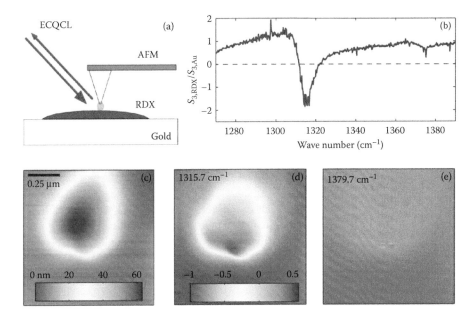

FIGURE 9.17 Nanospectroscopy and imaging of RDX particle on gold mirror. (a) Experimental configuration. (b) Near-field s-SNOM response of RDX particle. (c) AFM topographic data of RDX particle. (d) s-SNOM image with ECQCL wavelength at 1315.7 cm^{-1} (on-resonance). (e) s-SNOM image with ECQCL wavelength at 1379 cm^{-1} (off-resonance).

cases, the s-SNOM measurements can provide independent measurement of n and k [80,81]. However, analysis of spectra for materials with large dielectric functions, such as near the strong resonances for many explosives, remains an active area of research. Future research may include a method for determination of complex refractive indices of explosives for which the index is difficult or impossible to determine using far-field techniques such as spectroscopic ellipsometry [82]. In addition, the s-SNOM technique allows measurement of variation in complex refractive index at spatial scales far below optical diffraction limits, which may provide additional insight into the dependence of refractive index on crystalline form and orientation for materials such as RDX and PETN.

9.5 SPECTRAL-BASED DETECTION ALGORITHMS FOR STANDOFF EXPLOSIVES DETECTION

9.5.1 SPECTRAL ANALYSIS CONSIDERATIONS

While construction of instrumentation and collection of spectral or hyperspectral data are the first steps in standoff explosives detection, an equally important step is analysis of this data. The primary goal is to analyze measured spectral data to determine the presence or absence of various explosives compounds. This analysis should ultimately be performed in an automated (unsupervised) process, with a high probability of detection and low false-alarm rates. Many automated detection approaches seek to exploit libraries of reference spectra as well as data-extracted end-member spectra to determine the presence of target materials using various spectroscopic detection algorithms. In effect, these algorithms compare measured spectra against a set of reference spectra for the target materials, define a metric of similarity, and then make a decision based on this metric. Therefore, a critical part of spectral analysis is defining and measuring these reference, or library, spectra.

As shown in earlier examples, explosives detection using tunable ECQCLs provides a wealth of data concerning the transmission or reflectance of the explosives under test as a function of wavelength. For imaging systems, the target explosives particles usually represent a small portion of the total image and the remainder of the image consists of background materials or interferents, otherwise known as "clutter." The problem of identifying target materials in clutter based on optical spectra is similar to that encountered by users of passive imaging spectroscopy platforms used in remote sensing such as the airborne visible/infrared imaging spectrometer (AVIRIS) or the Hyperion sensor aboard the EO-1 spacecraft in which the ground sampling distance (GSD) is on the order of tens of meters. In both instances, image hypercubes are acquired for the field of view of the instrument at each wavelength step delivered by the dispersive sensor. As sensors evolved from multispectral (tens of bands) to hyperspectral (hundreds of bands), more sophisticated exploitation methods were sought, and these efforts provide a broad selection of methods to apply to standoff explosives detection using active wavelength-tunable illumination. The methods are typically divided into statistical (classification) and physics-based or spectral approaches. We will introduce examples of both methods and provide examples of their use in the standoff detection of explosives.

In this section, we discuss and apply various detection algorithms to analysis of hyperspectral images acquired with the ECQCL experiments described above. The first example uses a standoff specular reflectance configuration, as discussed in Section 9.4.3, applied to detection of RDX residue on a coin. The second example uses hyperspectral transmission microscopy, as described in Section 9.4.6, applied to analysis of PETN crystals. In both cases, we apply full-spectrum analysis techniques to utilize the complete data set acquired.

9.5.2 STATISTICS-BASED DETECTION METHODS

These methods involve classification of image pixels using either supervised or unsupervised approaches, of which there are many, described in great detail elsewhere [83]. Due to the large amount of information provided by imaging spectroscopy, a first look at the data using principal component analysis (PCA) using the Karhunen–Loève transform can give an initial indication of whether there is anything of spectral interest contained in the image hypercube. PCA is used to reduce data dimensionality, to segregate noise from signal components, and to produce uncorrelated output bands. The number of output bands after transformation can equal the number of input bands, and the output bands exhibit decreasing variance as the component index increases. Tools to perform the analysis are contained in many scientific analysis software packages or in freeware based on commercial software programs such as MATLAB [84] and ENVI [85].

9.5.3 SPECTRAL-BASED DETECTION METHODS

Due to the rich data set obtained using the active hyperspectral imaging approach made possible by wavelength-tuned QCL illumination, spectral identification methods pioneered by remote-sensing researchers can provide powerful and accurate identification of unknown explosives residues. These range from the relatively straightforward method of spectral feature fitting (SFF) to the more involved MTMF method. These methods will be explained briefly and illustrated with experimental results.

9.5.3.1 Spectral Feature Fitting (SFF)

SFF is essentially a least-squares approach to matching the spectra contained in each hypercube pixel with library spectra. In the implementation contained in ENVI, the data background is removed using a continuum removal approach, and the corrected spectrum is then compared to the reference spectra. The algorithm returns two images: a scale image, which shows absorption depth and material abundance, as well as a root mean square (RMS) image, which contains the fit errors. When a "fit" image is formed by dividing the scale image by the RMS image, bright pixels show a higher likelihood of the presence of the target explosive species. Figure 9.18 shows the fit image to highlight the areas of RDX residue on the coin surface, using

(a)   (b)

FIGURE 9.18 Fit image comparing pixel spectra to RDX library spectra along with grayscale image with overlay. The fit image in (a) is formed using results of the SFF algorithm by dividing the scale image obtained by the RMS or fit error image to show the location and abundance of RDX on the coin surface. Brighter pixels indicate a greater likelihood of RDX being present. In (b), pixels are selected using a 2D scatter plot of scale versus RMS images. Pixels are selected from the scatter plot with high scale values and low RMS values, and are shown in *red* on the coin surface.

the library spectrum shown in Figure 9.5a. In the Figure 9.18a image, the ratio of the scale image to the RMS image is displayed. In Figure 9.18b, a 2D scatter plot of scale versus RMS images is used to select pixels having high scale values and low RMS values, and these are overlaid in red on the grayscale coin image.

9.5.3.2 Matched Filtering

Matched filtering (MF) approaches arose from the need to solve the challenge of detecting small targets in clutter [86]. In airborne hyperspectral imaging, due to the instrument's GSD, which can range from ~15 to 30 m depending on the platform's altitude, it is not uncommon to collect spectral information from a mixture of objects contained within one pixel, which results in a "mixed pixel." In hyperspectral imaging, mixed pixels content can be considered clutter, and MF can test for the presence of a target material while suppressing the response of the composite unknown background. In explosives detection, pure pixels are equally improbable, since the residue resides on a surface having other contaminants present, and there may also be phenomena caused by diffuse scattering or varying contributions of surface reflections and volume absorption, which can vary the spectral signature of the unknown explosive relative to a library spectrum for the "pure" material. Farrand [86] describes the MF as a type of constrained energy minimization algorithm initially described by Harsanyi [87] and also Farrand and Harsanyi [88] in which a vector operator is found that suppresses the unknown background and enhances the target spectrum.

9.5.3.3 Mixture-Tuned Matched Filtering

MTMF is an extension of the matched filtering algorithm with initial data conditioning that applies an MNF transform [89] on the data to segregate good data from noisy data, reduce the amount of computation needed (noisy data are discarded), and reduce the dimensionality of the data set [90]. The MNF transform (also called the noise-adjusted PCA transform) provides a method to select new components that maximize signal-to-noise rather than maximizing variance as the PCA transform does. The matched filtering component suppresses the background while maximizing the target detection. MTMF, however, adds an infeasibility image in addition to the target abundance image, in order to reject false positives more effectively than the matched filter algorithm alone. According to Boardman [91], "MTMF combines the best parts of the Linear Spectral Mixing model and the statistical Matched Filter model while avoiding the drawbacks of each parent method." The MTMF algorithm consists of three steps: (1) preprocessing, (2) matched filtering, and (3) mixture tuning to reject false positives. Mixture tuning measures the feasibility of each pixel as a mixture of the target spectra and background, which minimizes false positives.

9.5.4 END-MEMBER EXTRACTION

The examples shown of the coin thus far were analyzed using a laboratory-acquired absorption spectrum for RDX. This approach is tractable when the measurement can ensure that all recorded spectra for RDX match the library spectrum, such as in a specular reflection geometry. When a mixture of specular and scattered light is received, or when the explosive is crystalline and has different spectra depending on its orientation due to its interaction with the strongly polarized laser illumination, signatures of explosives can differ significantly from transmission spectra often used for explosives and general chemical detection. In these cases, methods of extracting spectra or end-members from the data can provide a method to identify all of the pixels of interest in a hyperspectral image. As an initial example, the spectral maximum angle convex cone (SMACC) [92] end-member extraction function can be applied in ENVI, and the appropriate end-members that are found can be used for spectral identification using MTMF. In linear algebra, a cone is a subset of a vector space that is closed under multiplication by positive scalars. A convex cone is a subset of a vector space over an ordered field that is closed under linear combinations with positive coefficients. SMACC produces a list of spectra meeting certain uniqueness criteria and also produces abundance images to assist the selection of the most appropriate end-members for further processing. In SMACC, (also called residual minimization), extreme vectors are found in the data set and these extreme vectors are used as end-members. According to Gruninger et al. [92], "The extreme vectors or endmembers form a convex cone that contains the remaining data vectors. The convex cone provides a linear mixing model for the data vectors, with the positive coefficients being identified with the abundance of the endmember in the mixture model of a data vector."

For the coin with RDX residue previously shown, the SMACC algorithm was used to extract all unique end-members and corresponding abundance images. By examining the abundance images, three end-members were identified within portions of the RDX film. These are shown in Figure 9.19, where they are also linearly combined, scaled, and compared with a library spectrum for RDX. The variations in spectra could be due to the different crystalline forms of RDX, as observed in Figures 9.11 and 9.16, although the different spectral region for the results shown in Figure 9.19 prevents a direct comparison. The end-members shown in Figure 9.19 were used in the MTMF algorithm, where only the initial 15 eigenvalues were retained, which resulted in three pairs of score and infeasibility images, which were then examined using the scatter plot method to select those pixels having a high score and low infeasibility. The results are shown in Figure 9.20. Only a fraction of the pixels in the coin are a faithful match to the library spectrum, due to varying film

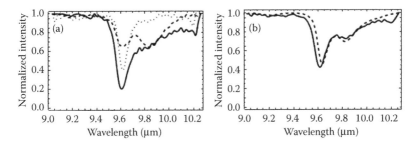

FIGURE 9.19   End-member spectra found using SMACC and selected after viewing abundance images for the coin with RDX residue (a). When the three spectra are combined linearly (b, *solid trace*) they compare favorably with a library spectrum of RDX (b, *dashed trace*).

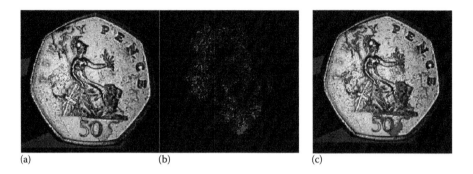

(a)   (b)   (c)

FIGURE 9.20   Reference image of coin in (a). In (b), SMACC spectra (*solid*, *dashed*, and *dotted* traces from previous figure) are mapped into pixel locations with *red*, *green*, and *blue* colors using MTMF spectral analysis. After applying the MTMF approach to spectral feature identification with end-members extracted from the coin image using the SMACC algorithm, the resultant classified image is shown in which pixels having abundance greater than zero and low infeasibility are colored *red* (c).

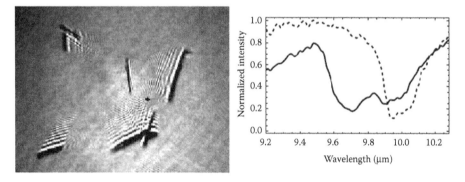

FIGURE 9.21   Microscopy image of PETN crystals (*left*) and two end-members extracted from the image (*right*) using the SMACC algorithm. In this example, the *solid black* spectrum includes those crystals near the horizontal, while the *dashed* spectrum is found in those crystals nearer the vertical orientation.

thickness, variations in the surface roughness of the coin, and local variations in the reflectance of the coin surface at IR wavelengths.

As was observed experimentally in Figure 9.15, the absorption spectrum of PETN is anisotropic, depending on the relative orientation of crystals and light polarization direction. For further investigation of this effect, we examine a transmission microscopy image of PETN, in which crystals of PETN were placed on a transparent IR window and imaged in transmission using the microscope configuration described in Section 9.4.6. Since the particles were randomly oriented, their response to polarized light differs depending on their orientation.

The image in Figure 9.21 shows the IR micrograph along with two end-members extracted from the PETN image using SMACC. The hypercube is composed of 100 bands of data from 1099.12 to 946.51 cm^{-1} (9.1 to 10.6 μm), and the image is intensity

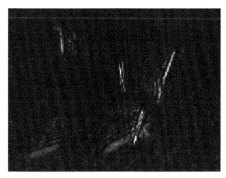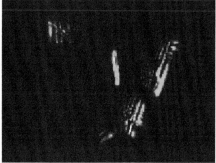

FIGURE 9.22 MTMF results for the extracted end-members shown in Figure 9.21. Those crystals nearer horizontal are well matched by the solid black spectrum (Figure 9.21 left), while those that are vertical better match the dashed black spectrum (Figure 9.21 right). In both cases the matched filter results are used to stretch the image histograms to highlight material abundances (MF results greater than zero), while the score and infeasibility images resulting from the MTMF method are used to highlight those pixels with scores >0 and with low infeasibility values.

normalized across the bands using the internal average relative (IAR) reflectance method, in which the average spectrum is found for the image and used as a reference spectrum to normalize the image reflectance. The spectrum depicted using a solid black trace is associated with those crystals nearer the horizontal orientation, while the spectrum shown in the dashed black trace is found in those crystals near vertically oriented. Due to the coherence of the illumination laser, edge diffraction from the crystal edges is evident in the ringing artifacts. No attempt was made to minimize this artifact, but its effect could be eliminated cosmetically by rapidly dithering the illumination beam, if desired. When the two end-members shown in Figure 9.21 are used for spectral identification by applying the MTMF method, we obtain the two images shown in Figure 9.22. The images were formed by first using the matched filter abundance images and histogram stretched to those pixel values greater than zero in the abundance images in order to highlight the location of pixels associated with each end-member and to suppress the image background. Then, as was done for the coin image, the MTMF abundance and infeasibility images for each member were examined on a 2D scatter plot to select those pixels with abundance scores greater than zero and low infeasibility values. The pixels found using this method are classified by color for each end-member. In this example, the crystalline nature of PETN gives rise to different spectra for polarized sources depending on the orientation of the crystal with respect to the illumination beam's polarization orientation. Library spectra taken using thermal sources would provide a superposition of both for an average spectrum that would have difficulty identifying all pixels containing PETN. Similarly, using only a single library spectrum would neglect the anisotropy and prevent detection of all pixels containing PETN.

The detection of explosives using active ECQCL illumination provides an effective method to identify unknown residue without swabbing or any contact with the surface of interest. In the examples shown, image hypercubes with up to 100 images were acquired as quickly as our microbolometer camera could be interrogated. Due to the rich features provided by military as well as homemade explosives, spectral identification methods developed for passive hyperspectral imaging can be exploited to reliably identify the presence of explosives on surfaces. Examples were provided for neat samples, both in reflection and transmission, using library spectra and methods that automatically extracted end-members from the data set. As illumination and imaging angles move from collection of specularly reflected light (or transmitted light) to predominantly scattering geometries, the amount of light collected by the imaging system will decrease, but, more importantly, the spectra of the imaged pixels will deviate from both transmission spectra of thin samples and reflection spectra acquired at single geometry illumination conditions. As active imaging spectroscopy matures, the study of the phenomenology of spectra obtained from solid materials under real-world conditions will become essential to ensure reliable and rapid identification of explosive threats.

9.6 CONCLUSIONS AND OUTLOOK

In this chapter, we have presented an overview of standoff reflectance spectroscopy of explosives. Theoretical considerations for reflectance spectroscopy lead to a phenomenological basis for the variability in reflectance spectra from thin films, residues, or particles of explosives on surfaces. We have presented a summary of recent work we have performed using ECQCLs to measure IR spectra of explosives particles and residues. These measurements included both standoff techniques and microscopy techniques, ranging from length scales of meters to nanometers. While our initial work was directed at developing systems for standoff detection, this research raised a number of other questions related to reflectance spectroscopy. The microscopy results presented here clearly demonstrate the high sensitivity of IR spectroscopy for the detection of trace amounts of explosives; the absorption strength of the explosives themselves is rarely the limiting factor in standoff detection using reflectance spectroscopy. More important is the imaging system itself, in two ways. First, the imaging system must collect enough light scattered from the target to perform spectroscopy with a high enough SNR for positive detection. Due to the low scattering efficiency of many surfaces in the MIR, as shown in Section 9.4.4, high laser power and sensitive detectors are often required. Second, the imaging system must have enough magnification to resolve the particles or spatial features of the explosives residues. In other words, the explosives particles must fill a large enough area of the imaging pixels that the spectral response of a small particle is detectable when averaged with the spectral response of the background material within the pixel.

As the instrumentation for standoff explosives detection using QCLs has evolved, as highlighted by the references discussed in Section 9.3, many of the radiometric and imaging problems have been solved, and instrumentation can now be constructed to detect small quantities of explosives at meter-scale distances. However, a fundamental challenge remaining is development of robust spectral analysis algorithms to maximize detection probability and minimize false alarms. As we and other researchers have observed, measured spectra of explosives (and many other solid materials for that matter) show a high degree of variability based on both measurement geometry and sample properties such as morphology. To understand the source of this variability, microscopy techniques are particularly valuable as they allow isolation of individual particles or crystalline forms, or even identification of particle orientation. The ECQCL-based microscopy techniques we have demonstrated are valuable tools for characterization of spectra from explosives. In addition, the hyperspectral microscopy techniques demonstrate both high sensitivity and rapid measurement speed and may find use in analytical or forensics applications.

Once the spectral responses of individual particles have been measured and understood, it is possible to simulate the spectral response of an ensemble of particles. While the differentiation of α- versus β-crystalline forms of RDX may or may not be directly relevant for detection of explosives formulations such as C-4, it serves as an example of how one molecular species may exhibit different spectral responses when in a solid form. Understanding and characterizing the expected spectral variability of the target explosives is crucial for spectroscopic based detection algorithms. Once the materials are characterized fully, it may be possible to identify a minimum basis set of library spectra that capture the spectral variability; however, care must be taken to limit the overall size of the spectral library to avoid overconstraining the problem and introducing false detections.

Material identification methods were described that were adopted from the remote sensing community, where they were used to identify natural resources for monitoring and extraction, and the earth sciences, primarily for VNIR/SWIR passive dispersive sensors. In some ways, active IR illumination is simpler than passive methods, due to its shorter standoff distance and freedom from the necessity to remove the effects of downwelling solar radiance and atmospheric absorbance features. Certainly, due to the high spectral radiance of ECQCLs and application of reflectance spectroscopy methods, identification of explosive materials on substrates with emissivity comparable to the explosive samples is more straightforward and amenable to the methods used to exploit passive reflectance hypercubes. It was shown that full spectral coverage is essential for accurate identification of explosive threats, due to perturbations in the spectral shapes because of sample conditions and scattering geometry that would not yield to simple methods such as band ratio methods. The methods of spectral identification presented take full advantage of the rich reflectance spectra possible with active illumination of tunable ECQCL sources, but often require human intervention. Practical systems will require methods of automated threat recognition (ATR), which is a vibrant area of research in hyperspectral imaging and should be applied to the spectroscopy of explosives detection and identification.

Future work is needed to develop better physical models for diffuse reflectance spectroscopy as it applies to detection of residues of explosives. For these models, quantitative complex refractive index data (n, k) are required. However, only limited (n, k) data are currently available for explosives and many potential substrates, especially in the MIR spectral region. Part of this is due to the difficulty of measurement of complex refractive indices. Spectroscopic ellipsometry [82] provides a robust technique for measurement of complex refractive index; however, it requires the use of smooth, homogeneous surfaces. Many explosives cannot be conveniently produced in this form. Other techniques may be useful for determination of the complex refractive indexes of particulates of explosives [16,17,68]. However, as we have observed and presented here, many crystalline explosives change their optical properties with crystalline form or crystal orientation or are highly dependent on sample preparation. Developing methods for accurate determination of complex refractive index in the MIR for these materials remains a challenge, but is a high priority for input into modeling studies.

In summary, we have shown how ECQCL-based spectroscopy can be used to detect and identify particles and residues on surfaces over a wide range of length scales. The high spectral radiance, broad tuning range, rapid wavelength tuning, high scan reproducibility, and low noise of the ECQCL source permits measurements in configurations that would be either impractical or overly time-consuming with broadband, incoherent IR sources. In this chapter, we presented standoff hyperspectral reflectance imaging, quantitative measurements of diffuse reflectance spectra, reflection–absorption IR spectroscopy, microscopic imaging and spectroscopy, and nanoscale imaging and spectroscopy with ECQCLs. The results demonstrate the potential for ECQCL-based spectroscopy to perform rapid measurements and to have high detection sensitivity, both of which are important for standoff explosives detection. While radiometric challenges remain, due to low scattering efficiency of materials in the MIR, future developments in high-power ECQCLs and sensitive detectors are expected to mitigate these issues and allow standoff detection at larger distances. Optical models for diffuse reflectance targeted specifically for spectroscopic detection of explosives and better understanding of the variability of reflectance spectra of explosives are required, but future research is expected to develop more robust spectral analysis algorithms to improve performance of standoff explosives detection systems.

ACKNOWLEDGMENTS

The authors gratefully acknowledge the contributions of numerous colleagues at PNNL who assisted with the experiments described above or performed related experiments not shown. Specifically, the authors acknowledge Jonathan Suter, Ian Craig, Matthew Taubman, Thomas Blake, and Timothy Johnson. The research described in this paper was conducted under the Laboratory Directed Research and Development Program at PNNL. Development of the ECQCL was supported by the DOE/NNSA Office of Nonproliferation and Verification Research and Development (NA-22). PNNL is operated for the US Department of Energy (DOE) by the Battelle Memorial Institute under Contract No. DE-AC05-76RL01830.

REFERENCES

1. C. Bauer, A. K. Sharma, U. Willer, J. Burgmeier, B. Braunschweig, W. Schade, S. Blaser, L. Hvozdara, A. Muller, and G. Holl, Potentials and limits of mid-infrared laser spectroscopy for the detection of explosives. *Applied Physics B: Lasers and Optics* **92**, 327 (2008).
2. S. Wallin, A. Pettersson, H. Ostmark, and A. Hobro, Laser-based standoff detection of explosives: A critical review. *Analytical and Bioanalytical Chemistry* **395**, 259 (2009).

3. J. Faist, F. Capasso, D. L. Sivco, C. Sirtori, A. L. Hutchinson, and A. Y. Cho, Quantum cascade laser. *Science* **264**, 553 (1994).

4. Y. Bai, S. Slivken, S. Kuboya, S. R. Darvish, and M. Razeghi, Quantum cascade lasers that emit more light than heat. *Nature Photonics* **4**, 99 (2010).

5. A. Hugi, R. Maulini, and J. Faist, External cavity quantum cascade laser. *Semiconductor Science and Technology* **25**, 083001 (2010).

6. A. Mukherjee, S. Von der Porten, and C. K. N. Patel, Standoff detection of explosive substances at distances of up to 150 m. *Applied Optics* **49**, 2072 (2010).

7. R. Furstenberg, C. A. Kendziora, J. Stepnowski, S. V. Stepnowski, M. Rake, M. R. Papantonakis, V. Nguyen, G. K. Hubler, and R. A. McGill, Stand-off detection of trace explosives via resonant infrared photothermal imaging. *Applied Physics Letters* **93**, 224103 (2008).

8. M. R. Papantonakis, C. A. Kendziora, R. Furstenberg, J. Stepnowski, M. Rake, S. V. Stepnowski, and R. A. McGill, Stand-off detection of trace explosives by infrared photothermal imaging. *Proceedings of SPIE* **7304**, 730418 (2009).

9. L. A. Skvortsov and E. M. Maksimov, Application of laser photothermal spectroscopy for standoff detection of trace explosive residues on surfaces. *Quantum Electronics* **40**, 565 (2010).

10. X. Chen, D. K. Guo, F. S. Choa, C. C. Wang, S. Trivedi, A. P. Snyder, G. Y. Ru, and J. Y. Fan, Standoff photoacoustic detection of explosives using quantum cascade laser and an ultrasensitive microphone. *Applied Optics* **52**, 2626 (2013).

11. C. A. Kendziora, R. Furstenberg, M. Papantonakis, V. Nguyen, J. Borchert, J. Byers, and R. A. McGill, Infrared photothermal imaging of trace explosives on relevant substrates. *Proceedings of SPIE* **8709**, 87090O (2013).

12. P. S. Cho, R. M. Jones, T. Shuman, D. Scoglietti, and G. Harston, Investigation of stand-off explosives detection via photothermal/photoacoustic interferometry. *Proceedings of SPIE* **8018**, 80181T (2011).

13. R. H. Farahi, A. Passian, L. Tetard, and T. Thundat, Pump-probe photothermal spectroscopy using quantum cascade lasers. *Journal of Physics D: Applied Physics* **45**, 7 (2012).

14. J. M. Bennet and L. Mattsson, *Introduction to Surface Roughness and Scattering*. Optical Society of America, Washington, DC (1999).

15. A. K. Goyal, M. Spencer, M. Kelly, J. Costa, M. DiLiberto, E. Meyer, and T. Jeys, Active infrared multispectral imaging of chemicals on surfaces. *Proceedings of SPIE* **8018**, 80180N (2011).

16. R. A. Isbell and M. Q. Brewster, Optical properties of RDX and HMX. *Materials Research Society Symposium Proceedings* **418**, 85 (1996).

17. R. A. Isbell and M. Q. Brewster, Optical properties of energetic materials: RDX, HMX, AP, NC/NG, and HTPB. *Propellants Explosives Pyrotechnics* **23**, 218 (1998).

18. J. D. Jackson, *Classical Electrodynamics*. Wiley, Hoboken, NJ (2009).

19. E. Hecht, *Optics*. Addison-Wesley (2002).

20. E. H. Korte and A. Roseler, Infrared reststrahlen revisited: Commonly disregarded optical details related to $n < 1$. *Analytical Bioanalytical Chemistry* **382**, 1987 (2005).

21. A. E. Miroshnichenko, S. Flach, and Y. S. Kivshar, Fano resonances in nanoscale structures. *Reviews of Modern Physics* **82**, 2257 (2010).

22. E. O. Heavens, *Optical Properties of Thin Solid Films*. Butterworths Scientific Publications, London (1955).

23. M. C. Phillips, J. D. Suter, B. E. Bernacki, and T. J. Johnson, Challenges of infrared reflective spectroscopy of solid-phase explosives and chemicals on surfaces. *Proceedings of SPIE* **8358**, 83580T (2012).

24. J. W. Salisbury, Mid-infrared spectroscopy: Laboratory data, in C. M. Pieters and A. J. Englert (eds), *Remote Geochemical Analysis*, p. 79. Cambridge University Press, New York (1993).

25. B. Philips-Invernizzi and D. Dupont, Bibliographical review for reflectance of diffusing media. *Optical Engineering* **40**, 1082 (2001).
26. C. W. Van Neste, L. R. Senesac, and T. Thundat, Standoff photoacoustic spectroscopy. *Applied Physics Letters* **92**, 234102 (2008).
27. C. W. Van Neste, L. R. Senesac, and T. Thundat, Standoff spectroscopy of surface adsorbed chemicals. *Analytical Chemistry* **81**, 1952 (2009).
28. R. C. Sharma, D. Kumar, N. Bhardwaj, S. Gupta, H. Chandra, and A. K. Maini, Portable detection system for standoff sensing of explosives and hazardous materials. *Optics Communications* **309**, 44 (2013).
29. X. C. Liu, C. W. Van Neste, M. Gupta, Y. Y. Tsui, S. Kim, and T. Thundat, Standoff reflection–absorption spectra of surface adsorbed explosives measured with pulsed quantum cascade lasers. *Sensors and Actuators B: Chemical* **191**, 450 (2014).
30. B. E. Bernacki and M. C. Phillips, Standoff hyperspectral imaging of explosives residues using broadly tunable external cavity quantum cascade laser illumination. *Proceedings of SPIE* **7665**, 76650I (2010).
31. B. Hinkov, F. Fuchs, J. M. Kaster, Q. Yang, W. Bronner, R. Aidam, and K. Kohler, Broad band tunable quantum cascade lasers for stand-off detection of explosives. *Proceedings of SPIE* **7484**, 748406 (2009).
32. J. R. Castro-Suarez, L. C. Pacheco-Londono, W. Ortiz-Rivera, M. Velez-Reyes, M. Diem, and S. P. Hernandez-Rivera, Open path FTIR detection of threat chemicals in air and on surfaces. *Proceedings of SPIE* **8012**, 801209 (2011).
33. T. A. Blake, J. F. Kelly, N. B. Gallagher, P. L. Gassman, and T. J. Johnson, Passive stand-off detection of RDX residues on metal surfaces via infrared hyperspectral imaging. *Analytical and Bioanalytical Chemistry* **395**, 337 (2009).
34. L. C. Pacheco-Londono, W. Ortiz-Rivera, O. M. Primera-Pedrozo, and S. P. Hernandez-Rivera, Vibrational spectroscopy standoff detection of explosives. *Analytical and Bioanalytical Chemistry* **395**, 323 (2009).
35. M. Ortolani and U. Schade, Fourier-transform far-infrared spectroscopic ellipsometry for standoff material identification. *Nuclear Instruments and Methods in Physics Research Section A: Accelerators, Spectrometers, Detectors and Associated Equipment* **623**, 791 (2010).
36. J. Suter, B. Bernacki, and M. Phillips, Spectral and angular dependence of mid-infrared diffuse scattering from explosives residues for standoff detection using external cavity quantum cascade lasers. *Applied Physics B: Lasers and Optics* **108**, 965 (2012).
37. J. D. Suter, B. E. Bernacki, and M. C. Phillips, Angle-resolved scattering spectroscopy of explosives using an external cavity quantum cascade laser. *Proceedings of SPIE* **8268**, 82681O (2012).
38. L. C. Pacheco-Londono, J. R. Castro-Suarez, J. Aparicio-Bolanos, and S. P. Hernandez-Rivera, Angular dependence of source-target-detector in active mode standoff infrared detection. *Proceedings of SPIE* **8711**, 871108 (2013).
39. C. A. Ortega-Zuñiga, N. Y. Galán-Freyle, J. R. Castro-Suarez, J. Aparicio-Bolaño, L. C. Pacheco-Londoño, and S. P. Hernández-Rivera, Dependence of detection limits on angular alignment, substrate type and surface concentration in active mode standoff IR. *Proceedings of SPIE* **8734**, 87340R (2013).
40. M. E. Morales-Rodriguez, L. R. Senesac, T. Thundat, M. K. Rafailov, and P. G. Datskos, Standoff imaging of chemicals using IR spectroscopy. *Proceedings of SPIE* **8031**, 80312D (2011).
41. C. R. Schwarze, E. Schundler, R. Vaillancourt, S. Newbry, and R. Benedict-Gill, Risley prism scan-based approach to standoff trace explosive detection. *Optical Engineering* **53**, 021110 (2014).

42. F. Fuchs, B. Hinkov, S. Hugger, J. M. Kaster, R. Aidam, W. Bronner, K. Kohler, et al., Imaging stand-off detection of explosives using tunable MIR quantum cascade lasers. *Proceedings of SPIE* **7608**, 760809 (2010).

43. F. Fuchs, S. Hugger, M. Kinzer, R. Aidam, W. Bronner, R. Losch, and Q. Yang, Imaging standoff detection of explosives using widely tunable midinfrared quantum cascade lasers. *Optical Engineering* **49**, 111127 (2010).

44. F. Fuchs, S. Hugger, J. Jarvis, V. Blattmann, M. Kinzer, Q. K. Yang, R. Ostendorf, et al., Infrared hyperspectral standoff detection of explosives. *Chemical, Biological, Radiological, Nuclear, and Explosives (CBRNE) Sensing XIV* **8710**, 8 (2013).

45. S. Hugger, F. Fuchs, J. Jarvis, M. Kinzer, Q. K. Yang, W. Bronner, R. Driad, R. Aidam, K. Degreif, and F. Schnurer, Broadband tunable external cavity quantum cascade lasers for standoff detection of explosives. *Proceedings of SPIE* **8373**, 83732G (2012).

46. F. Fuchs, S. Hugger, M. Kinzer, Q. K. Yang, W. Bronner, R. Aidam, K. Degreif, S. Rademacher, F. Schnurer, and W. Schweikert, Standoff detection of explosives with broad band tunable external cavity quantum cascade lasers. *Proceedings of SPIE* **8268**, 82681N (2012).

47. K. Degreif, S. Rademacher, P. Dasheva, F. Fuchs, S. Hugger, F. Schnurer, and W. Schweikert, Stand-off explosive detection on surfaces using multispectral MIR-Imaging. *Proceedings of SPIE* **7945**, 79450P (2011).

48. S. Hugger, F. Fuchs, J. Jarvis, M. Kinzer, Q. K. Yang, R. Driad, R. Aidam, and J. Wagner, Broadband-tunable external-cavity quantum cascade lasers for the spectroscopic detection of hazardous substances. *Proceedings of SPIE* **8631**, 86312I (2013).

49. A. Goyal, T. Myers, C. A. Wang, M. Kelly, B. Tyrrell, B. Gokden, A. Sanchez, G. Turner, and F. Capasso, Active hyperspectral imaging using a quantum cascade laser (QCL) array and digital-pixel focal plane array (DFPA) camera. *Optics Express* **22**, 14392 (2014).

50. G. P. Luo, C. Peng, H. Q. Le, S. S. Pei, W. Y. Hwang, B. Ishaug, J. Um, J. N. Baillargeon, and C. H. Lin, Grating-tuned external-cavity quantum-cascade semiconductor lasers. *Applied Physics Letters* **78**, 2834 (2001).

51. R. Maulini, A. Mohan, M. Giovannini, J. Faist, and E. Gini, External cavity quantum-cascade laser tunable from 8.2 to 10.4 µm using a gain element with a heterogeneous cascade. *Applied Physics Letters* **88**, 201113 (2006).

52. C. Peng, G. P. Luo, and H. Q. Le, Broadband, continuous, and fine-tune properties of external-cavity thermoelectric-stabilized mid-infrared quantum-cascade lasers. *Applied Optics* **42**, 4877 (2003).

53. M. C. Phillips, T. L. Myers, M. D. Wojcik, and B. D. Cannon, External cavity quantum cascade laser for quartz tuning fork photoacoustic spectroscopy of broad absorption features. *Optics Letters* **32**, 1177 (2007).

54. G. Totschnig, F. Winter, V. Pustogov, J. Faist, and A. Muller, Mid-infrared external-cavity quantum-cascade laser. *Optics Letters* **27**, 1788 (2002).

55. G. Wysocki, R. F. Curl, F. K. Tittel, R. Maulini, J. M. Bulliard, and J. Faist, Widely tunable mode-hop free external cavity quantum cascade laser for high resolution spectroscopic applications. *Applied Physics B: Lasers and Optics* **81**, 769 (2005).

56. M. Pushkarsky, A. Tsekoun, I. G. Dunayevskiy, R. Go, and C. K. N. Patel, Sub-parts-per-billion level detection of NO_2 using room-temperature quantum cascade lasers. *Proceedings of the National Academy of Sciences of the United States of America* **103**, 10846 (2006).

57. G. Wysocki, R. Lewicki, R. F. Curl, F. K. Tittel, L. Diehl, F. Capasso, M. Troccoli, et al., Widely tunable mode-hop free external cavity quantum cascade lasers for high resolution spectroscopy and chemical sensing. *Applied Physics B: Lasers and Optics* **92**, 305 (2008).

58. M. C. Phillips, M. S. Taubman, B. E. Bernacki, B. D. Cannon, J. T. Schiffern, and T. L. Myers, Design and performance of a sensor system for detection of multiple chemicals using an external cavity quantum cascade laser. *Proceedings of SPIE* **7608**, 76080D (2010).

59. M. C. Phillips, M. S. Taubman, B. E. Bernacki, B. D. Cannon, R. D. Stahl, J. T. Schiffern, and T. L. Myers, Real-time trace gas sensing of fluorocarbons using a swept-wavelength external cavity quantum cascade laser. *Analyst* **139**, 2047 (2014).

60. S. Riedi, A. Hugi, A. Bismuto, M. Beck, and J. Faist, Broadband external cavity tuning in the 3–4 µm window. *Applied Physics Letters* **103**, 4 (2013).

61. A. Hugi, R. Terazzi, Y. Bonetti, A. Wittmann, M. Fischer, M. Beck, J. Faist, and E. Gini, External cavity quantum cascade laser tunable from 7.6 to 11.4 µm. *Applied Physics Letters* **95**, 3 (2009).

62. E. R. Deutsch, F. G. Haibach, and A. Mazurenko, Detection and quantification of explosives and CWAs using a handheld widely-tunable quantum cascade laser. *Proceedings of SPIE* **8374**, 83740M (2012).

63. M. G. Littman and H. J. Metcalf, Spectrally narrow pulsed dye-laser without beam expander. *Applied Optics* **17**, 2224 (1978).

64. M. C. Phillips and N. Ho, Infrared hyperspectral imaging using a broadly tunable external cavity quantum cascade laser and microbolometer focal plane array. *Optics Express* **16**, 1836 (2008).

65. M. S. Taubman, T. L. Myers, B. E. Bernacki, R. D. Stahl, B. D. Cannon, J. T. Schiffern, and M. C. Phillips, A modular architecture for multi-channel external cavity quantum cascade laser-based chemical sensors: A systems approach. *Proceedings of SPIE* **8268**, 82682G (2012).

66. M. C. Phillips and B. E. Bernacki, Hyperspectral microscopy of explosives particles using an external cavity quantum cascade laser. *Optical Engineering* **52**, 061302 (2012).

67. S. K. Andersson and G. A. Niklasson, Influence of surface roughness on the infrared reststrahlen band. *Journal of Physics: Condensed Matter* **7**, 7173 (1995).

68. R. A. Isbell and M. Q. Brewster, Optical properties of energetic materials from infrared spectroscopy. *Journal of Thermophysics and Heat Transfer* **11**, 65 (1997).

69. R. Maulini, I. Dunayevskiy, A. Lyakh, A. Tsekoun, C. K. N. Patel, L. Diehl, C. Pflugl, and F. Capaso, Widely tunable high-power external cavity quantum cascade laser operating at continuous-wave at room temperature. *Electronics Letters* **45**, 107 (2009).

70. American National Standards Institute, American National Standard for Safe Use of Lasers. Laser Institute of America, Orlando, FL, ANSI Z136. 1-2000, revision of ANSI. 1-1993 (2000)

71. C. Weitkamp, *Lidar: Range-Resolved Optical Remote Sensing of the Atmosphere.* Springer, New York (2005).

72. M. C. Phillips, I. M. Craig, and T. A. Blake, Reflection-absorption infrared spectroscopy of thin films using an external cavity quantum cascade laser. *Proceedings of SPIE* **8631**, 86310C (2013).

73. W. G. Golden, D. D. Saperstein, M. W. Severson, and J. Overend, Infrared reflection–absorption spectroscopy of surface species: A comparison of Fourier transform and dispersion methods. *The Journal of Physical Chemistry* **88**, 574 (1984).

74. R. G. Greenler, Infrared study of adsorbed molecules on metal surfaces by reflection techniques. *Journal of Chemical Physics* **44**, 310 (1966).

75. M. Trenary, Reflection absorption infrared spectroscopy and the structure of molecular adsorbates on metal surfaces. *Annual Review of Physical Chemistry* **51**, 381 (2000).

76. T. Buffeteau, B. Desbat, and J. M. Turlet, Polarization modulation FT-IR spectroscopy of surfaces and ultra-thin films: Experimental procedure and quantitative analysis. *Applied Spectroscopy* **45**, 380 (1991).

77. P. Torres, L. Mercado, I. Cotte, S. P. Hernández, N. Mina, A. Santana, R. T. Chamberlain, R. Lareau, and M. E. Castro, Vibrational spectroscopy study of β and α RDX deposits. *The Journal of Physical Chemistry B* **108**, 8799 (2004).

78. R. J. Karpowicz and T. B. Brill, Comparison of the molecular structure of hexahydro-1,3,5-trinitro-s-triazine in the vapor, solution, and solid phases. *Journal of Physical Chemistry* **88**, 348 (1984).

79. I. M. Craig, M. S. Taubman, A. S. Lea, M. C. Phillips, E. E. Josberger, and M. B. Raschke, Infrared near-field spectroscopy of trace explosives using an external cavity quantum cascade laser. *Optics Express* **21**, 30401 (2013).

80. A. A. Govyadinov, I. Amenabar, F. Huth, P. S. Carney, and R. Hillenbrand, Quantitative measurement of local infrared absorption and dielectric function with tip-enhanced near-field microscopy. *Journal of Physical Chemistry Letters* **4**, 1526 (2013).

81. F. Huth, A. Govyadinov, S. Amarie, W. Nuansing, F. Keilmann, and R. Hilenbrand, Nano-FTIR absorption spectroscopy of molecular fingerprints at 20 nm spatial resolution. *Nano Letters* **12**, 3973 (2012).

82. H. G. Tompkins and W. A. McGahan, *Spectroscopic Ellipsometry and Reflectometry: A User's Guide*. Wiley-Interscience, New York (1999).

83. J. A. Richards and X. Jia, *Remote Sensing Digital Image Analysis: An Introduction*, 4th edn. Springer-Verlag, Berlin (2006).

84. E. Arzuaga-Cruz, L. O. Jimenez-Rodriguez, M. Velez-Reyes, D. Kaeli, E. Rodriguez-Diaz, H. T. Velazquez-Santana, A. Castrodad-Carrau, L. E. Santos-Campis, and C. Santiago, A MATLAB toolbox for hyperspectral image analysis, in *Proceedings of the 2004 IEEE International Geoscience and Remote Sensing Symposium, IGARSS*, vol. 7, p. 4839–4842.

85. Exelis Visual Information Solutions, version 5.1, Boulder, CO. http://www.exelisvis.com/envi.

86. W. H. Farrand, Analysis of AVIRIS data: A comparison of performance of commercial software with published algorithms. JPL Publication 02-1, AVIRIS Workshop (2001).

87. J. C. Harsanyi, Detection and classification of subpixel spectral signatures in hyperspectral image sequences. PhD dissertation, University of Maryland, Baltimore County (1993).

88. W. H. Farrand and J. C. Harsanyi, Mapping the distribution of mine tailings in the Coeur d'Alene River Valley, Idaho, through the use of a constrained energy minimization technique. *Remote Sensing of Environment* **59**, 64 (1997).

89. A. A. Green, M. Berman, P. Switzer, and M. D. Craig, A transformation for ordering multispectral data in terms of image quality with implications for noise removal. *IEEE Transactions on Geoscience and Remote Sensing* **26**, 65 (1988).

90. J. W. Boardman and F. A. Kruse, Analysis of imaging spectrometer data using N-dimensional geometry and a mixture-tuned matched filtering approach. *IEEE Transactions on Geoscience and Remote Sensing* **49**, 4138 (2011).

91. J. W. Boardman, Leveraging the high dimensionality of AVIRIS data for improved subpixel target unmixing and rejection of false positives: Mixture-tuned matched filtering. In *Summaries of Seventh Annual JPL Earth Science Workshop*, vol. 1, p. 55. Pasadena, CA, JPL Publication 98-94 (1998).

92. J. H. Gruninger, A. J. Ratkowski, and M. L. Hoke, The sequential maximum angle convex cone (SMACC) endmember model. *Proceedings of SPIE* **5425**, 1 (2004).

10 Photothermal Methods for Laser-Based Detection of Explosives

Christopher A. Kendziora, Robert Furstenberg,
Michael R. Papantonakis, Viet K. Nguyen,
Jeff M. Byers, and R. Andrew McGill

CONTENTS

10.1 INTRODUCTION

Detection and identification of explosives has been a highly active area of research and development, particularly since the outset of the conflicts in Afghanistan (2001) and Iraq (2003). This chapter emphasizes recent developments in the laser-based detection of explosives exploiting photothermal (PT) spectroscopy. Successful applications of a variety of photothermal methods in gases and condensed matter have resulted in a substantial body of literature. The following discussion summarizes these methods and the experimental components and arrangements for PT detection, as well as its use for explosive sensing. Applications of photothermal spectroscopy, specifically laser-based sensing schemes, are reviewed, along with standoff detection techniques. It is not our intention to discuss all photothermal methods in great detail, as not all of these have been applied to explosives detection. Instead, we hope to provide the reader with a general overview of contemporary photothermal methods used for explosives detection.

We have benefited from consulting several previous reviews. For additional information regarding general methods for photothermal analysis,[1,2] explosives detection,[3,4] optical methods for standoff detection,[5-7] photo-acoustic techniques,[8-10] backscatter imaging,[11] and standoff photothermal detection,[12] we recommend these thorough review articles and many of the references therein.

10.2 PHOTOTHERMAL PHENOMENA

10.2.1 Traditional Photothermal Methods

Photothermal (PT) phenomena are the basis for a wide number of experimental techniques for testing, characterization, and detection of materials. The common thread in all these techniques is the measurement and analysis of the signals produced when light interacts with and subsequently heats matter. This photothermal heat is then either directly measured as in, for example, photothermal radiometry, or indirectly, through induced changes in material properties (e.g., index of refraction) or mechanical expansion of the material (thermoelastic effect), as exploited in photo-acoustic and photothermal deflection techniques. While these phenomena can occur in all phases of matter, we will restrict our discussion to solids as, under conditions typically found in the field, most explosives of interest are present in their solid form. Furthermore, owing to the low vapor pressure of many explosives, their vapor signatures are exceedingly small[13] and the predominant photothermal detection signature is associated with light interacting with solid explosives mostly found in the form of micron-sized particulates adhered to surfaces.[14]

Arguably, the first major study featuring the photothermal effect was published by John Tyndall in 1872, in which he describes measuring heat absorbed by gases.[15] Subsequently, it was Alexander Graham Bell in 1880 describing his "photophone" and thereby laying the foundation for photo-acoustics.[16] These were remarkable experimental efforts given the lack of appropriate instrumentation available at the time. In light of this long history (over 140 years), it would be beyond the scope of this chapter to give an exhaustive review of photothermal and photo-acoustic

TABLE 10.1

Summary of Photothermal-Based Techniques and Useful References

Technique	Measured Property	Probe beam	Detector Type	References
Photothermal radiometry (PTR)	Thermal emission (integrated)	No	IR detector	17
Photothermal imaging spectroscopy (PT-IRIS)	Spectrally resolved thermal emission	No	IR camera	18,19
Photothermal deflection spectroscopy (PTDS)	Change in index of refraction at or above sample, sample movement	Yes (visible laser)	Photodetector	20
Photothermal phase-shift spectroscopy (PTPS)	Change in index of refraction through sample	Yes (visible laser)	Interferometer setup with photodetector	21,22
Photothermal interferometry (PTI)	Change in index of refraction at or above sample, sample movement	Yes (visible laser)	Interferometer setup with photodetector	23 (for liquids)
Photothermal lensing spectroscopy (PTLS)	Change in index of refraction	Yes (visible laser)	Photodetector	24
Photo-acoustic spectroscopy (PAS)	Heat-induced sound waves	No	Microphone	25

phenomena. Instead, the main categories of techniques (along with brief descriptions and relevant citations) are given in Table 10.1. For a detailed description and theory for each of these techniques, the reader is directed to several good reviews in the literature. Also, Section 10.5 provides a more detailed description for some of these techniques that have been used in explosives detection. For each technique in Table 10.1, we list the physical property being measured, detector type used to measure it, and whether a probe beam is required for the given measurement. Techniques that do not measure the photothermal effect directly (by measuring the heat or sound generated) require a separate probe beam, most often a visible-light laser. We also provide an early literature citation for each.

The basic photothermal instrumentation consists of a light source to heat the sample and a detector to measure a subsequent signal change. Light sources traditionally used are halogen lamps, xenon arc, and visible or near-infrared (IR) lasers, but any other radiation source (e.g., microwave, x-ray) capable of heating a sample can excite the photothermal effect.[2] Illumination is usually modulated or delivered in a single pulse or a series of pulses. Recently, with advances in IR lasers, most notably high-power, wavelength-tunable quantum cascade lasers (QCLs),[26] it is possible to perform photothermal experiments in the IR range, which significantly improves the spectroscopic detection capabilities of photothermal techniques.[10,18,27–29] It is important to point out that earlier photothermal experiments typically used a fixed wavelength (visible or near-IR) light source. Also, the samples were often coated with a black paint to enhance absorption.

10.2.2   LASER SOURCES EMPLOYED FOR PHOTOTHERMAL METHODS

As this review is focused on laser-based approaches, we restrict our discussion of photothermal sources to lasers. In general, lasers are highly advantageous for PT excitation due to their high brightness and coherence. In fact, the vast majority of PT studies have used lasers since their invention in the 1960s.[1] As PT phenomena derive from the absorption properties of the target, PT signals increase (in some cases more than linearly) with photon flux. The coherence of laser sources allows them to be focused to small spots or to maintain high flux when directed toward remote targets for standoff applications. Coherence also enables PT interferometry approaches, especially for the probe beams.

Traditional laser sources for PT studies include those at visible (e.g., neodymium-doped:yttrium aluminum garnet (Nd:YAG), 355–1064 nm) and near-IR (1550 nm)[29] wavelengths, as well as optical parametric oscillators (OPOs)[29] and carbon dioxide (CO_2)[19,30] lasers in the IR. Visible wavelength lasers have the advantages of being convenient to locate and focus on a target, offer very high powers (or pulse energies), and typically represent durable, cost-effective, mature technology. However, visible lasers have much lower maximum permissible exposure (MPE) levels for eye and skin safety[31] (refer to Chapter 2) and may alert the target (if a person), and visible wavelengths offer poor molecular selectivity by absorption cross sections. Near-IR lasers have the advantages of being invisible to the unaided eye (and therefore stealthy), allowing much higher (up to >1000×, depending on the wavelength) power/energy levels at MPE, and being able to exploit the mature diode technology developed for the telecom industry. However, near-IR wavelengths offer poor molecular selectivity compared to the mid-IR wavelengths, and alignment of invisible beams can be challenging. Finally, mid-IR lasers offer the advantages of coupling to fundamental vibrational modes in molecules of interest, higher MPE, lower susceptibility to scattering by dust (due to their longer wavelength), potentially very high power or peak energy (in the case of CO_2),[32] and stealth. However, traditional mid-IR lasers are often bulky, can have very low efficiency at the longer wavelengths (in the case of OPOs), and are limited (in the case of CO_2 lasers) to certain discrete wavelengths.[32]

Since their invention in 1994,[33] QCLs have emerged as useful tools for IR measurements,[26,34] including PT-based detection of explosives.[10,18,27–29,35–38] Figure 10.1 shows a photograph of a QCL on a copper thermal chuck. Compared to traditional IR lasers, QCLs (which are typically fabricated by molecular beam epitaxy or MBE) have the advantages of being compact, tailored to a specific wavelength range of interest, and continually tunable. At specific wavelengths, QCLs can emit multiple watts of continuous-wave (CW) power,[39] although 10 to >100 mW is more typical from tunable devices today (2014). Challenges for QCL technology include light conversion efficiency at room temperature (and therefore total power output) and tuning range for a given chip, but these parameters continue to improve rapidly as QCLs are developed for the commercial market.

FIGURE 10.1 Photograph of a quantum cascade laser (QCL). The arrow on the right illustrates the direction of the laser beam emission.

10.2.3 Detectors Employed for Photothermal Methods

The sensors used for PT detection of explosives include microphones,[8] cantilevers,[40] tuning forks,[41,42] bolometers,[18] mercury cadmium telluride (MCT) single-channel IR detectors,[18] MCT arrays,[19,43] and even full spectrometers.[30] Because these instruments are common to many other types of experiments, their detailed principles of operation and performance characteristics are outside the scope of this review. However, specific detectors will be discussed with respect to highlighted technologies and apparatus.

Table 10.2 lists a comparison of sensors used in photothermal technologies for explosives detection. The table lists the types of applications the sensors are used for, their operating frequency ranges, their relative sensitivity, relative cost, whether cooling is required, and examples of PT technologies that employ them. For additional detail on IR sensors,[44–46] focal plane arrays,[47] and photo-acoustic transducers,[48] the reader may also consult the examples given.

10.3 THEORY AND MODELING OF PHOTOTHERMAL HEATING OF SOLID PARTICLES ON SUBSTRATES

10.3.1 Photothermal Heating

The photothermal effect can be thought of as a combination of two phenomena: laser–matter interaction resulting in heating of the sample, and the subsequent (or concurrent) thermal heat transfer processes. Laser–matter interaction (as described by the amount of absorbed heat by the sample) is mostly responsible for the spectral absorption features of the photothermal signal, while the thermal processes are mostly responsible for the temporal emission signatures.[1] These two phenomena can be described/modeled separately (as uncoupled equations) when the amount of

TABLE 10.2

Summary of Sensors Used in Photothermal Technologies for Explosives Detection

Sensor	Utility	Frequency/ Wavelength	Sensitivity	Cost	Cooling	Examples
MCT	Mid-LWIR PT	2–20 μm	High	Med.	TE, LN$_2$	Ref. 18
MCT array	Mid-LWIR PT	2–11 μm	(NETD)	High	LN$_2$, Stirling	Refs. 19,43
Bolometer array	LWIR PT	7–14 μm	(NETD)	Med.	None	Ref. 18
FTIR spectrometer	Mid-LWIR PT	2–20 μm	Low	High	LN$_2$	Ref. 30
Microphone	Photo-acoustic	20–30,000 Hz	Medium	Low	None	Refs. 8,37
Tuning fork	Photo-acoustic/ photothermal	Resonant v~33 kHz	Very high at specific frequency	Very low	None	Refs. 40,41
Cantilever	Photo-acoustic/ photothermal	Resonant v~100 Hz	High	Low	None	Refs. 40,49

Med., medium; TE, thermoelectric; LN$_2$, liquid nitrogen dewar; Stirling, Stirling closed-cycle refrigerator; NETD; noise equivalent temperature difference; LWIR, long-wave infrared; FTIR, Fourier transform infrared.

heating of the sample is small enough not to alter the thermophysical properties of the sample, and, conversely, the heating of the sample will not alter the optical properties and thus will not influence the laser–matter interaction.

There is extensive literature on the interaction of light with matter and also on the relaxation mechanisms (thermal and nonthermal) within materials that have been excited above thermal equilibrium.[1,2] However, most of the fundamental theoretical and experimental considerations have been performed for bulk materials or uniformly thick films. Since the relevant case for explosives detection tends to be small solid particles on surfaces,[14] approximations and formulations made for bulk or film may not be applicable. In this section, the specific case of solid particles on substrates is examined. By approximating that the optical and thermal phenomena may be decoupled, the light–matter interactions and heat transport are considered separately.[50,51]

10.3.2 HEATING OF SOLID PARTICLES ON SUBSTRATES

As the size and geometry of the sample govern the choice of applicable theories and methods, we will restrict our analysis of the photothermal signal to the relevant geometry of a collection of small particles of explosives materials on a solid substrate. Furthermore, we will assume the particles are contacting the substrate over a limited surface area. This is in line with the properties of explosives traces generated through fingerprint transfer or other means of contact.[52] Figure 10.2a illustrates

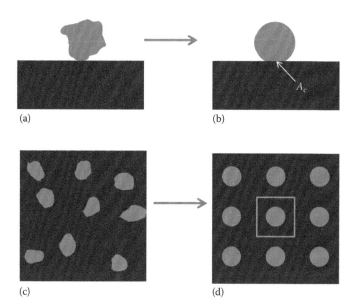

FIGURE 10.2 A schematic drawing of an explosives particle on a flat substrate (a) and its equivalent sphere approximation used in numerical modeling (b). Ensembles of particles (c) are treated as regular arrays for modeling purposes (d).

the basic geometry of the problem for a single isolated particle (of explosives). The particle is irregular in shape as it is a product of fingerprint transfers during which it is crushed.[14] Figure 10.2b shows an approximation where the particle is replaced by a sphere of equivalent volume, touching the substrate at a single point with a certain contact area. For higher spatial concentration of analyte (as in Figure 10.2c), an array of particles was considered, as shown in Figure 10.2d.

10.3.3 LASER–MATTER INTERACTION

In order to solve the heat transfer problem, we first need to know the amount of heat generated when the sample is illuminated by a laser (most often an IR laser). The amount of laser power absorbed by the semi-infinite substrate (away from the particle, per unit volume) is straightforward and in accordance with the Beer–Lambert law[53]; it is given by Equation 10.1:

$$q_{sub} = \alpha I (1 - R) e^{-\alpha z},\qquad (10.1)$$

where:
z = depth into the substrate
α = laser wavelength-dependent absorption coefficient of the substrate
R = reflectance at the air–substrate interface
I = laser irradiance (in W/m^2)

Estimating the light absorbed (or scattered) by the particles themselves is much more challenging, and we are not aware of a single theory of laser absorption and scattering in particles that accounts for all the complexities involved: irregular particle shape, micron-scale particle size, particle and substrate surface roughness, sparse particle coverage, presence of substrate, complex substrates (e.g., fabric or multilayered car paints), and possible interferents surrounding the explosives particle as well as those mixed in the particles themselves (e.g., plasticizers, additives).

Several notable attempts have been made to solve this problem, at least for the case of laser (back)-scattering. In one approach, by Hinkov et al.,[54] the particles were approximated as a thin film on a substrate. This approximation was successfully used to explain some spectral changes observed on explosives deposited on different substrates. Another notable approach is the "transflection" theory by Phillips et al.[55] in which they further demonstrate the importance of substrate on particle spectra (see Chapters 8 and 9). Mukherjee et al. discuss an example of the thin-film approximation in photothermal theory of explosives on substrates.[19]

Due to the size of the particles of interest for forensic applications, which is on the order of magnitude of the wavelength of the IR laser, it is important to consider relevant scattering theories that address the original particle geometry (shape and size). As a first approximation, we have applied the Mie scattering theory[53,56] to the spherical approximation in Figure 10.2b, with the exception that the substrate was ignored, which may be justified for certain (but not all) substrates. It was shown that particle size alone can be responsible for spectral changes in addition to any changes due to the presence of substrate as discussed above.[50,51]

The amount of laser power absorbed by a particle is then simply[53]

$$q_{part} = IQ_{abs}R^2\pi, \tag{10.2}$$

where:

Q_{abs} = absorption cross-section efficiency
R = particle radius

An example of how the particle size can affect the spectral response is illustrated in Figure 10.3. Figure 10.3a shows the absorption cross-section efficiency (Q_{abs}) given by the Mie theory for a series of 2,4-dinitrotoluene (DNT) (a precursor of the explosive 2,4,6-trinitrotoluene (TNT)) spheres with diameters ranging from 1 to 25 μm. Figure 10.3b shows the *reflection* cross section (Q_r) for the same DNT spheres. It can be seen that for the relevant particle size range (5–10 μm), the absorption cross section is consistent with the absorption coefficient and absorbance commonly found in Fourier transform infrared (FTIR) spectral libraries for bulk materials, while the reflection cross section appears to be more particle size–dependent.

10.3.4 HEAT TRANSPORT

It has been shown[50,51] that the only relevant heat transfer mechanism for small particles is conduction. Both radiative heat transfer and convection can be ignored due to

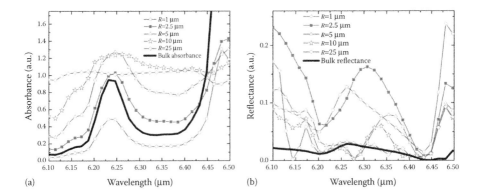

(a)   (b)

FIGURE 10.3 Mie scattering efficiencies for absorbance (a) and reflectance (b) as a function of particle size, along with bulk values. The figure illustrates that particle size affects measured spectra, especially for reflectance. Absorption efficiencies appear less dependent on particle size.

the combination of small particle sizes and small induced temperature differences. The main conduction mechanism found was that between the particle and air. This is somewhat counterintuitive, as one would expect the major effect to be the cooling of particles at the contact with the substrate. However, contact with the substrate only contributes roughly 20% to the overall heat transport from a *loosely* bound particle.

For these reasons, a pure heat conduction model is sufficient:

$$\frac{\partial T_i}{\partial t} - \alpha_i \Delta T_i = \frac{q_i}{\rho_i \cdot c_{P,i}}, \tag{10.3}$$

where:

T = three-dimensional temperature field within the domain
Δ = Laplace operator
α = thermal diffusivity
ρ = density
c_P = specific heat capacity at constant pressure
i = a, p, s is a label indicating the spatial domain (air, particle, substrate)
q_i = heat sources modeling the laser heating of the respective domains

Initial and boundary conditions are generally

$$T_i = T_0, \quad t = 0 \tag{10.4}$$

$$T_i \rightarrow T_0, |\mathbf{x}| \rightarrow \infty.$$

Solving the above set of equations, we can assume the laser heating is uniform throughout the particle due to the low Biot number of the particle that ensures that this heat, while nonuniformly deposited, is very rapidly uniformly redistributed.[50]

There are no simple analytical solutions to this problem, but by using the lumped-capacitance method, we obtained approximate analytical expressions for the thermal time constant and maximum temperature of a particle subject to a laser pulse.[50] This modeling and simulation method and results are discussed in Section 10.3.5.

10.3.5 MODELING AND SIMULATION

Simulations were performed using the software package COMSOL Multiphysics,[50,51] modeling the particles as spheres in contact with the substrate, as shown in Figure 10.2c. The thermal resistance of the contact can vary depending on the material, its shape, and how it sits on the substrate. Previously, we have measured the contact conductance (H) for polyethylene spheres.[50] We apply the same contact conductance here as well: $H = 3500 R_p^2$ W/K, where R_p is the particle radius. The size of the substrate and air region above the sphere is the same as for the flat solid described above. The laser beam is assumed to be uniformly absorbed in the top hemisphere to simulate resonant heating, which yields the highest temperature. We refer to this absorbed power as P_{abs}. The average temperature of the top hemisphere is then determined.

A series of over 3000 simulations were performed while varying the particle size (R_p from 0.5 to 50 µm), and thermophysical properties of both the particle (p) and substrate (s): thermal conductivity K from 0.05 to 400 W/m/K, density from 100 to 10,000 kg/m³ and heat capacity from 100 to 5,000 J/kg/K. The duration of the laser pulse ($P_{abs}=0.1$ mW) was five times the estimated thermal time constant for the given sample. For each simulation run, we obtained ΔT_{max} and τ by fitting the particle temperature rise to Equation 10.5:[57]

$$\Delta T(t) = \Delta T_{max}\left(1 - \exp(-t/\tau)\right). \tag{10.5}$$

The results of a typical simulation are shown in Figure 10.4. Figure 10.4a is the temperature map of a 10 µm diameter TNT particle (and surrounding air) on glass at the end of the laser pulse (5 ms). The temperature rise of the particle is shown in Figure 10.4b, along with the fit to Equation 10.6.

It was found that ΔT_{max} has a near linear dependence on $1/\sqrt{K}$. Also, it was found that $\Delta T_{max} \sim R$. Combining these two along with $\Delta T_{max} \sim I_{laser}$, we fit the simulation data points to the parameterized formula in Equation 10.6:[57]

$$\Delta T_{max} = I_{laser}Q_{abs}R_p\pi\left(-0.04573 + \frac{0.2616}{\sqrt{K_p}} + \frac{0.4135}{\sqrt{K_s}} + \frac{0.6296}{K_p} + \frac{0.06960}{K_s}\right.$$
$$\left. - \frac{0.04070}{K_p\sqrt{K_s}} - \frac{0.1025}{K_s\sqrt{K_p}} + \frac{0.02704}{K_pK_s} - \frac{0.2406}{K_p^{3/2}} + \frac{0.026425}{K_p^2}\right). \tag{10.6}$$

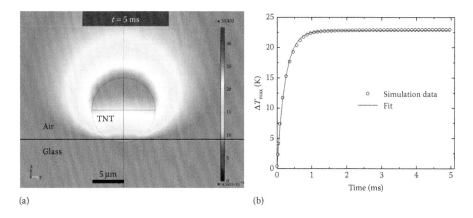

(a)   (b)

FIGURE 10.4 (a) Simulated temperature ΔT (K) map of a 10 μm diameter TNT particle on a soda-lime glass slide surrounded by air and illuminated with a Gaussian beam ($P_0=0.1$ mW, $r_0=5$ μm) at the end of a 5 ms laser pulse. (b) The temperature rise of the particle along with the fit to Equation 10.6.

For the thermal time constant, we get Equation 10.7:

$$\tau = \rho_p c_p R_P^2 \left(-0.1413 + \frac{1.0948}{\sqrt{K_p}} + \frac{1.9135}{\sqrt{K_s}} + \frac{1.6356}{K_p} + \frac{0.4321}{K_s} \right.$$

$$\left. - \frac{0.2079}{K_p\sqrt{K_s}} - \frac{0.4338}{K_s\sqrt{K_p}} + \frac{0.1178}{K_p K_s} - \frac{0.6653}{K_p^{3/2}} + \frac{0.0745}{K_p^2} \right). \quad (10.7)$$

These parameterized formulas should be applicable to a wide number of samples where the thermal properties of the particle and substrate materials are independently known. The R (particle size) dependence of these formulas is extremely close to R and R^2, respectively, and should be valid well outside the 0.5–50 μm range tested.

By combining Equations 10.7 and 10.8 with Equation 10.6, one can estimate the ΔT and τ for any sample where the thermal properties of the particle and substrate materials are independently known. In the case of realistic explosive particle traces, ΔT is usually a fraction of a degree and the time constants are on the order of 1 ms. We have also measured ΔT and τ as functions of R for polyethylene microspheres in air and in vacuum.[50] The size range (~10–100 μm) was slightly larger than typical explosive particles found in the field. Results are shown in Figure 10.5. In vacuum, the heat loss through air is minimized, which results in an (~5×) increase in ΔT. By comparing vacuum and in-air values, we conclude that heat loss through air is the dominant heat loss mechanism for loosely coupled particles and is responsible for up to 80% of the total heat loss. The anomalies observed for smaller particles (~10 μm) are caused by increased agglomeration of particles and higher areal coverage.

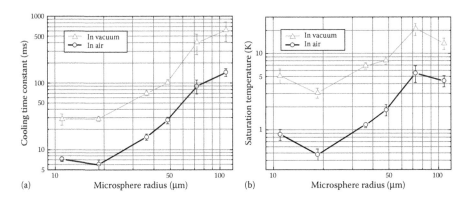

(a)   Microsphere radius (μm)   (b)   Microsphere radius (μm)

FIGURE 10.5 Experimentally measured cooling time constants (a) and temperature increases (b) for blackened polyethylene microspheres on a polished copper surface. Values are given for both air (normal pressure) and vacuum. (From Großer, J. et al., *International Journal of Heat and Mass Transfer*, 55, 8038–8050, 2012.)

10.4 INFRARED SIGNATURES: SPECTROSCOPIC BASIS FOR CHEMICAL IDENTIFICATION

It has been shown that PT phenomena derive from the absorption properties of the illuminated material.[1] Therefore, the absorption spectra of explosives such as those shown in Figure 10.6 provide the spectroscopic basis for their PT-derived chemical identification. Figure 10.6 shows the ambient temperature IR absorbance spectra of several analytes of interest: ammonium nitrate (AN), TNT, and 1,3,5-trinitroper-hydro-1,3,5-triazine (RDX). The spectra were measured using a standard FTIR with an attenuated total reflection (ATR) crystal. Peaks in the curves represent characteristic molecular absorption bands, which can be considered as identification "signatures" for each analyte. In practice, PT techniques primarily exploit these absorption bands to resonantly couple energy into the target analyte, but by

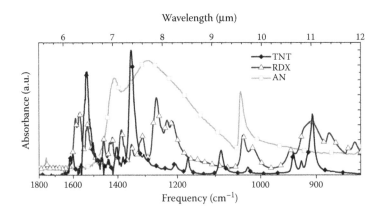

FIGURE 10.6 FTIR-ATR absorbance spectra of TNT, RDX, and AN.

Kirchhoff's law, the analytes' emissivities also exhibit the same spectral properties. This characteristic emission property forms the basis of some types of passive IR hyperspectral imaging (HSI)[58] (refer to Chapter 8). For standoff applications, the absorption due to water vapor in air is a vital consideration. Although there are many strong (but narrow) absorption bands in the 5–8 μm region,[18] in practice, the narrow linewidth of QCL lasers permits wavelength selections that avoid these features for propagation through the air to the target with minimal attenuation. More importantly from an identification standpoint, emitted long-wave IR (LWIR) (8–12 μm) photothermal signals transmit without much attenuation even through a humid atmosphere. This preserves the spectral content of the emission and enables significant standoff of the detector from the target surface.

In general, PT phenomenology offers "multidimensional" paths to enhancing both *sensitivity* to trace quantities and *selectivity* to distinguish threat analytes from benign substrates and interferents. These "dimensions" primarily comprise (1) multiwavelength excitation to spectrally resolve various *absorption* features of interest; (2) selection of spectral bandpass filters that optimize collection of explosives *emission* relative to substrates or interferents; (3) *spatial contrast* between the particles on the surface and the surrounding substrate; and (4) differences in the *time dependence* of thermal emissions from substrates and surface particles.

For photothermal detection in general, the fundamental limits to standoff distance depend on the laser power and divergence, as well as the collection efficiency and detector sensitivity. The American National Standard Institute (ANSI) standard[31] MPE for eye safety in this IR wavelength range is 0.1 W/cm^2 for CW illumination. For intermittent or pulsed illumination, somewhat higher peak powers can be safely used. A beam that starts out at the eye-safe level will decrease in intensity due to divergence and atmospheric absorption. At the eye-safe MPE, surface particles heat by up to a few degrees Celsius. While a few degrees of change can be readily observed at standoff, it becomes difficult to resolve small spatial features. For example, using a 100 mm focal length lens, at 26.3 m standoff, a typical 40 μm pixel within a focal plane array will image to a 1 cm diameter spot on the target. As thermal emission can be considered isotropic, the collected thermal signal should decrease with distance by $1/R^2$, where R is the standoff distance.

For PT-based approaches, further complications introduced by typical explosives-bearing surfaces arise from the combination of their sheer physical thickness and continuum absorbing nature; polymeric materials tend to absorb (and emit) throughout the IR. Thus, whether the target substrate is a multilayer modern vehicle coating or the fibers that constitute synthetic and natural textiles that typify clothing, physical thicknesses on the order of >10 μm are common. Consequently, even weakly absorbed laser wavelengths will effectively be completely absorbed at some depth within these materials. They are therefore best characterized as being "optically thick," in the sense that they absorb all IR wavelengths at some point within their depth. These universal considerations introduced by real-world surfaces complicate the detection and identification of explosives materials, including by PT-based methods.

10.5 EXAMPLES OF EXPERIMENTAL ARRANGEMENTS AND APPLICATIONS FOR EXPLOSIVES DETECTION

10.5.1 CALORIMETRY AND THERMAL PHASE CHANGE

Calorimetry is characterized by the direct measurement of the temperature increase due to PT excitation. Calorimetry with nonphotothermal energy sources is commonly used to analyze and forensically detect explosives due to their exothermic properties.[59] However, due to the implicit requirement of contact with the explosive to be detected, photothermal calorimetry is not widely used for explosives, particularly not with laser-based approaches.

However, as a material heats, it can undergo phase changes such as from solid to liquid or vapor. For the purposes of this chapter, we neglect chemical changes, such as photo fragmentation, which take place on timescales shorter than the thermal properties of the targets. Because many explosives of interest have very low vapor pressures at room temperature,[13] they can be very difficult to detect by mature vapor-sensing technologies such as ion mobility spectroscopy (IMS) or mass spectrometry.[60] Nonetheless, photothermal vaporization can enable coupling to vapor sensing detection technology (optical or other types such as IMS) for noncontact explosives detection.[28,35,61–67] It should be noted that while high-energy photons such as those at visible wavelengths are known to decompose explosives,[61] IR photons can be used to generate the vapor phase without breaking the molecular bonds critical to many chemical signatures for identification.[35,68–70]

An example configuration for resonant IR laser-induced PT vaporization was described by Furstenberg et al.[28] In this case, the vapor was collected and analyzed using a commercial IMS explosives detector. A specific advantage of coupling the photothermal vaporization to the mature IMS trace detection technology is the robustness of the chemical identification due to orthogonal (IR absorption vs. ion mobility) physical characteristics. The advantage of this approach over traditional "swabbing" techniques for sampling trace explosives is that it does not require contact with the target surface, thereby avoiding potential cross-contamination. The advantage to using a resonant IR wavelength coupled to specific absorption bands of interest is the increased signal-to-noise ratio (SNR) of analyte vapors relative to background or interferent vapors. The disadvantages of laser-induced vaporization are: (1) that it typically requires high (non-eye-safe) laser power levels; (2) that the vapor intake is typically required to be fairly close to the target surface; and (3) that the target analyte should typically absorb the wavelength of light used. For a wide variety of analytes, this could require multiple wavelengths or even multiple laser sources.

10.5.2 PHOTOTHERMAL INTERFEROMETRY

Traditional photothermal interferometry (PTI) is a pump-probe method where the pump source is used to generate a refractive index change in the target, which is then probed with a second beam.[1,71] These index changes cause intensity and/or phase shifts in the probe beam, which can be measured interferometrically. This has been shown to enhance sensitivity by many orders of magnitude.[72] However, traditional implementations of PTI are based on a transmission geometry where the probe source and detector are on opposite sides of the target, which is not practical for most applications of explosives detection.

Recently, PTI has been adapted to a backscattering geometry for stand-off detection of explosives.[29] A schematic of the setup is shown in Figure 10.7. A pulsed tunable IR QCL beam is directed toward the target at 5 m standoff distance. A second, CW, beam is overlapped with the pulsed beam as a probe. Photothermal/photo-acoustic impulses generated by resonant absorption in the target shift the amplitude and/or phase of the reflected CW beam, which is mixed interferometrically with the probe source. As can be seen in Figure 10.8, the PTI spectra measured at 5 m standoff match well with the FTIR spectra from octa-hydro-1,3,5,7-tetranitro-1,3,5,7-tetrazocine (HMX) and RDX target samples. A distinct advantage of this method is that the probe beam wavelength is 1550 nm, in the peak of the "eye-safe" band. This allows use of strong probe intensity to amplify the interferometer signal. Another advantage of this technique is that the processing of the interferometric quadrature signals generates signal with very small background noise. These advantages enable the experiments to demonstrate a good match to the target explosives absorption spectra as the QCL is tuned from 8.8–10.2 µm. From a practical point of view, one disadvantage of this approach is that it requires that two beams of different wavelengths from different sources be overlapped at standoff distance. This can be a challenge due to wavelength-dependent refraction in the atmosphere. Another disadvantage could be when the target is far off normal angle to the probe source/collector system, resulting in a very low backscattered probe intensity, which would be difficult to measure interferometrically.

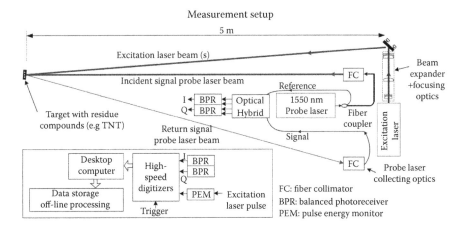

FIGURE 10.7 Photothermal interferometry (PTI) schematic. (From Cho, P. S., Jones, R. M., Shuman, T., Scoglietti, D. and Harston, G. *Proceedings of SPIE*, 8018, 80181T, 2011. With permission.)

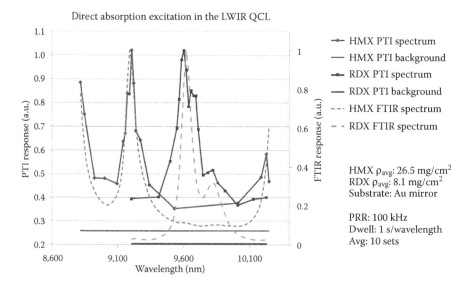

FIGURE 10.8 The photothermal interference (PTI) signal from RDX and HMX explosives measured at 5 m standoff. The FTIR spectra are shown for comparison. (From Cho, P. S., Jones, R. M., Shuman, T., Scoglietti, D. and Harston, G. *Proceedings of SPIE*, 8018, 80181T, 2011. With permission.)

10.5.3 PHOTOTHERMAL DEFLECTION

Traditional photothermal deflection (together with refraction and diffraction) approaches are based on the *mirage effect*, where the photothermal heating of a target indirectly heats the surrounding gas, thereby changing its index of refraction.[1,73] A probe beam passing through this heated gas is deflected, thereby changing the signal intensity at a given sensor position. Because such deflections are typically at small angles for low analyte concentrations, these approaches favor transmission geometries not practical for many explosives detection applications.

A recent series of experiments exploits another form of photothermal deflection based on the sensitive motion of micromechanical cantilevers.[40,49] The experimental configuration is based on a microelectromechanical systems (MEMS) cantilever where one side of the cantilever arm is used to sorb explosive trace vapors. The opposite side of the cantilever is probed with a laser beam, which is reflected onto a position-sensitive detector. The photothermal sensing aspect of this arrangement is implemented by a continuously tunable monochromatic IR source used to illuminate the sorbed vapors, thereby slightly heating the cantilever, causing it to bend. The cantilever deflection measured by the position-sensitive detector is a function of the wavelength-dependent absorption, allowing the analyte spectrum to be collected as the IR source is tuned. The explosives detection is achieved by matching the spectrum to the known spectra of analytes of interest. Early versions of this approach were based on broadband IR sources passed through a monochromator. Later versions used a tunable QCL for IR excitation. Further variations of this approach replaced the laser-based deflection measurement with shifts in the resonant frequency of the cantilever itself. For increased sensitivity, the cantilevers were replaced with quartz crystal tuning forks, which feature very high quality (Q) factors.[41] Advantages to these approaches include the sensitivity to nanogram quantities of analyte in the gas phase and the high fidelity of the PT spectroscopic match to the reference spectrum. However, disadvantages to these approaches include the low vapor pressure of most explosives and the challenge of collecting enough analyte on the small cantilever surface. While itself not directly applicable to standoff detection of explosives, this cantilever-based methodology was successfully leveraged for standoff applications based on IR reflectivity and backscattering approaches.[74]

Due to the diffraction limit, IR wavelengths present a challenge to spatially resolving trace analysis of individual particles of explosives with diameters of 1–20 μm by IR spectroscopy. However, PT deflection approaches offer the potential for higher spatial resolution by exploiting shorter (visible)-wavelength probes in a diffraction-limited spot.[75,76] An example experimental configuration is shown in Figure 10.9. A mid-IR QCL PT pump beam is focused to a diffraction-limited spot with diameter ~10 μm. A visible laser probe beam is folded collinearly and focused to a diffraction-limited spot (diameter ~0.5 μm) coincident with the IR. Upon wavelength-dependent PT excitation of the target, the particles on the surface exhibit thermo-optic or thermoelastic deflection of the probe beam. To minimize the extent of the thermal diffusion (and therefore maintain the spatial resolution), as well as to enhance the detection sensitivity by lock-in amplification, the pump beam is pulsed or chopped as shown in Figure 10.9. An advantage of this approach is that it allows IR microscopy to compete

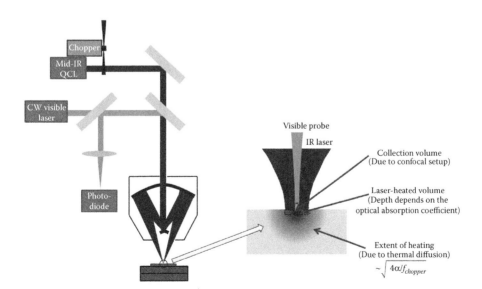

FIGURE 10.9 Experimental configuration for high-spatial-resolution photothermal microscopy. The spatial extent of heating is a function of the thermal diffusivity (α) and the chopping frequency ($f_{chopper}$). (From Furstenberg, R., Kendziora, C. A., Papantonakis, M. R., Viet, N. and McGill, R. A., *Proceedings of SPIE*, 8374, 837411, 2012.)

with (or even be combined with) Raman spectroscopy (refer to Chapter 5) for chemical identification of individual particles of explosives. IR has advantages over Raman for samples with large luminescence or Raman-inactive vibrational bands. Another advantage of microscopy is that the small spot sizes enable dramatically higher laser flux, resulting in much higher target temperatures and therefore higher PT signals. A disadvantage of PT IR microscopy is that it is limited to point detection (although it can be rastered to generate images).[75,76] Another potential disadvantage of the high laser flux is thermal decomposition or vaporization of the target particles.

More recently, an adapted photothermal deflection method has been developed for standoff application.[77] The schematic arrangement for this approach is shown in Figure 10.10.

In this standoff photothermal deflection approach, the pump QCL selectively heats the surface, causing a deflection of the probe beam from the position-sensitive detector. This deflection is measured as a function of pump laser wavelength to generate spectral information on the interrogated spot. By raster scanning over the surface area, this point detection approach can generate an image of the target. Advantages of this method include the abilities for standoff spectroscopic mapping of the target without requiring complex or expensive components such as telescopes, cooled detectors, or imaging arrays. Disadvantages of this approach include the requirement to maintain coalignment of the pump and probe beams as a function of standoff distance, the time and position stability necessary to collect spectra from adjacent spots to create an image, and the angular dependence of the reflected beam on the target surface flatness.

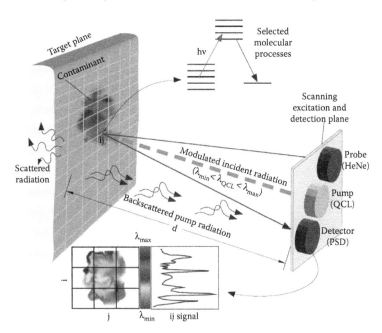

FIGURE 10.10 Pump-probe standoff hyperspectral imaging based on photothermal deflection. (From Farahi, R. H., Passian, A., Tetard, L. and Thundat, T., *Journal of Physics D: Applied Physics*, **45**, 125101, 2012.)

10.5.4 PHOTO-ACOUSTICS

Traditional photo-acoustic methods utilize pressure-sensitive detectors, such as microphones, to detect signals generated by probing the target with light pulsed or chopped at acoustic frequencies.[1,48] Photo-acoustic detection of explosives has been the subject of previous reviews.[8,10,78] The preponderance of these studies investigate the analyte in the gas phase, typically within a dedicated photo-acoustic cell. Recently, an apparatus has been built and tested for standoff photo-acoustic detection of solid-phase explosives.[42] A schematic diagram of the instrument is shown in Figure 10.11. Similar approaches (including multiple QCLs) have been used elsewhere to detect explosives on surfaces.[27,74] The configuration shown in Figure 10.11 features a 45° mirror to fold the PT excitation beam collinear with the receiving mirror, which facilitates coregistration at different standoff distances.

This setup leverages the photothermal response of the target surface to QCL beams pulsed at acoustic frequencies. The frequency is chosen to match the resonance of a sensitive quartz crystal tuning fork (QCTF). Example spectra collected at 5 m and 10 m standoff are shown in Figure 10.12. The spectra measured with the quartz-enhanced photo-acoustic spectroscopy (QEPAS) system match the reference FTIR spectrum of the DNT analyte. In addition to standoff and spectroscopic capability, advantages of this approach include the very low total energy deposited on the target (due to the pulsed beam) and the high sensitivity/low cost performance of the QCTF. Disadvantages of this approach include the limitation to point detection and the requirement that the target be near normal incidence. Point detection approaches in general present challenges for background subtraction of signal due to the substrate by requiring *a priori* knowledge of an uncontaminated substrate area to measure for comparison.

Recently, a similar photo-acoustic approach was taken using an ultrasensitive microphone for signal transduction.[37] The sensitivity and standoff distance are further enhanced by using a 2 ft. diameter parabolic reflector to collect sound waves and focus them on the microphone. Using this apparatus, TNT powder could be detected

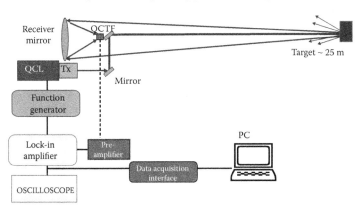

FIGURE 10.11 Experimental setup for standoff laser photo-acoustic detection. The reflected signal is transduced by a quartz crystal tuning fork (QCTF). (From Sharma, R. C. et al., *Optics Communications*, **309**, 44–49, 2013. With permission.)

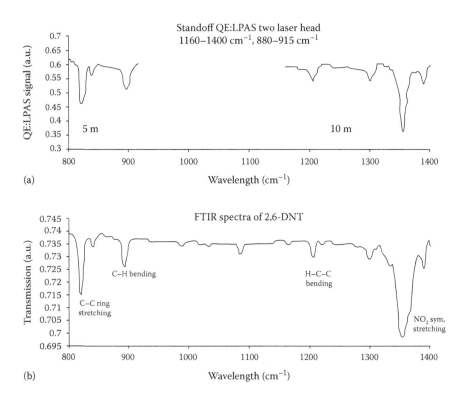

FIGURE 10.12   The photo-acoustic spectrum measured at 5 m and 10 m standoff (a) compared with the reference FTIR spectrum of 2,6-DNT (b). (From Sharma, R. C. et al., *Optics Communications*, 309, 44–49, 2013. With permission.)

by photo-acoustic response to a QCL operating at 7.35 μm, close to an absorption feature of TNT. In principle, such a method could be extended using a tunable QCL to extract a wavelength-dependent response for chemical identification. A potential advantage to measuring the sound waves directly (as opposed to measuring the backscattered light and transducing it acoustically with a QCTF) is the retention of additional frequency components, and the potentially larger relative signal for oblique target surfaces, since acoustic signals scale with $1/R$, where R is the standoff distance.[37] Potential disadvantages to measuring the sound waves directly include the need to filter out ambient noise, the relatively slower speed of sound versus light through air, and the relatively inefficient transduction of light to sound at the target/air interface.

10.5.5   PHOTOTHERMAL RADIOMETRY

Traditional photothermal radiometry utilizes a light source to heat the target while collecting the emitted Planck radiation to measure the temperature.[1,17,79] Furthermore, because by Kirchhoff's law the wavelength-dependent emission of a material matches its absorption spectrum, laser-induced thermal emission (LITE) has been used to capture the spectral signature of the target.[80] In principle, the excitation can be at any wavelength absorbed by the sample, but the stronger the absorption, the warmer the sample and the brighter the emission signal. For this reason, recent PT radiometry techniques have exploited the resonant heating accessible with IR laser sources.[18,19,30,75,76,81]

A recent application of PT radiometry to explosives detection of individual trace particles of explosives is based on a confocal microscope design.[75,76] A schematic of the PT confocal microscope is shown in Figure 10.13. In this design, the coherence properties of the QCL laser are exploited to focus the IR beam to a near-diffraction-limited beam spot diameter.

To demonstrate the chemical imaging capability of the microscope, a small crystal of 2,4-DNT on a glass slide was measured by raster scanning. The spatial raster scan consisted of a 12 × 12 array of points with a 15 μm step between each point. The laser was tuned to 35 discrete wavelengths in the 6–6.6 μm spectral region. The raster scan was repeated for each wavelength. The dwell time at each point was ~300 ms, which was the minimum time required to move the stage between points. Figure 10.14a shows the optical image of the crystal, while Figure 10.14b shows the confocal microscope image. Figures 10.14c through 10.14e show images generated by the PT IR probe, with the laser tuned to three different wavelengths. The wavelength in Figure 10.14e is near an absorption peak of DNT. Advantages of PT radiometric microscopy include the ability to detect and identify individual trace particles of explosives without damaging them. Also, IR microscopy has advantages over Raman where the target analytes exhibit fluorescence, are damaged by high photon energies, or are not Raman active. Disadvantages to this approach include

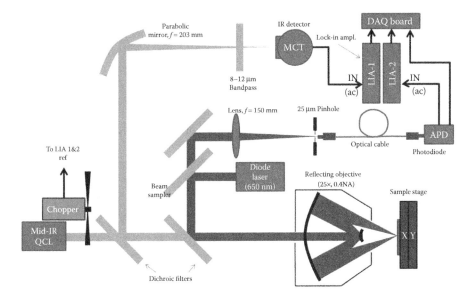

FIGURE 10.13 Schematic of the photothermal confocal microscope. (From Furstenberg, R., Kendziora, C. A., Papantonakis, M. R., Nguyen, V. and McGill, R. A., *MRS Proceedings Fall 2011*, **1415**, 2012.)

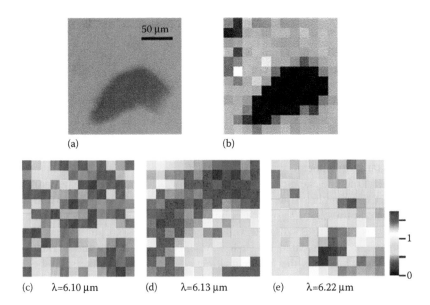

(a)   (b)

(c)   λ=6.10 μm   (d)   λ=6.13 μm   (e)   λ=6.22 μm

FIGURE 10.14   Photothermal imaging of a DNT crystal on glass. (From Furstenberg, R., Kendziora, C. A., Papantonakis, M. R., Nguyen, V. and McGill, R. A., *MRS Proceedings Fall 2011*, **1415**, 2012.)

the low spatial resolution due to the diffraction limit, and the need to position the target very close to the objective as well as to keep the target very still during the entire raster scan.

While PT radiometry was traditionally applied in close proximity, recently, LITE has been applied to explosives at standoff distances.[30] A schematic of the experimental setup is shown in Figure 10.15. In this configuration, a high-power (4–36 W) CO_2 laser was used to heat a target at a standoff distance of 4 m. The beam was on for up to 1 min to heat the sample. Advantages of this approach include the collection of the full FTIR spectrum for chemical identification and the potential for long standoff distances. Example spectra from explosives samples are shown in Figure 10.16. Disadvantages of this approach include the very high laser (non-eye-safe) power needed, the long times needed to heat the sample, potential vaporization or decomposition of the sample at high temperatures, and the point detection limitation.

10.5.6   Photothermal Infrared Imaging Spectroscopy (PT-IRIS)

Photothermal infrared imaging spectroscopy (PT-IRIS) is an active approach under development by the authors for standoff detection using IR QCLs for surface illumination and IR focal plane array (FPA) detectors for imaging.[18,63–66] A simplified block schematic diagram is shown in Figure 10.17.

In PT-IRIS, an IR source—usually a QCL—is used to illuminate a surface potentially contaminated with residues of interest. If the wavelength of the light is resonant with absorption features of surface residues, the analyte heats by ~1°C. This thermal

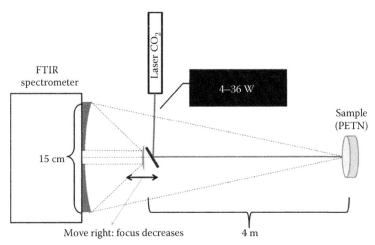

FIGURE 10.15 Laser-induced thermal emission (LITE) instrument setup. (From Galan-Freyle, N. Y., Pacheco-Londono, L. C., Figueroa-Navedo, A. and Hernandez-Rivera, S. P. *Proceedings of SPIE*, 8705, 870508, 2013. With permission.)

emission is observed using a LWIR camera, which collects images at a video frame rate (>30 Hz). Differential imaging (i.e., subtracting the image with the laser "off" from the frame with the laser "on") removes the time-independent background and reveals the trace residues. The IR excitation wavelength is tuned to couple to a wide range of absorption bands within the target analytes. PT-IRIS has been previously demonstrated[18,28,35,43,51,82–85] at different standoff distances for a range of analytes and substrates, including fabrics[83] and painted steel car panels.[84,85] Figure 10.18 shows a cart-based platform used to demonstrate the PT-IRIS approach during outdoor testing.

Figure 10.19 illustrates the PT-IRIS method using TNT and RDX analytes deposited on an IR-transparent polypropylene film.[18] As the QCL wavelength is tuned in four steps across the nitrogen–oxygen (N–O) absorption band, the analytes are selectively heated. The spot size, illustrated by the dashed circle, is ~10 mm. This sequence of images demonstrates the selectivity of the photothermal technique, even between types of explosives. Advantages of the PT-IRIS approach include efficacy below the MPE, imaging capability, and the "two-dimensional" spectroscopic advantage of tuning both the excitation wavelength and the bandpass filter region. Disadvantages include the competing metrics of area coverage versus signal intensity for a given laser power, and the computational requirements associated with this type of hyperspectral imaging.

A related approach has been taken, although using a high-power CO_2 laser and telescope optics, for very long standoffs.[19] The schematic diagram for the remote optothermal sensor (ROSE) instrument is shown in Figure 10.20. A notch-blocking ($10.3 > \lambda > 10.9$ μm) filter (not shown) is used in front of the camera to prevent back-scattering from the laser. The CO_2 laser is collimated/focused at standoff using a

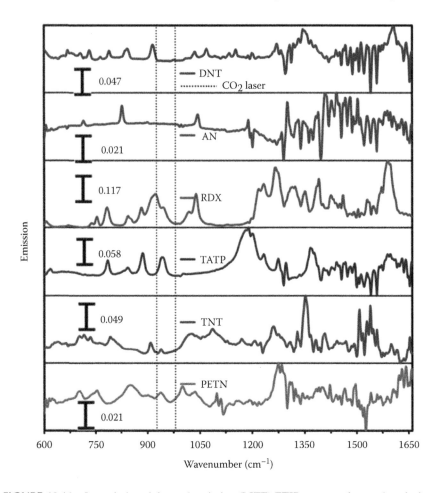

FIGURE 10.16 Laser-induced thermal emission (LITE) FTIR spectra of several explosives measured at 4 m standoff. (From Galan-Freyle, N. Y., Pacheco-Londono, L. C., Figueroa-Navedo, A. and Hernandez-Rivera, S. P. *Proceedings of SPIE*, 8705, 870508, 2013. With permission.)

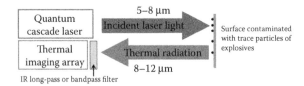

FIGURE 10.17 Block schematic diagram for photothermal infrared imaging spectroscopy (PT-IRIS).

FIGURE 10.18 The PT-IRIS cart system during outdoor testing. The sample tripod is positioned at 10 m standoff distance. The dark arrow (top center of photo) illustrates the approximate path of the invisible IR beam.

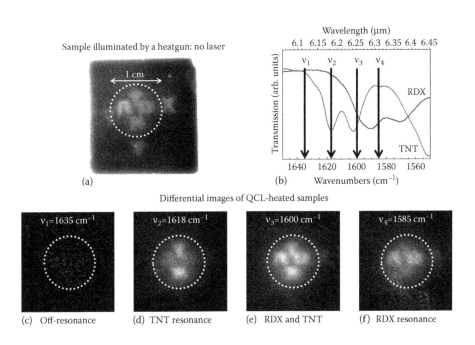

FIGURE 10.19 Illustration of resonant photothermal heating. Letters "RDX" and "TNT" were written on a polypropylene film using solution of the respective explosives. (a) Result of nonselective heating using a heatgun. (b) IR spectrum of RDX and TNT showing resonances. (c) Laser heating with off-resonance wavelength. (d) Laser heating on resonance with TNT only. (e) Laser heating on resonance with RDX and TNT. (f) Laser heating on resonance with RDX only. (From Furstenberg, R. et al., *Applied Physics Letters*, **93**, 224103, 2008.)

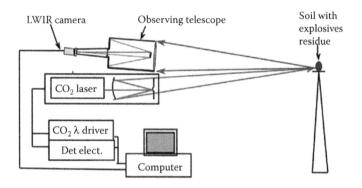

FIGURE 10.20 Schematic of the remote optothermal sensor (ROSE). (From Mukherjee, A., Von der Porten, S. and Patel, C. K. N., *Applied Optics*, **49**, 2072–2078, 2010.)

12 in. reflecting telescope. The target is observed and imaged onto the LWIR (cooled MCT) camera using a 24 in. reflecting telescope. A photograph of the truck-mounted ROSE system during field-testing at the Naval Surface Warfare Center (NSWC) at China Lake, California, is shown in Figure 10.21. The CO_2 laser can be tuned to different lines within a band of ~10.5–10.8 μm to generate a spectrum for target chemical identification. Figure 10.22 shows typical data from the ROSE instrument taken at 150 m standoff. The spectral response from a contaminated dirt sample measured across 11 discrete excitation wavelengths clearly shows a peak at 10.65 μm, near the reference absorption peak for TNT. Clean soil shows a much smaller signal with no discernable wavelength dependence. Advantages to the ROSE approach include most of the advantages of the PT-IRIS method (e.g., imaging, standoff), but also

FIGURE 10.21 Photograph of the remote optothermal sensor (ROSE) during testing at the NSWC, China Lake. (From Mukherjee, A., Von der Porten, S. and Patel, C. K. N., *Applied Optics*, **49**, 2072–2078, 2010.)

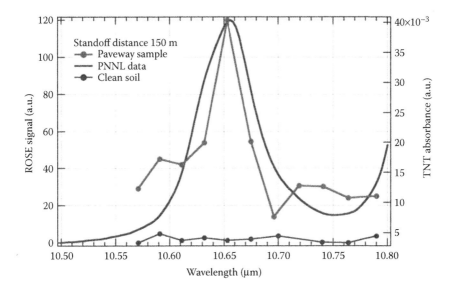

FIGURE 10.22 Comparison of the signal from the remote optothermal sensor (ROSE) for an explosives contaminated dirt sample and comparison with the reference TNT absorbance at 10.65 μm. (From Mukherjee, A., Von der Porten, S. and Patel, C. K. N., *Applied Optics*, **49**, 2072–2078, 2010.)

include the very high power available from the CO_2 laser and the very long standoff capability provided by the 24 in. collection aperture. Disadvantages include the non-eye-safe laser and the overall size and cost of the instrument.

10.6 CONCLUSIONS AND OUTLOOK

Photothermal spectroscopy has been developed over more than 100 years as a sensitive, noncontact way to measure material parameters such as wavelength-dependent absorptivity, which can be challenging to determine under typical reflection or transmission conditions. More recently, photothermal methods originally developed for laboratory characterization of fundamental materials properties have been successfully adapted to the sensing and identification of explosives. This has been demonstrated for gas-phase analytes, as well as for the case of trace particles on surfaces. For the gas phase, photo-acoustic approaches have been highly effective. For particulates, a variety of approaches based on PT phase changes, PT radiometry, PT deflection, and PT interference, as well as photo-acoustic methods, have been successful at sensing explosives. Whereas most of these techniques have been developed in the laboratory, several have also been tested at standoff distances, demonstrating their relevance to the challenging but crucial goal of standoff detection of explosives.

As the instruments used for PT investigations (such as QCLs, IR detectors and focal plane arrays, and quartz crystal tuning forks) continue to improve, so will the ability of these methods to detect ever smaller quantities of explosives at ever longer standoff distances.

ACKNOWLEDGMENTS

This work was sponsored by NRL/ONR core funds and by the US Army REDCOM CERDEC NVESD. Previous support was provided by the Rapid Reaction Technology Office of the Office of the Secretary of Defense.

REFERENCES

1. Bialkowski, S. E. *Photothermal Spectroscopy Methods for Chemical Analysis*, Volume 134. John Wiley & Sons, Hoboken, NJ (1996).
2. Mandelis, A. *Principles and Perspectives of Photothermal and Photoacoustic Phenomena*, Volume 1. Elsevier, Amsterdam (1992).
3. Colton, R. J. and Russell, J. N. Making the world a safer place. *Science* **299**, 1324–1325 (2003).
4. Council, N. R. Determining core capabilities in chemical and biological defense science and technology. Report No. 9780309265355, The National Academies Press, Washington, DC (2012).
5. Petryk, M. W. P. Promising spectroscopic techniques for the portable detection of condensed-phase contaminants on surfaces. *Applied Spectroscopy Reviews* **42**, 287–343 (2007).
6. Wallin, S., Pettersson, A., Ostmark, H. and Hobro, A. Laser-based standoff detection of explosives: A critical review. *Analytical and Bioanalytical Chemistry* **395**, 259–274 (2009).
7. López-López, M. and García-Ruiz, C. Infrared and Raman spectroscopy techniques applied to identification of explosives. *Trends in Analytical Chemistry* **54**, 36–44 (2014).
8. Holthoff, E. L. and Pellegrino, P. M. Sensing applications using photoacoustic spectroscopy. In K. Iniewski (ed.), *Optical, Acoustic, Magnetic, and Mechanical Sensor Technologies*. CRC Press, Boca Raton, FL (2012).
9. Willer, U. and Schade, W. Photonic sensor devices for explosive detection. *Analytical and Bioanalytical Chemistry* **395**, 275–282 (2009).
10. Patel, C. K. N. Laser photoacoustic spectroscopy helps fight terrorism: High sensitivity detection of chemical warfare agent and explosives. *European Physical Journal: Special Topics* **153**, 1–18 (2008).
11. Fuchs, F. et al. Broadband-tunable external-cavity quantum cascade lasers for spectroscopy and standoff detection. In M. Razeghi, L. Esaki, and K. von Klitzing (eds), *The Wonder of Nanotechnology: Quantum Optoelectronic Devices and Applications*. SPIE, Bellingham, WA (2013).
12. Skvortsov, L. A. and Maksimov, E. M. Application of laser photothermal spectroscopy for standoff detection of trace explosive residues on surfaces. *Quantum Electronics* **40**, 565–578 (2010).
13. Ostmark, H., Wallin, S. and Ang, H. G. Vapor pressure of explosives: A critical review. *Propellants Explosives Pyrotechnics* **37**, 12–23 (2012).
14. Verkouteren, J. R. Particle characteristics of trace high explosives: RDX and PETN. *Journal of Forensic Sciences* **52**, 335–340 (2007).
15. Tyndall, J. *Contributions to Molecular Physics in the Domain of Radiant Heat*. D. Appelton, New York (1873).
16. Bell, A. G. On the production and reproduction of sound by light. *American Journal of Science* **20**, 305–324 (1880).
17. Nordal, P. E. and Kanstad, S. O. Photothermal Radiometry. *Physica Scripta* **20**, 659–662 (1979).
18. Furstenberg, R. et al. Stand-off detection of trace explosives via resonant infrared photothermal imaging. *Applied Physics Letters* **93**, 224103 (2008).

19. Mukherjee, A., Von der Porten, S. and Patel, C. K. N. Standoff detection of explosive substances at distances of up to 150 m. *Applied Optics* **49**, 2072–2078 (2010).
20. Jackson, W. B., Amer, N. M., Boccara, A. C. and Fournier, D. Photothermal deflection spectroscopy and detection. *Applied Optics* **20**, 1333–1344 (1981).
21. Stone, J. Thermooptical technique for the measurement of absorption loss spectrum in liquids. *Applied Optics* **12**, 1828–1830 (1973).
22. Davis, C. C. Trace detection in gases using phase fluctuation optical heterodyne spectroscopy. *Applied Physics Letters* **36**, 515–518 (1980).
23. Stone, J. Measurements of absorption of light in low-loss liquids. *Journal of the Optical Society of America* **62**, 327333 (1972).
24. Gordon, J. P., Leite, R. C. C., Moore, R. S., Porto, S. P. S. and Whinnery, J. R. Long-transient effects in lasers with inserted liquid samples. *Journal of Applied Physics* **36**, 38 (1965).
25. Rosencwaig, A. Photoacoustic spectroscopy of solids. *Optics Communications* **7**, 305–308 (1973).
26. Takeuchi, E., Thomas, K. and Day, T., Quantum-cascade lasers: Applications multiply for external-cavity QCLs in *Laser Focus World*, Volume 45, 83–86 (2009).
27. Van Neste, C. W., Senesac, L. R. and Thundat, T. Standoff Spectroscopy of Surface Adsorbed Chemicals. *Analytical Chemistry* **81**, 1952–1956 (2009).
28. Furstenberg, R., Papantonakis, M., Kendziora, C. A., Bubb, D. M., Corgan, J., and McGill, R. A. Laser vaporization of trace explosives for enhanced non-contact detection. *Proceedings of SPIE* **7665**, 76650Q (2010).
29. Cho, P. S., Jones, R. M., Shuman, T., Scoglietti, D. and Harston, G. Investigation of standoff explosives detection via photothermal/photoacoustic interferometry. *Proceedings of SPIE* **8018**, 80181T (2011).
30. Galan-Freyle, N. Y., Pacheco-Londono, L. C., Figueroa-Navedo, A. and Hernandez-Rivera, S. P. Standoff laser-induced thermal emission of explosives. *Proceedings of SPIE* **8705**, 870508 (2013).
31. American National Standards Institute. American National Standard for Safe Use of Lasers. Vol. Z136.1. ANSI, Orlando, FL (2007).
32. Patel, C. K. N. Continuous-wave laser action on vibrational-rotational transitions of CO_2. *Physical Review* **136**, A1187–A1193 (1964).
33. Faist, J. et al. Quantum cascade laser. *Science* **264**, 553–556 (1994).
34. Gmachl, C., Capasso, F., Sivco, D. L. and Cho, A. Y. Recent progress in quantum cascade lasers and applications. *Reports on Progress in Physics* **64**, 1533–1601 (2001).
35. Papantonakis, M. R., Kendziora, C., Furstenberg, R., Stepnowski, S. V., Rake, M., Stepnowski, J. and McGill, R. A. Stand-off detection of trace explosives by infrared photothermal imaging. *Proceedings of SPIE* **7304**, 730418 (2009).
36. Pushkarsky, M. B. et al. High-sensitivity detection of TNT. *Proceedings of the National Academy of Sciences of the United States of America* **103**, 19630–19634 (2006).
37. Chen, X. et al. Standoff photoacoustic detection of explosives using quantum cascade laser and an ultrasensitive microphone. *Applied Optics* **52**, 2626–2632 (2013).
38. Bauer, C., Willer, U. and Schade, W. Use of quantum cascade lasers for detection of explosives: progress and challenges. *Optical Engineering* **49**, 111–126 (2010).
39. Bai, Y. B., Slivken, S., Kuboya, S., Darvish, S. R. and Razeghi, M. Quantum cascade lasers that emit more light than heat. *Nature Photonics* **4**, 99–102, (2010).
40. Thundat, T., Van Neste, C. W., Senesac, L. R. and Krause, A. R. Photothermal sensing of chemical vapors using microcantilevers. In A. Korkin, P. S. Krstic, and J. C. Wells (eds), *Nanotechnology for Electronics, Photonics, and Renewable Energy*, 183–191. Springer, New York (2010).

41. Van Neste, C. W., Morales-Rodriguez, M. E., Senesac, L. R., Mahajan, S. M. and Thundat, T. Quartz crystal tuning fork photoacoustic point sensing. *Sensors and Actuators B: Chemical* **150**, 402–405 (2010).

42. Sharma, R. C. et al. Portable detection system for standoff sensing of explosives and hazardous materials. *Optics Communications* **309**, 44–49 (2013).

43. Christopher A. Kendziora, Robert Furstenberg, Michael Papantonakis, Viet Nguyen, Jennifer Stepnowski, R. Andrew McGill. Advances in standoff detection of trace explosives by infrared photo-thermal imaging, *Proceedings of SPIE* **7664**, 76641J (2010).

44. Corsi, C. History highlights and future trends of infrared sensors. *Journal of Modern Optics* **57**, 1663–1686 (2010).

45. Talghader, J. J., Gawarikar, A. S. and Shea, R. P. Spectral selectivity in infrared thermal detection. *Light-Science and Applications* **1**, e24 (2012).

46. Downs, C. and Vandervelde, T. E. Progress in infrared photodetectors since 2000. *Sensors* **13**, 5054–5098 (2013).

47. Rogalski, A. Progress in focal plane array technologies. *Progress in Quantum Electronics* **36**, 342–473 (2012).

48. Li, J. S., Chen, W. D. and Yu, B. L. Recent progress on infrared photoacoustic spectroscopy techniques. *Applied Spectroscopy Reviews* **46**, 440–471 (2011).

49. Krause, A. R., Van Neste, C., Senesac, L., Thundat, T. and Finot, E. Trace explosive detection using photothermal deflection spectroscopy. *Journal of Applied Physics* **103**, 094906 (2008).

50. Großer, J. et al. Modeling of the heat transfer in laser-heated small particles on surfaces. *International Journal of Heat and Mass Transfer* **55**, 8038–8050 (2012).

51. Furstenberg, R., Großer, J., Kendziora, C. A., Papantonakis, M. R., Nguyen, V. and McGill, R. A. Modeling of laser-analyte-substrate interaction in photo-thermal infrared imaging and laser trace vaporization. *Proceedings of the SPIE* **8013**, 801318 (2011).

52. Verkouteren, J. R., Coleman, J. L. and Cho, I. Automated mapping of explosives particles in composition C-4 fingerprints. *Journal of Forensic Sciences* **55**, 334–340 (2010).

53. Bohren, C. F. and Huffman, D. R. *Absorption and Scattering of Light by Small Particles*. Wiley-VCH, Weinheim, Germany (1983).

54. Hinkov, B., Fuchs, F., Kaster, J. M., Yang, Q., Bronner, W., Aidam, R. and Köhler, K. Broad band tunable quantum cascade lasers for stand-off detection of explosives. *Proceedings of SPIE* **7484**, 748406 (2009).

55. Phillips, M. C., Suter, J. D., Bernacki, B. E. and Johnson, T. J. Challenges of infrared reflective spectroscopy of solid-phase explosives and chemicals on surfaces. *Proceedings of SPIE* **8358**, 83580T (2012).

56. Mie, G. Articles on the optical characteristics of turbid tubes, especially colloidal metal solutions. *Annalen Der Physik* **25**, 377–445 (1908).

57. Furstenberg, R., Kendziora, C., Papantonakis, M., Viet, N. and McGill, A. Advances in photo-thermal infrared imaging microspectroscopy. *Proceedings of SPIE* **8729**, 87290H (2013).

58. Blake, T. A., Kelly, J. F., Gallagher, N. B., Gassman, P. L. and Johnson, T. J. Passive standoff detection of RDX residues on metal surfaces via infrared hyperspectral imaging. *Analytical and Bioanalytical Chemistry* **395**, 337–348 (2009).

59. Liu, Y.-S., Ugaz, V. M., North, S. W., Rogers, W. J. and Mannan, M. S. Development of a miniature calorimeter for identification and detection of explosives and other energetic compounds. *Journal of Hazardous Materials* **142**, 662–668 (2007).

60. Caygill, J. S., Davis, F. and Higson, S. P. J. Current trends in explosive detection techniques. *Talanta* **88**, 14–29 (2012).

61. Monat, J. E. and Gump, J. C. Shock compression of condensed matter, in M. L. Elert et al. (eds), *AIP Conference Proceedings*. 2009, Part 1 and 2. Volume 1195, 1309–1312, 28 June–3 July, Nashville, TN (2009).

62. Papantonakis, M. et al. Laser trace vaporization of trace explosive materials. *MRS Proceedings Fall 2011* **1405**, (2012).

63. McGill, R. A. et al. Analyte detection with infrared light. US Patent, US8421017 B2 (2013).
64. McGill, R. A. et al. Detection of chemicals with Infrared Light. US Patent, US8222604 B2 (2012).
65. McGill, R. A. et al. Detection of chemicals with infrared light. US Patent, US8101915 B2 (2012).
66. McGill, R. A. et al. Detection of chemicals with infrared light. US Patent, US8421018 B2 (2013).
67. Belyakov, V. V. and Bunker, S. N. Flash vapor sampling for a trace chemical detector. US Patent, US7098672 B2 (2006).
68. Bubb, D. M. et al. Resonant infrared pulsed-laser deposition of polymer films using a free-electron laser. *Journal of Vacuum Science & Technology A: Vacuum Surfaces and Films* **19**, 2698–2702 (2001).
69. Haglund, Jr., R. F., Belmont, R. J., Bubb, D. M., Dygert, N. L., Johnson, Jr., S. L. and Schriver, K. E. Mechanism of resonant infrared laser vaporization of intact polymers. *Proceedings of SPIE* **6261**, 62610V (2006).
70. Chrisey, D. B. et al. Laser deposition of polymer and biomaterial films. *Chemical Reviews* **103**, 553–576, (2003).
71. Longaker, P. R. and Litvak, M. M. Pertubation of refractive index of absorbing media by a pulsed laser beam. *Journal of Applied Physics* **40**, 4033–4041 (1969).
72. Davis, C. C. and Petuchowski, S. J. Phase fluctuation optical heterodyne spectroscopy of gases. *Applied Optics* **20**, 2539–2554 (1981).
73. Pichon, C., Leliboux, M., Fournier, D. and Boccara, A. C. Variable-temperature photoacoustic effect—Application to phase-transition. *Applied Physics Letters* **35**, 435–437 (1979).
74. Van Neste, C. W., Senesac, L. R. and Thundat, T. Standoff photoacoustic spectroscopy. *Applied Physics Letters* **92**, 134102 (2008).
75. Furstenberg, R., Kendziora, C. A., Papantonakis, M. R., Nguyen, V. and McGill, R. A. Photo-thermal spectroscopic imaging of MEMS structures with sub-micron spatial resolution. *MRS Proceedings Fall 2011* **1415**, (2012).
76. Furstenberg, R., Kendziora, C. A., Papantonakis, M. R., Viet, N. and McGill, R. A. Chemical imaging using infrared photothermal microspectroscopy. *Proceedings of SPIE* **8374**, 837411 (2012)
77. Farahi, R. H., Passian, A., Tetard, L. and Thundat, T. Pump-probe photothermal spectroscopy using quantum cascade lasers. *Journal of Physics D: Applied Physics* **45**, 125101 (2012).
78. Hildenbrand, J., Herbst, J., Wollenstein, J. and Lambrecht, A. Explosive detection using infrared laser spectroscopy. *Quantum Sensing and Nanophotonic Devices VI* **7222**, 72220B (2009).
79. Tam, A. C. Pulsed Photothermal Radiometry for Noncontact Spectroscopy, Material Testing and Inspection Measurements. *Infrared Physics* **25**, 305–313 (1985).
80. Lin, L. T., Archibald, D. D. and Honigs, D. E. Preliminary studies of laser-induced thermal emission-spectroscopy of condensed phases. *Applied Spectroscopy* **42**, 477–483 (1988).
81. Plutov, D. V. and Killinger, D. K. Modeling of spectral emission-based lidar remote sensing. *Proceedings of SPIE* **7665**, 76650D (2010).
82. Furstenberg, R. et al. Stand-off detection of trace explosives by infrared photo-thermal spectroscopy. *2009 IEEE Conference on Technologies for Homeland Security (HST)*, 465–471 (2009).
83. Kendziora, C. A. et al. Remote explosives detection (RED) by Infrared photothermal imaging. *MRS Proceedings Fall 2011* **1405**, (2012).
84. Kendziora, C. A., Furstenberg, R., Papantonakis, M., Nguyen, V., Borchert, J., Byers, J. and McGill, R. A. Infrared photothermal imaging of trace explosives on relevant substrates. *Proceedings of SPIE* **8709**, 870900 (2013).
85. Kendziora, C. A., Jones, R. M., Furstenberg, R., Papantonakis, M., Nguyen, V. and McGill, R. A. Infrared photothermal imaging for standoff detection applications. *Proceedings of SPIE* **8373**, 83732H (2012).

.

11 Detecting Explosives and Chemical Weapons Using Cavity-Enhanced Absorption Spectrometry

J. Brian Leen, Manish Gupta, and Douglas S. Baer

CONTENTS

11.1 INTRODUCTION

The rapid and accurate detection of vapor-phase explosives and chemical warfare agents (CWAs) requires extremely sensitive analytical technologies. The low vapor pressure of common explosives means that exceedingly small concentrations are likely to be found in the ambient atmosphere even under the best of conditions; when the explosive is intentionally concealed, detection becomes nearly impossible.[1] For this reason, safety specialists desire the lowest possible detection limits for explosives. The required detection limits for CWAs are more favorable, because their use involves intentional dispersal at high parts-per-billion (ppb) concentrations or more, and these compounds typically exist in the vapor phase. However, CWA detection must be performed rapidly (within minutes), because affected personnel need to don personal protective equipment as quickly as possible. From an operational standpoint, both explosive and CWA measurement methods must not produce frequent false positives due to the high cost of responding to false alarms.

The absorption of light by molecules is an attractive method for detection of explosives and CWAs. Foremost, optical absorption measurements are generally quantitative, because a measurement of transmitted optical intensity is directly proportional to the concentration of target analytes via Beer's law.[2] Additionally, frequency-resolved absorption spectroscopy yields extensive information about molecular structure.[3] Together, these qualities allow spectroscopists to identify and quantify small concentrations of target analytes in complicated background gas matrices. This selectivity is particularly effective in the mid-infrared (MIR) optical spectral region (between 5 and 12 µm), a spectral window known as the *molecular fingerprint* region.

Unfortunately, conventional optical absorption spectrometry methods are not capable of detecting very low analyte concentrations. Thus, a method of increasing the optical absorption is required to provide a measurable signal. Cavity-enhanced absorption spectroscopy (CEAS) serves this purpose by passing the light beam through the gas sample multiple times, thus enabling measurements of very weak molecular absorptions that are not possible to measure with conventional absorption spectroscopy (CAS). Although many configurations of CEAS have been devised and successfully implemented, the most widely used is cavity ring-down spectroscopy (CRDS), which is the focus in this chapter. (Note that the term CEAS has sometimes been applied only to a subset of cavity-enhanced methods.[4] Here, the term is used in the most general sense: to denote all absorption spectroscopies that use a high-finesse optical cavity to enhance the effective optical pathlength.)

CRDS was first used by O'Keefe and Deacon to measure weak, "forbidden" transitions of O_2 and the conclusion of that work[5]—that high-finesse optical cavities can be used to improve the sensitivity of absorption spectroscopy—has resulted in over a thousand publications and the development of numerous alternative CEAS configurations. Applications for CEAS include fundamental analytical chemistry,[6,7] astrochemistry,[8] atmospheric chemistry,[9–12] isotopic studies,[13,14] industrial process monitoring,[15] medical diagnostics,[16–19] explosives detection,[20–24] and chemical weapon agents detection.[25] CEAS has been performed with a myriad of light sources: pulsed and continuous lasers, lamps,[26] light-emitting diodes (LEDs),[27] and super-continuum sources.[28] The unifying feature of CEAS is a high-finesse optical cavity that serves to trap the probe light so that it passes through the sample gas multiple times as it is reflected by the cavity mirrors. The resulting effective optical pathlength is very long, increasing the probability of optical absorption in the gas.

Although CEAS can be used to measure very small concentrations (down to parts per quadrillion[29]), it cannot generally match the sensitivity of mass spectrometry. However, CEAS has a more robust experimental setup (e.g., no high vacuum) and measures absolute absorptions, thus allowing a straightforward and accurate quantification of analyte concentrations without extensive calibration. Additionally, it can be applied to nonionizing species and can differentiate between isomers (e.g., m,o,p-xylene) and isotopomers (e.g., the nitrogen (N) isotopomers $^{15}N^{\alpha}$ and $^{15}N^{\beta}$ of nitrous oxide (N_2O)[30]).

This chapter first covers the principles of conventional optical absorption spectroscopy and the basic equations governing CRDS. The remainder of the chapter focuses on using CEAS for explosives and CWA detection, with a special emphasis on multimode CRDS (MM-CRDS).

11.2   CONVENTIONAL ABSORPTION SPECTROSCOPY (CAS)

Due to their specific electronic structure, molecules absorb light only at very specific frequencies. Almost all covalently bonded molecules with an electric dipole moment will absorb light in the infrared (IR) portion of the spectrum. This MIR optical absorption typically excites a vibrational mode of the molecule. Importantly, MIR spectroscopic measurements are inherently selective, with even closely related molecules exhibiting entirely unique optical absorption spectra. Thus, these spectra can be used to "fingerprint" molecules in a complicated mixture.

CAS describes an experimental setup where light is directed through an absorbing medium and the intensity of the light is measured before and after passing through the medium to determine the amount of light absorbed. Frequency-resolved measurements provide an absorption spectrum that can be interpreted to yield the concentrations of multiple species in the mixture. The absorption process is described by the Beer–Lambert–Bougher law (or more commonly Beer's law), which states that light passing through an absorbing medium experiences an exponential decrease in the transmitted optical intensity with the absorbing pathlength:[2]

$$I = I_0 e^{-\alpha(\nu)l}, \tag{11.1}$$

where:

I = measured optical intensity after transmission through the medium
I_0 = incident intensity
l = absorption pathlength
$\alpha(\nu)$ = absorption coefficient of the medium at frequency ν

The absorption coefficient can be expressed as $\alpha(\nu) = S\chi(\nu)p_a$, where S is the transition line strength, $\chi(\nu)$ is the area-normalized lineshape function at frequency ν, and p_a

is the partial pressure of the absorber. If multiple absorbers are present, the total absorption coefficient, $\alpha_t(\nu)$, is the sum of all the absorbers: $\alpha_t(\nu)+\alpha_2(\nu)+...+\alpha_i(\nu)$. This additive property of absorption is important in complicated mixtures, where the absorption of a target analyte is likely to overlap with that of other constituents (see Section 11.4.1.3).

11.2.1 COMMON TYPES OF CONVENTIONAL ABSORPTION SPECTROSCOPY

There are many common measurement methods that use Beer's law.[31] For example, nondispersive infrared (NDIR) spectroscopy involves using a filtered lamp that provides a fixed wavelength and measuring the optical absorption after propagation though a gas sample. Any change in transmitted intensity is attributed to the target specie, and the concentration is calculated using Beer's law.[32] Multiple species can be measured by changing the lamp filter to provide a different probe wavelength.

In tunable diode laser absorption spectroscopy (TDLAS),[33–35] a laser is tuned over the absorption features of a target molecule to yield a high-resolution spectrum. The measurement of a spectrum allows very accurate speciation in mixtures of gases and is frequently used in industrial process control, where an analyte is present in a mixture of other absorbers.[36] TDLAS is sometimes combined with frequency modulation (e.g., wavelength modulation spectroscopy) to avoid 1/f noise and thus improve the sensitivity by one or two orders of magnitude.[37–39]

Finally, Fourier transform infrared (FTIR) spectroscopy is widely used and has been applied to the detection of CWAs[40–42] and explosives forensics.[43] FTIR utilizes a broadband light source (e.g., lamps, thermal bars) that typically operates over the spectral range of 400–4000 cm^{-1}. As this is a large wavelength range that spans the entire MIR molecular fingerprint region, FTIR excels at identifying analytes in complex, real-world mixtures, where there is substantial overlap of absorption features from different gases. Analyte concentration is typically extracted from FTIR spectra using least-squares or chemometric fitting algorithms that are very similar to those used in Sections 2.4 and 2.5.[44]

11.2.2 LONG-PATH CONVENTIONAL ABSORPTION SPECTROSCOPY

The simplest way to improve the sensitivity of CAS is to extend the absorption pathlength by increasing the distance between the light source and detector. Due to physical constraints, this is a limited approach, and, for portable or benchtop instruments, the pathlength rarely exceeds 1 m. However, for some outdoor applications, the detector can be placed far from the light source, allowing the system to probe kilometers of air. This method has been used with FTIR for the monitoring of fence lines[45,46] and atmospheric constituents.[44,47,48] Whereas this technique can be used to detect broadly distributed, toxic compounds,[41] it is not suited to the detection of discrete point sources (e.g., explosives). Additionally, the improvement in sensitivity achieved with open-path methods is always partially negated by the increased noise caused by atmospheric turbulence and transient scattering sources such as water vapor,[49] smoke,[50] dust, pollen, insects, etc., which can change the baseline absorption in unpredictable ways.[51]

Another method of increasing the effective optical pathlength is to reflect the light beam through the sample volume multiple times using a set of mirrors. Herriott[52] and White[53] cells offer pathlengths of hundreds of meters in a benchtop instrument, with excellent sensitivity.[54,55] Other configurations, such as ring reflectors, also exist.[56] However, instruments with hundreds of passes require precise alignment and are highly sensitive to vibration and temperature fluctuations;[57] thus they cannot attain the sensitivity possible with the best cavity-enhanced methods.[58]

11.2.3 CONVENTIONAL ABSORPTION SPECTROSCOPY SENSITIVITY LIMITS

Because S and $\chi(\nu)$ are physical properties of each analyte, trace concentrations (i.e., low p_a) result in very small absorptions. Clearly, if $\alpha(\nu)$ of Equation 11.1 is small, I will be very close to I_0 and will thus be undetectable due to measurement noise (e.g., detector noise, light source fluctuations, etc.). As a result, the sensitivity of CAS is limited by the ability to measure very small changes in light intensity $(\Delta I = I - I_0)$ and is often described by the minimum resolvable ΔI. For well-designed, field-deployable systems, $\Delta I_{min}/I_0$ is 1×10^{-5}, which corresponds to an αl product of 1×10^{-5}.[51,59,60] Unfortunately, sensitivity in many systems is limited by etalons (or fringes) caused by interference within and between optical elements, often placing ΔI between 1×10^{-3} and 1×10^{-4}.[61,62] The $\Delta I/I_0$ metric is useful for CAS systems because the optical pathlength is usually defined by the deployment location. Paldus and others have outlined the minimum detectable absorption loss (MDAL) as a common metric for comparing the sensitivity of various absorption spectroscopies.[59] For CAS, the MDAL is given by Equation 11.2:

$$\alpha_{min} \cong \frac{\Delta I_{min}}{I_0 l}. \tag{11.2}$$

Thus, a conventional absorption spectrometer with $\Delta I/I_0 = 1 \times 10^{-5}$ and a 1 m pathlength will have an MDAL of 1×10^{-7} cm^{-1} at a signal-to-noise-ratio (SNR) of 1:1. For those wishing to detect threat-level quantities of explosives and CWAs in the vapor phase, the MDALs of conventional absorption spectroscopy are inadequate (see Sections 11.4 and 11.5); a more sensitive method is required that dramatically increases the effective optical pathlength.

11.3 CAVITY-ENHANCED ABSORPTION SPECTROSCOPY

CEAS uses highly reflective dielectric mirrors to reflect probe light through a sample many times, much like White and Herriott cells. However, by passing the light through the mirror instead of through holes drilled in the mirror, CEAS provides thousands of passes and effective optical pathlengths of several kilometers in a benchtop instrument.

CRDS was the first CEAS. Subsequent implementations have been developed that have resolved certain limitations of CRDS. There are many excellent reviews covering the theory and applications of CRDS more extensively.[4,6,9,63–71]

CRDS originated in the early 1980s from the need to quantitatively measure dielectric mirrors with reflectivities (R) near unity (e.g., $R > 99.99\%$).[72,73] By measuring the rate at which light escapes from a high-reflectivity optical cavity, it is possible to accurately determine the optical loss within the cavity and thus calculate the very small losses due to the mirror transmission and absorption. A simplistic model of this, sometimes referred to as the photon bullet model, is shown in Figure 11.1.

The rate at which light decays out of the cavity can be derived using the same method used to obtain Beer's law. The loss through the mirrors for one round trip (assuming they are identical) is shown in Equation 11.3:[74,75]

$$2(1 - R) = 2(T + A + S), \tag{11.3}$$

where:
R = mirror reflectivity
T = mirror transmission
A = absorption within the cavity
S = scattering from the mirror surface.

This loss occurs over the round trip time of the cavity, which is given by Equation 11.4:

$$2Ln / c, \tag{11.4}$$

where:
L = cavity length
n = refractive index within the cavity
c = speed of light

The differential loss is then given by Equation 11.5:

$$\frac{dI}{dt} = -\frac{c(1 - R)}{nL} I = -\frac{c(T + A + S)}{nL} I, \tag{11.5}$$

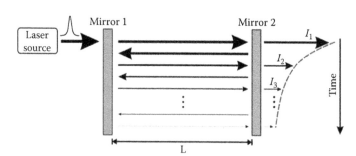

FIGURE 11.1 Schematic representation of a cavity ring-down event. A laser pulse is injected into the cavity. The first light to pass out of the cavity to the right has a high intensity; each subsequent reflection loses intensity via transmission through the mirror and absorption in the cavity, resulting in an exponentially decreasing amplitude over time.

where I is the power in the optical cavity.

The solution for Equation 11.5 is

$$I(t) = I_0 e^{-\frac{c}{nL}(1-R)t} ,$$ (11.6)

where the empty cavity decay time or ring-down time is given by

$$\tau_0 = \frac{nL}{c(1-R)} .$$ (11.7)

The optical loss upon reflection is frequently described as a portion of the light lost, for example, if 100 of every 1 million photons are transmitted, absorbed, or scattered, then the mirror is said to have a loss of 100 parts per million. A typical ring-down cavity will have a length of 10 cm^{-1} m and mirror loss $(1 - R$ or $T+A+S)$ of 10–1000 ppm (R=99.999–99.9%), although mirror losses of 1 ppm (R=99.9999%) have been reported.[76] As an example, a 30 cm cell with 50 ppm mirrors will have a ring-down time of 20 μs. A sample ring-down trace taken in the MIR, where the mirror losses are typically much higher due to material absorptions in the mirror dielectric stack (200–1000 ppm), is shown in Figure 11.2. This cavity is 80 cm long, and the measured ring-down is ~3.3 μs, which corresponds to a mirror loss of roughly 800 ppm.

O'Keefe and Deacon realized that if optically absorbing species were placed within the cavity, the additional optical losses could be measurable with extreme sensitivity and that this optical enhancement could be used as a spectroscopic tool for the measurement of minute absorbers.[5] With a uniformly distributed, linear absorber (i.e., nonsaturated) in the cavity, the rate at which light leaks out is now described by

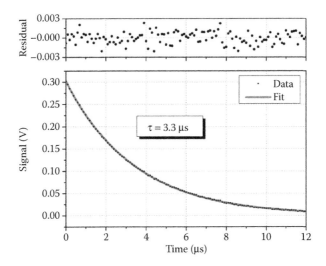

FIGURE 11.2 A sample average of 10 ring-down events measured at $v = 960$ cm^{-1}. The measured voltage is fit to a single exponential, and the fit residual is shown in the top panel.

$$\frac{dI}{dt} = -\frac{c}{nL}\left[(1-R)+\alpha l\right]I,$$ (11.8)

where:
  α = same per-length absorption found in Equation 11.1
  l = length of the absorber within the cavity (which does not need to equal the cavity length, L)

The solution to Equation 1.8 describes the cavity lifetime with an additional absorber:

$$I(t) = I_0 e^{-\frac{c}{nL}\left[(1-R)+\alpha l\right]t}.$$ (11.9)

The ring-down time is then

$$\tau = \frac{nL}{c\left[(1-R)+\alpha l\right]}.$$ (11.10)

In the case of CWAs and explosives, there will be other absorbers in the spectrum and, as before, α can be represented as a sum of the absorbers at the probe wavelength. Additionally, the mirror reflectivity varies with wavelength, and the ring-down time can be written as

$$\tau(\nu) = \frac{nL}{c\left[(1-R(\nu))+\sum_i \alpha_i(\nu)l\right]}.$$ (11.11)

Experimentally, the value of τ is obtained by fitting the measured ring-down voltage to a decaying exponential using the Levenberg-Marquardt (LM) nonlinear minimization algorithm.[77] In the past, less computationally intensive fitting methods have been explored; however, the LM algorithm has been shown to be optimal for ring-down fitting,[78] and computing power is now sufficient for nearly all applications (e.g., an optimized LM fitting algorithm runs at 10 kHz in the author's lab on a standard PC with a processor clocked at 2.6 GHz). Thus, LM fitting is now the method of choice for most applications.

The total absorption attributable to the intracavity absorber ($\alpha_t(\nu) = \sum_i \alpha i(\nu)$) is then calculated according to Equation 11.12:

$$\alpha_t(\nu) = \frac{L}{cl}\left[\frac{1}{\tau} - \frac{1}{\tau_0}\right],$$ (11.12)

where τ_0 is the ringdown time of the cavity without optical absorbers ($\alpha = 0$).

The differential detection limit is again defined as $\Delta\tau/\tau_0 = (\tau_0 - \tau)/\tau_0$, which is typically measurable to better than 1% and down to 0.001%.[79] The MDAL for CRDS can be expressed as

$$\alpha_{min}(\nu) = \frac{1 - R(\nu)}{l}\left(\frac{\Delta\tau}{\tau_0}\right)_{min}. \qquad (11.13)$$

For the example MIR system at 960 cm^{-1} with an 80 cm cell, $1 - R = 800$ ppm, and assuming a $(\Delta\tau/\tau_0)_{min} = 0.1\%$, $\alpha_{min}(\nu)$ is 1×10^{-8} cm^{-1}. For comparison, for a 30 cm cell with 50 ppm mirrors (which is more representative of near-infrared (NIR) CRDS systems), the same $(\Delta\tau/\tau_0)_{min}$ yields an $\alpha_{min}(\nu)$ of 1×10^{-9} cm^{-1}. Even at the modest example $(\Delta\tau/\tau_0)_{min}$ of 0.1%, both of these MDALs are significantly better than the optimized CAS system with an MDAL of 1×10^{-7} cm^{-1} given in Section 11.2.3. Note that to facilitate direct comparison between methods, the $\sqrt{N}$ improvement with increased averaging is usually incorporated into the MDAL, which is given in units of cm^{-1}/$\sqrt{Hz}$.

The primary reason for the increased sensitivity is the long effective optical pathlength of the cavity L_{eff}, which is given by

$$L_{eff} = \tau c. \qquad (11.14)$$

In the MIR case ($\tau = 3.3$ μs), $L_{eff} = 990$ m, while in the NIR case ($\tau = 20$ μs), $L_{eff} = 6$ km! These very long effective CEAS pathlengths provide impressive sensitivity in a benchtop instrument. The best reported CEAS sensitivity to date is 10^{-14} cm^{-1}/$\sqrt{Hz}$,[80] and sensitivities of 10^{-10}– 10^{-11} cm^{-1}/$\sqrt{Hz}$ are routinely achieved.[59]

Another potential benefit of CRDS is that the measurement is fundamentally time based, and the measurement of τ does not depend on the laser intensity. Thus, CRDS was first successfully used with pulsed lasers, whose highly variable pulse amplitudes render CAS measurements almost impossible. In current research, this advantage is exploited with recently introduced broadly tunable, external-cavity quantum cascade lasers (EC-QCL[81–84]). Although these MIR lasers have a pulse-to-pulse amplitude stability of only about 1%, they produce average powers in the milliwatt range (with pulse peak powers of hundreds of milliwatts) and are capable of tuning over a wavelength range of >250 cm^{-1} with a single gain chip. They are routinely multiplexed to operate continuously from 5 to 14 μm. These lasers are especially well suited for detecting CWAs, whose absorption features are typically tens of wavenumbers wide, which in the past were only measureable by spectrally dim broadband sources (e.g., lamps) or by specialized optical parametric oscillators (OPOs),[20] which are generally not viable in field instruments. The extraordinary tuning range of EC-QCLs has made it feasible to speciate CWAs and explosives in real-world mixtures with the extraordinary MDALs of CEAS.

Despite its ability to measure very small absorptions, the dynamic range of CRDS systems is fairly limited. As the absorption within the cavity increases, the ring-down time decreases and eventually becomes unmeasurably short (referred to as dark-out) due to the laser turn-off time, response time of the detector, or data digitization rate.

Notably, cavity-enhanced transmission methods such as integrated cavity output spectroscopy (ICOS) do not suffer from this problem and thus have a much larger dynamic range. For all CEAS, the sensitivity to small absorbers decreases when superimposed on large absorbers, because the large absorber decreases the effective optical pathlength. This often occurs in real sensing situations where background components and the limited dynamic range of CRDS could potentially be used to mask the presence of contraband material. For instance, acetone has overlapping absorptions with TATP near 1200 cm^{-1}, and a small, leaky container of acetone (e.g., nail polish remover) could blind a CRDS sensor in this region. Fortunately, this is not a fatal weakness, but the CRDS spectrometer must measure over a large wavelength range to record multiple absorption bands of target analytes to avoid crude blinding.

11.3.1 CAVITY TRANSMISSION AND MODE FORMATION

Coupling of light to the cavity is determined by the cavity's mode structure.[85] The longitudinal and transverse cavity modes have significant ramifications for CEAS instruments. Mode formation in optical resonators is discussed extensively in [75] and, with respect to CRDS, in chapter 4 of [70] and chapter 2 of [71]. Its properties and importance are only outlined here.

11.3.1.1 Longitudinal Modes

The first condition for light admission into the cavity is constructive interference along the length of the cavity. These resonances are called longitudinal modes and occur at regularly spaced intervals when the cavity length is equal to an integer number of half wavelengths of the incoming light. Peaks in transmission through the cavity due to these modes will appear as either the cavity length or laser wavelength is changed, causing overlap between the cavity mode and the laser frequency (Figure 11.3, overlapped with mode q00). The separation between modes is called the cavity-free spectral range (FSR) and is given by the inverse of the cavity round-trip time:

$$\Delta v_{FSR} = \frac{c}{2nL} . \qquad (11.15)$$

11.3.1.2 Transverse Modes

Stable optical resonators also support specific transverse modes. The lowest-order mode has a Gaussian cross section and is referred to as the TEM$_{00}$ mode. Higher-order modes are typically expressed in the Hermite-Gaussian basis set and designated as TEM$_{nm}$. Critically for CEAS, different transverse modes travel slightly different pathlengths within the cavity, incorporating more off-axis propagation with increasing mode order, or, more precisely, they accumulate additional phase as they travel through the cavity's focus (see Guoy's phase in Ref. 75). The different phase accumulation manifests as frequency displacement for higher-order transverse modes, as shown in Figure 11.3, and is described by Equation 11.16:

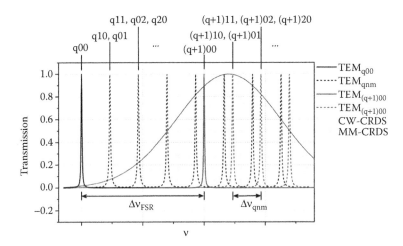

FIGURE 11.3 Schematic representation of CEAS mode structure. Note that lower q, higher n,m modes can overlap with higher q, lower n,m modes to produce a very dense mode structure. The laser bandwidth of conventional mode-matching is shown in green under q00. The laser bandwidth of MM-CRDS is shown as the broad blue envelope and covers many cavity modes.

$$v_{qnm} = \frac{c}{2L}\left[q + \frac{n+m+1}{\pi}\arccos\left(\sqrt{g_1 g_2}\right)\right], \tag{11.16}$$

where:

q = longitudinal cavity mode

n and m specify the transverse mode number

g_1 and g_2 = cavity g-parameters of mirrors 1 and 2, respectively (again, see[75])

g-parameters are commonly used in the analysis of laser mode structure and are defined as

$$g = 1 - \frac{L}{R} \tag{11.17}$$

where:

L = cavity length

R = radius of curvature of the mirror

The intertransverse mode spacing is determined by the cavity g-parameters and can overlap with higher q modes. Notably, a cavity with a $g_1 g_2$ product of 1 (e.g., concentric) will have a transverse mode spacing that is degenerate with the TEM_{00} modes. Because of the differing phase accumulation, excitation of multiple transverse modes results in mode beating, causing a superimposed oscillation on the ring down and making it difficult to accurately determine τ.[86] If these higher-order modes are to be avoided, the spatial profile and wavefront curvature of the light incident on the cavity must match the cavity's Gaussian TEM_{00} profile—referred to as mode-matching. Off-axis integrated cavity output spectroscopy (OA-ICOS) and MM-CRDS (see Section 11.4.1) are two important methods that *do not* require mode-matching. Instead, many transverse modes are intentionally excited by properly selecting g_1g_2 and introducing the laser at an angle and off the central axis of the cavity. This configuration produces a very narrow longitudinal mode spacing relative to the laser bandwidth, resulting in uniform cavity transmission and minimal mode beating.[87–89] The two approaches are represented in Figure 11.3.

11.3.2 Types of CEAS

There are many types of CEAS, and dedicated review articles, with more detailed treatments, exist for many of them.[65] The first and experimentally simplest type of CEAS is pulsed-CRDS (P-CRDS), where a laser pulse is introduced into the cavity and the ring-down is measured.[5] If the pulse duration is much greater than the cavity round-trip time, then a continuous ring down is observed. Efforts are typically made to match the source laser to the transverse electromagnetic (TEM_{00}) cavity mode, although MM-CRDS is an exception. To date, P-CRDS has achieved a maximum sensitivity of $\sim 10^{-10}\,cm^{-1}/\sqrt{Hz}$.[59] Continuous-wave CRDS (CW-CRDS) takes advantage of the improved laser stability and narrow linewidth of CW laser operation. It achieves higher sensitivities and higher spectral resolution than P-CRDS.[90] With CW-CRDS, the laser power is generally turned off using external modulation (e.g., acousto-optic or electro-optic modulators) to observe the ring-down event. The best sensitivities are achieved when the laser is locked to a single longitudinal mode, and this technique can reach a sensitivity of $\sim 10^{-12}\,cm^{-1}/\sqrt{Hz}$[91]. Noise-immune cavity-enhanced optical heterodyne molecular spectroscopy (NICE-OHMS) uses a cavity plus frequency modulation to suppress laser frequency noise resulting in a sensitivity of $\sim 10^{-14}\,cm^{-1}/\sqrt{Hz}$, the best of any CEAS technique.[80] Fourier transform-CRDS (FT-CRDS) pairs the enhanced sensitivity of CRDS with the spectral multiplexing of FTIR.[92] There are several variations of broadband CRDS (BB-CRDS)[93] using lamps,[26] LEDs,[94] super-continuum sources,[28,95,96] and millimeter waves.[97] Cavity-attenuated phase-shift spectroscopy (CAPS) measures the phase shift imparted on the transmitted signal to estimate the ring-down time using only a lock-in amplifier.[72,98] Finally, OA-ICOS takes the approach of measuring the transmitted intensity of the laser through the cavity.[88,99] The change in intensity in the presence of absorbing species is given by Equation 11.18:

$$\frac{\Delta I}{I_0} = \frac{GA}{1+GA}. \tag{11.18}$$

where:

A = single-pass absorption, $A = 1 - e^{-\alpha(v)l}$
G = gain factor, $G = R/(1-R)$

By introducing the laser beam off-axis, many transverse modes are excited, which leads to limited but continuous uniformly transmitted power that is robust to vibrations and temperature and pressure fluctuations. OA-ICOS has achieved a measurement sensitivity of 2.4×10^{-11} cm^{-1}/$\sqrt{\text{Hz}}$.[100]

11.4   DETECTION OF CWAs WITH CEAS

The spectroscopic detection of CWAs and explosives is especially challenging due to the very low required detection limits. For CWAs, the Centers for Disease Control and Prevention (CDC) lists a concentration of 15 ppb (0.1 mg/m^3) for GA (tabun) and 17.3 ppb (0.1 mg/m^3) for GB (sarin) as immediately dangerous to life and health (IDLH) for continuous exposure over a 30-min period (US Army recommendations from 1988 are double these values).[101] VX is much more potent, and the CDC IDLH recommendation is 9.1 ppb (3 μg/m^3). The short-term exposure limit (STEL) for these compounds is three orders of magnitude lower. Noticeable health effects occur at lower concentrations and have long-term consequences for victims.[102] For example, sarin causes miosis at 5.8 ppb (33.3 μg/m^3)[103] and thus requires a detector capable of sensing concentrations well below this value.

Using the MDAL of Equations 11.2 and 11.13 in conjunction with a target analyte's absorption cross section, it is straightforward to calculate a best-case limit of detection (LOD), typically expressed at an SNR ratio of 3:1. Peak absorption values ((α_i (ν)) of Equation 11.11) for CWAs in the MIR are approximately 3×10^{-5} [μmol/mol·cm]$^{-1}$,[104] which corresponds to a LOD of about 9 ppb for CAS (MDAL = 10^{-7} cm^{-1}/$\sqrt{\text{Hz}}$). This detection limit is too high when compared to the IDLH values (9–17 ppb) to provide a low false-positive detection rate. For our example, the MIR CRDS system has an MDAL of 10^{-8} cm^{-1}/$\sqrt{\text{Hz}}$ and a LOD of 0.9 ppb.

Although there have been several publications detailing the detection of CWAs with photo-acoustic spectroscopy,[103,105–107] there have been very few publications describing instruments designed to detect CWAs using CEAS. This is because the strongest absorption features for most chemical weapons are broad compared to the tuning range of distributed feedback lasers (DFB) and occur between 8 and 12 μm, where lasers are difficult to manufacture. Without the ability to tune beyond the target absorber, it is impossible to differentiate between targets, common atmospheric constituents, and innumerable interferents (e.g., gasoline, cleaning products, perfume, bug spray, etc.). However, with newly available ECQCLs capable of tuning over many microns in the MIR wavelength region, the detection of chemical weapons in real-world air at levels needed to protect civilians and military personnel (<5 ppb) has become a possibility. Primarily, these lasers have been used for standoff or long-path CAS detection,[108–110] but they are well suited for use in CRDS. Qu et al. have demonstrated the use of EC-QCLs in conjunction with cavity ring-down spectroscopy for the detection of the nerve agent simulant dimethyl methylphosphonate (DMMP) with a modest MDAL of 2×10^{-7} cm^{-1} and a detection limit of 77 ppb.[25]

11.4.1   MULTIMODE CAVITY RING-DOWN SPECTROSCOPY (MM-CRDS)

Figure 11.4 shows a schematic diagram of an MM-CRDS system developed at Los Gatos Research for the ultrasensitive, real-time detection of toxic volatile organic compounds (VOC), CWAs, explosives, and other MIR-absorbing molecules. The ECQCL operates in pulsed mode with pulse widths and peak powers of 500 ns and 10–250 mW, respectively. It is tunable from 7–14 μm with a spectral bandwidth of 0.5 cm⁻¹. The usable spectral range is presently limited by the mirror reflectivity, and the system incorporates custom broadband dielectric coated mirrors. Two sets of these mirrors have been tested and determined to have reflectivities of 99.89%–99.94% from 7.9 to 10.4 μm, and 99.67%–99.92% from 9.4 to 11.6 μm. Above 11.6 μm and at higher temperatures (>100°C), the mirror absorption losses become too large for CEAS.[111] A step-and-measure scheme is used to collect CEAS spectra, requiring 4–7 min to scan the entire mirror band. The cavity is 80 cm long, which corresponds to a $\Delta\nu_{FSR}$ of 187 MHz (0.006 cm⁻¹). As this is well below the 0.5 cm⁻¹ laser pulse width, longitudinal mode-matching is impossible and many cavity modes are excited by each laser pulse. The resulting, highly multiexponential ring-down provides an accurate measurement of optical loss.[66] In the absence of resonant mode-matching, >99% of the light is reflected from the front mirror. In order to capture some of this lost energy, a mirror with a small hole drilled off-center is used to reinject the reflected beam and thus increase the power circulating within the cavity.[112,113] The power transmitted through the exit mirror is detected with a liquid nitrogen–cooled mercury cadmium telluride (MCT) detector. Thermoelectrically cooled MCT detectors are also available to improve the fieldability; however, they are roughly 10 times less sensitive, resulting in substantially degraded performance.

In order to increase the vapor pressure of the measured compounds, the system is divided into three temperature zones. The first temperature zone (T_1) contains a sample oven containing the specie to be measured at the saturated vapor pressure. The second zone (T_2) contains transfer tubing, a flow restricting orifice, and a three-way valve switching between sample gas and zero air (used for the measurement

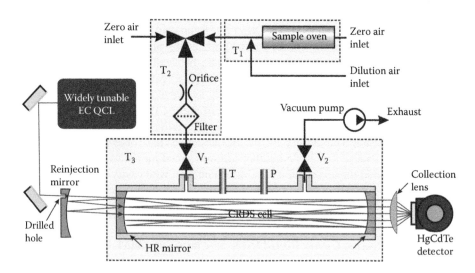

FIGURE 11.4 A diagram of the MM-CRDS instrument.

of τ_0). The third zone (T_3) consists of the wetted cavity surfaces and is held at a temperature greater than T_1 to prevent condensation. Gas is admitted into the cavity through V_1 and spectral measurement is performed after isolating the sample gas by closing V_1 and V_2.

11.4.1.1 Multimode Cavity Ring-Down Spectroscopy Minimum Detectable Absorption Loss (MM-CRDS MDAL)

The MDAL of the MM-CRDS system is obtained by calculating the precision of repeated measurements of τ_0 as shown in Figure 11.5. Over most of the mirror band, these measured values exceed $1 \times 10^{-8} \, \mathrm{cm^{-1}/\sqrt{Hz}}$ and approach $1 \times 10^{-9} \, \mathrm{cm^{-1}/\sqrt{Hz}}$ in the band center. The absence of mode-matching results in lower power on the detector and the system is limited in part by amplifier noise. Additionally, the MDAL is impacted by the irreproducibility of the laser tuning; the MDAL measured at a single wavelength (1100 cm^{-1}) is $2 \times 10^{-10} \, \mathrm{cm^{-1}/\sqrt{Hz}}$, an order of magnitude better than when the laser is tuned, indicating that improvements to the laser tuning mechanism will dramatically improve the instrument precision.

DMMP and diisopropyl methylphosphonate (DIMP) are commonly used, low-toxicity nerve agent simulants for G-agents[114,115] due to their similar physical properties. The optical absorption spectra of DMMP and DIMP are comparable in intensity to G-agents ($\sim 3 \times 10^{-5}$ [μmol/mol·cm]$^{-1}$) and have similar spectroscopic structures and locations.[104] The MM-CRDS absorption spectra of these two simulants are shown in Figure 11.6 and closely match previously measured spectra.[105]

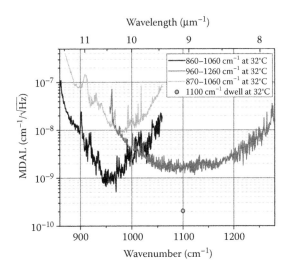

FIGURE 11.5 The measured MDAL of the Los Gatos Research MM-CRDS system calculated by repeated measurements of τ_0 for two broadband mirror sets. The MDAL of the long wave mirror is also measured at 100°C (light gray), where the increased optical losses in the dielectric mirror cause an increase of almost 10× in MDAL. The MDAL measured by dwelling at 1100 cm^{-1} is shown for comparison (gray dot).

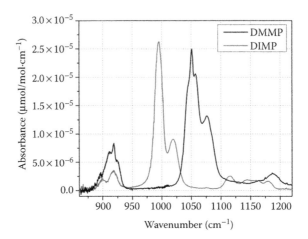

FIGURE 11.6   Measured absorption spectra for the CWA simulants DMMP and DIMP in ultra-zero air at 600 Torr. The spectra are composites taken with two different mirror sets, separately covering 860–1020 cm^{-1} and 1020–1220 cm^{-1}. The absolute concentrations were 360 ppb of DMMP and 410 ppb of DIMP.

11.4.1.2   Analyzing MM-CRDS Spectra

Most CEAS techniques target small molecules (2–4 atoms) with NIR absorption spectra that are well described by a few absorption peaks and can be fit using Voigt functions.[10,88] For larger molecules in the MIR, the rovibrational transitions become closely spaced, causing the absorption features to merge into a continuous absorption band covering many wavenumbers, as in Figure 11.6. Some of the individual line structures can be recovered by operating at low pressures to decrease collisional broadening, but this sacrifices sensitivity by reducing the total number of light-absorbing molecules in the cavity. In order to quantitatively extract the concentration from broadband absorption spectra, it is necessary to fit the measured spectra with a linear combination of known spectra from a library. These premeasured spectra are commonly called basis sets and many libraries exist for FTIR spectral analysis.[115] For example, Pacific Northwest National Laboratories (PNNL) maintains a high-resolution library (0.1 cm^{-1}) with over 400 common compounds (http://nwir.pnl. gov).[117,118] A constrained, classical least-squares minimization can be used to identify the linear combination of library spectra that best fits the measured spectrum of MM- or broadband CRDS. The best fit is found by minimizing the set of equations produced by Equation 11.19:

$$\alpha(\nu) = \sum_{l}^{N} X_{l}\alpha_{l}(\nu).$$
(11.19)

where:
$\alpha(\nu)$ = measured absorption at wavelength ν
N = number of species included

X_l = concentration coefficient of library spectrum l
$\alpha_l(\nu)$ = absorption coefficient of library spectrum l at wavelength ν

Temperature and pressure corrections according to the ideal gas law can also be applied, although, for large temperature and pressure ranges, multiple basis sets must be used.[118] The coefficients X_l produced from the minimization of Equation 11.19 directly yield the mole fraction of each library compound.

Chemometric fitting methods[120] can improve on the analysis of complicated MIR spectra by intelligently limiting the number of included basis sets and wavelengths used.[44,120] Mukherjee et al. showed that a manually selected spectrum subset that avoids nontarget absorption peaks improves the fitting results of MIR photo-acoustic spectra.[106]

An example of least-squares minimization fitting applied to the quantification of DMMP is shown in Figure 11.7. The sample is produced by diluting a saturated flow of DMMP with zero air. The measurement result is highly linear ($R^2 = 0.997$) and exhibits a 2% error in slope.

11.4.1.3   MM-CRDS of Multicomponent Mixtures and Outdoor Air

The basis set decomposition is also robust to interference caused by overlapping absorption peaks. Figure 11.8a shows spectra of samples containing a fixed 70 ppb of DMMP, plus varying concentrations of gasoline. California reformulated gasoline typically contains 10% ethanol,[122] whose absorption spectrum completely overlaps with that of DMMP. This test represents a worst-case scenario for interferences; regardless, because the two spectra have characteristic structures with distinct peaks, the basis set decomposition method is capable of extracting an accurate DMMP concentration with minimal interference (~10% of the DMMP concentration, Figure 11.8b). Note that, in the absence of this data analysis methodology, this is another case where common volatile compounds could have been used to mask the

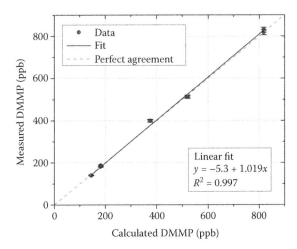

FIGURE 11.7   Measured vs. calculated DMMP concentration demonstrating the linearity of MM-CRDS.

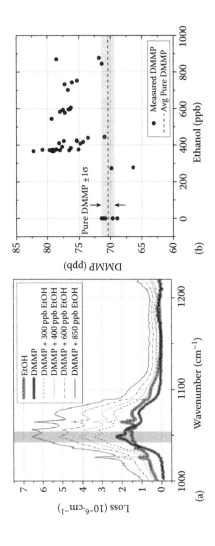

FIGURE 11.8 (a) Measured spectra with a constant 70 ppb of DMMP and variable concentrations of gasoline containing ~10% ethanol. Pure DMMP and ethanol spectra are shown with a halo. The gray band highlights the characteristic peak of DMMP, which is still visible even when superimposed on the largest ethanol peaks. (b) The measured DMMP concentration vs. the measured ethanol concentration. For this worst-case interference, the cross-talk is limited to ~10% of the measured DMMP value.

presence of dangerous nerve agents, since opening a bottle of distilled spirits near the inlet would blind the system to DMMP in this spectral region.

For CWA detection, it is critical that the CEAS system should provide early warning by offering real-time detection of single-digit ppb levels of CWAs in the presence of contaminated, urban, and battlefield air. The ability of the Los Gatos Research MM-CRDS system to detect DIMP in contaminated urban air has been tested by doping ambient air (Mountain View, CA) with a known concentration of DIMP. The primary spectral features in this case are water and carbon dioxide, with occasional contributions from industrial sources (e.g., refrigerants are often observed, particularly Freon-134a from air duster cans). A typical measured spectrum is shown in Figure 11.9a (black trace) along with the best fit (gray trace) obtained by minimizing Equation 11.19. The contributions to the best fit from water, carbon dioxide, and DIMP are also shown. As with the DMMP and ethanol example, the characteristic absorption features of each specie overlap and the large wavelength measurement range of the MM-CRDS instrument allows quantitative determination of the contributing species. Figure 11.9b shows the concentration of DIMP in Mountain View, California, air measured over a 45-h period for two different doping concentrations $(9.2 \pm 0.5$ and 18.4 ± 0.9 ppb). The respective measured values were 7.9 ± 1.1 and 19.3 ± 1.4 ppb, each within the convolved uncertainties of the calculated values and differing from the calculated value by 1.3 and 0.9 ppb for the low and high values, respectively.

Because CEAS produces direct absorption measurements, these concentrations are robust to changing ambient conditions (e.g., humidity) and can be calculated without any correction factors. Taking the standard deviation of the 8.2 ppb concentration ($\sigma = 1.1$ ppb), it is possible to calculate the probability of false positives (PFP) for various alarm thresholds (Figure 11.10). For example, the PFP for the 30 min miosis exposure threshold (5.8 ppb or 33 $\mu g/m^3$) is 7×10^{-8}, and, because the measurement period is ~4 min, it yields a false positive every ~5300 years! A commonly used maximum PFP is 1×10^{-6}, which occurs at a threshold of 5.2 ppb, and corresponds to a false positive on average once every 7.6 years.

11.5 DETECTION OF EXPLOSIVES WITH CEAS

Explosives present an even greater challenge because of their low vapor pressures (generally from 10^{-6} to 10^{-9} mbar[122]). For example, at 25°C, the vapor pressure of 2,4,6-trinitrotoluene (TNT) is 8 ppb, cyclotrimethylenetrinitramine (RDX) is 6 parts per trillion (ppt), pentaerythritol tetranitrate (PETN) is 20 ppt, and octogen (HMX) is 1×10^{-3} parts per quadrillion (ppq). Conversely, ethylene glycol dinitrate (EGDN) has a vapor pressure of 90 parts per million (ppm), triacetone triperoxide (TATP) (which has become popular with terrorists because of its ease of manufacture[1]) is 60 ppm, and 2,4-dinitrotoluene (DNT) (which is commonly found with TNT) is 0.3 ppm.[124] There are several comprehensive reviews on these and other challenges associated with detecting explosives.[1,60,125–129] Evaluation of the required detection limits is complicated by concealment, which is estimated to reduce the real-world concentrations by a factor of 1000,[130,131] with TNT concentrations in air above buried land mines reduced by as much as six orders of magnitude.[132]

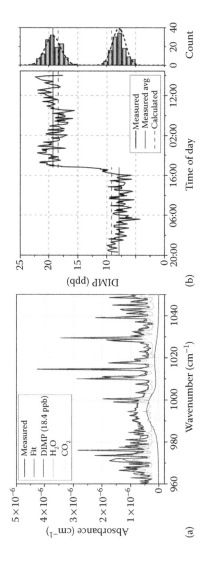

FIGURE 11.9 DIMP in Mountain View Air. (a) An example measured spectrum of Mountain View, California, air with 18.4 ppb of DIMP added (black). The constituent absorbers as determined by the fitting algorithm are also shown. (b) The concentration of DIMP as measured by the Los Gatos Research MM-CRDS instrument (black). The average measured value is shown as a solid line. The calculated doping level is shown as a dotted line. Each measured value agrees with the calculated value to within the convolved uncertainty.

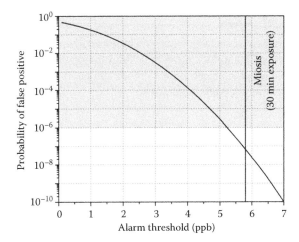

FIGURE 11.10 Probability of a false positive for various alarm thresholds, calculated using the 1σ found by measuring Mountain View, California, air doped with 8.4 ppb of DIMP. The PFP reaches 1×10^{-6} at 5.2 ppb (a false alarm every 7.6 years) and 7×10^{-8} at the 30 min threshold for miosis of 5.8 ppb (a false alarm every 5300 years).

Fortunately, absorption values for explosives in the MIR are similar to those of CWAs, although the largest absorption bands appear between 7 and 8 μm rather than 9–11 μm.[43,133–135] For example, the peak absorption of PETN between 7 and 8 μm is 6×10^{-4} [μmol/mol·cm]$^{-1}$, RDX is 2×10^{-5} [μmol/mol·cm]$^{-1}$, and TNT is 3.4×10^{-5} [μmol/mol·cm]$^{-1}$.[20] Banas et al. have tabulated the location and type of the major optical absorptions between 4000 and 400 cm^{-1} for C-4, PETN, and TNT.[43]

A high-performance CAS spectrometer (MDAL $= 1 \times 10^{-7}$ cm^{-1}/$\sqrt{\text{Hz}}$) will have a LOD for PETN of 50 ppb and for TNT of 9 ppb, which is insufficient when compared to the expected concentrations of most explosives.[136] The absorption cross sections are generally 10–100 times smaller in the NIR, where optical absorption is produced by overtones of the strong MIR ro-vibrational transitions. However, electronic transitions in the ultraviolet (UV) are characteristically larger than MIR vibrational transitions, particularly for aromatic molecules. For example, absorption cross sections for TNT, 2,4-DNT, RDX, PETN, and HMX are all about 100 times larger in the UV than in the MIR.[125] Unfortunately, the larger UV absorptions are accompanied by spectrally broad absorption spectra that are indistinguishable from one another and other common atmospheric constituents, making speciation impossible without a complementary separation technique.[21]

There are several existing techniques for addressing the low ambient concentrations of explosives. Most involve preconcentrating a large quantity of air onto a collection substrate (stainless steel, quartz, nickel, gold, platinum, copper, aluminum, and Teflon), followed by flash-heating to >150°C.[1,137,138] Preconcentration of the high-vapor-pressure explosives is relatively easy and effective, because they readily adhere to room temperature surfaces, while the most abundant ambient gases (nitrogen, oxygen, carbon dioxide, water) do not. The sampled air for preconcentration is either drawn from the atmosphere or, more commonly, an object suspected of harboring explosives (persons, parcels, airline baggage, etc.). The latter is placed in a portal and swept using high-volume air flows, air jets, air knives, or acoustic energy to dislodge residual explosive particles.[127,139] These particles can contain micrograms of material—orders of magnitude more molecules than even large volumes of saturated air.[140] In general, preconcentration methods can be used as front ends for multiple analytical techniques, including CEAS.

11.5.1 Measurements of Explosives Using CEAS

Unlike CWAs, there are several examples of explosives detection using CEAS. Perhaps the most heavily cited work is that of Todd et al., who used an MIR OPO coupled with CRDS to measure TATP, TNT, RDX, PETN, and tetryl.[20] The OPO provided a 200–300 µJ/pulse from 6–8 µm, and spectral measurements were reported from 7 to 8 µm. This instrument utilized a CRDS cell ($L=50$ cm) with dual temperature zones to thermally isolate the cavity mirrors; only the center of the cell ($l=10$ cm) was heated to 170°C. Although this configuration requires a complicated air purge between the sample gas in the center of the cell and the cavity mirrors, it was extremely effective, providing a TNT detection limit of ~8 ppb. Todd et al. also demonstrated two preconcentrators: an actively cooled and flash-heated, glass-coated, stainless steel tube and a dimethyl silicone membrane separator that improved the TNT detection limit to 1.2 ppb and 75 ppt, respectively.

Usachev et al. used CRDS to measure the optical properties of the strong, UV $n{\to}\pi^*$ electronic transitions of TNT in the vapor phase.[23] The tripled output of a neodymium-doped yttrium aluminum garnet (Nd:YAG) pumped-dye laser was used as the light source to produce radiation between 225 and 235 nm. This instrument used a 77.5 cm long cavity that incorporated a central sample section that was 24 cm long and heated to between 5°C and 110°C. They report a detection limit of <1 ppb of TNT.

Ramos and Dagdigian present a similar system for the measurement of 2,4-DNT, 2,6-DNT, 1,2-dinitrobenzene (DNB), 1,3-DNB, and 1,4-DNB.[21] This demonstration utilized a frequency-doubled OPO or dye laser to produce radiation around 240 nm. The cavity was 60 cm long, with a 35 cm long central sample section that was heated to 50°C. They report a detection limit of 100 ppt of TNT.

In each of the UV experiments listed above, the absorption spectra of explosives are broad and noncharacteristic.[21,141] Peroxyacetyl nitrate (PAN) and ozone are examples of interferents that possess UV absorptions that overlap those of target explosives and have highly variable concentrations in urban environments. The indistinct UV spectra of explosives make it impossible to positively identify them without a front-end analyte separation technique (e.g., gas chromatography).

Another approach is to thermally decompose the explosives and detect the decomposition products. Wojtas et al. were able to detect nitroglycerine, ammonium nitrate, TNT, PETN, RDX, and HMX by monitoring the thermal decomposition products (N_2O, nitric oxide (NO), and nitrogen dioxide (NO_2)), using three different CRDS systems operating at 4.53 µm, 5.26 µm, and 410 nm, respectively.[142,143] They also utilized several preconcentration methods. The NO_2-only system was field-tested in a Polish mine, where the authors demonstrated nonquantitative detection of nitroglycerine and ammonium nitrate emitted from passing miners. The N_2O and NO systems were able to detect 1 ng of TNT, PETN, RDX, and HMX. In theory, this triplicate detection of decomposition products could be used to provide some speciation between explosives, which could have different characteristic ratios of N_2O, NO, and NO_2, but this was not demonstrated.

Taha et al. detect only NO_2 emitted from catalytically converted explosives using a blue laser diode coupled to CRDS.[24] The explosive (1,3-DNB, 2,4-DNT, or 2,4,6-TNT) is catalytically converted to NO_2 using PtO_2 heated to 250°C. As NO_2 concentrations in urban and industrial environments fluctuate rapidly, the background NO_2 concentration was measured using a parallel reference cell without the catalytic converter, and the measured value was subtracted from the catalytically converted concentration. They report a detection limit of 500 ppt of TNT. Again, this technique does not provide speciation and is prone to interferences from other nitrogen-containing compounds that convert to NO_2 (e.g., PANs).

Snels et al. have demonstrated detection of TNT, 2,4-DNT, and 2,6-DNT with NIR CRDS. This system operated at 1560–1680 nm, using an external cavity diode laser (ECDL).[22] In this case, a solid sample was introduced into the vapor phase by flash-heating a 5 cm long sample tube located in the center of the cavity. This configuration mimics applications where solid samples are obtained from preconcentration, fingerprints, or other sources. Finally, rather than using a least-squares fit to known basis sets for quantification of spectral constituents, the authors apply a principle component analysis to demonstrate speciation of the three measured compounds (no other constituents are included in the analysis). They report a detection limit of 7.4 ppm.

Miller et al. have proposed a novel diffusion time-of-flight (DiTOF) chamber, coupled with CRDS for the detection of explosives and other compounds.[144] By observing changes in the ring-down time of two CRDS cells at a single wavelength, placed perpendicular to the drift direction and displaced from one another along the drift direction, the mixture constituents and their concentrations can be calculated. This system also forgoes the least-squares fit of spectra in favor of a Bayesian estimation of the constituent concentrations based on known diffusion rates.[144] They predict a detection limit for TNT of 12 ppt in a heavily contaminated sample background by assuming operation at 3.2 μm and an aggressive MDAL of 1.4×10^{-12} cm$^{-1}/\sqrt{Hz}$.

11.5.1.1 Detection of Explosives with MM-CRDS

The Los Gatos Research MM-CRDS system (described in Section 11.4.1) can also be used to detect explosives in the 9.4–11.5 μm region. For these measurements, T_2 was increased to 150°C to avoid adsorption of the explosives during transfer, and T_3 was elevated to 100°C to reduce adsorption onto the cell walls; 200 μg of explosive was deposited on a glass slide and placed in the sample oven (Figure 11.4). T_1 was set to 60°C for TNT or 100°C for PETN. The vapor within the sample oven was then drawn into the CRDS cell by a 300 sccm flow of zero air to produce a concentration of ~1 ppm of TNT or ~3 ppm of PETN in the optical cavity. The resulting MM-CRDS spectra, measured at 700 Torr, are shown in Figure 11.11. Many of the peaks identified in the solid-phase FTIR data of Banas et al. are present in Figure 11.11, but with a spectral shift (Table 11.1). An additional TNT peak that is

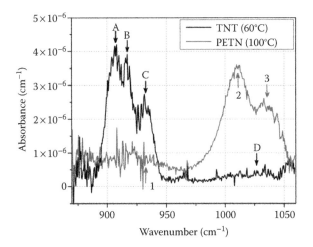

FIGURE 11.11 Measured absorption spectra of TNT and PETN measured with the Los Gatos Research MM-CRDS system. The measured concentration is approximately 1 ppm of TNT and 3 ppm of PETN. A description of the peaks labeled A–D and 1–3 is shown in Table 11.1.

not present in the FTIR data is resolvable at 917 cm^{-1}. Although this spectral region is not ideal for the detection of explosives, reasonable detection limits are still found; using the MDAL of a heated cell from Figure 11.5, LODs of 34 ppb for TNT and 29 ppb for PETN are achieved. With a dual temperature zone cell as in Todd et al., the MDAL is expected to increase by an order of magnitude (see Figure 11.5), resulting in detection limits of 2.9 ppb for PETN and 3.4 ppb for TNT.[46] Further

TABLE 11.1
MM-CRDS Absorption Peaks for TNT and PETN between 9.4 and 11.5 μm

| Explosive | Peak | Wavenumber (cm^{-1}) | | Description |
		FTIR[a]	Actual	
TNT	A	908	907	Carbon–nitrogen (C–N) stretch
	B	—	917	Unidentified
	C	939	932	C–N stretch
	D	1026	Ambiguous	Methyl (CH$_3$) deformation (def)
PETN	1	939	Ambiguous	Methylene bridge (CH$_2$) torsion + CCC def
	2	1002	1011	Carbon–oxygen (CO) stretch + CCC def
	3	1037	1035	CO stretch + C$_5$ skeletal + NO$_2$ rock

[a] Solid-phase FTIR as described in Banas et al. *Vib. Spectrosc.,* 51, 168–176, **2009**.

TABLE 11.2
Reported Limit of Detection without Preconcentration

Reference	Wavelength Region	Specific	MDAL	Explosive	LOD (3σ)
Todd[20]	7–8 μm	Yes	$9 \times 10^{-9}\,\text{cm}^{-1}/\sqrt{\text{Hz}}$	TNT	~8 ppb
Usachev[23]	225–235 nm	No	$3 \times 10^{-6}\,\text{cm}^{-1}/\sqrt{\text{Hz}}$	TNT	<1 ppb
Ramos[21]	235–250 nm	No	$2 \times 10^{-7}\,\text{cm}^{-1}$	2,4-DNT	100 ppt
Wojtas[141]	410 nm	No	Not given	NG, NH$_4$NO$_3$[a]	Nonquantitative
	4.53–5.26 μm	No	Not given	TNT, others	1 ng
Taha[24]	405 nm	No	Not given	TNT, others	500 ppt
Snels[22]	1560–1680 nm	Yes	$4 \times 10^{-8}\,\text{cm}^{-1}/\sqrt{\text{Hz}}$	TNT	7.4 ppm
Miller[143]	3.2 μm	Yes	$1.4 \times 10^{-12}\,\text{cm}^{-1}/\sqrt{\text{Hz}}$[b]	TNT	12 ppt[b]
This work	9.4–11.5 μm	Yes	$1 \times 10^{-7}\,\text{cm}^{-1}/\sqrt{\text{Hz}}$	TNT	34 (3.4[c]) ppb
			$2 \times 10^{-8}\,\text{cm}^{-1}/\sqrt{\text{Hz}}$	PETN	29 (2.9[c]) ppb

[a] Nitroglycerin (NG), ammonium nitrate (NH$_4$NO$_3$).
[b] Simulated.
[c] Projected for a dual temperature zone cell.

improvements are possible by measuring the larger absorption peaks between 7 and 8 μm and by preconcentration of the sample gas.

A summary of the published detection limits achieved *without preconcentration* is given in Table 11.2. The best-demonstrated detection limits are sub-ppb, indicating that the state-of-the-art analyzers are incapable of realistic detection of low-volatility explosives. Of the commonly tested explosives, only TATP, DNT, DNB, and TNT are likely detectable using the reported methods under real-world conditions.

However, with an achievable preconcentration factor of 1×10^3, PETN and RDX come within reach and, importantly, *concealed* TNT becomes detectable. Additionally, many of these systems have relatively poor reported MDALs because of the challenges associated with CEAS in the UV and MIR; further improvements in light sources, detectors, and coating technology will yield improved MDALs and proportionally decreased LODs.

11.6   CONCLUSION

By reflecting probe light through a sample many times, CEAS offers sensitivity that is orders of magnitude better than the best conventional absorption measurements. Absorptions sensitivities of $1 \times 10^{-10} - 1 \times 10^{-11}\,\text{cm}^{-1}$ are routinely achieved and sensitivities of $1 \times 10^{-14}\,\text{cm}^{-1}/\sqrt{\text{Hz}}$ have been demonstrated.

Despite this advantage, CEAS has seldom been used for CWA detection and the reported sensitivities are inadequate for safety monitoring applications. However, by using the Los Gatos Research MM-CRDS system in conjunction with a broadly tunable MIR EC-QCL, we have demonstrated the real-time detection of low ppb levels of CWA simulants in ambient air samples. This system is sufficient for early warning detection with a PFP rate of 1×10^{-6} (i.e., a false alarm once every 7.6 years).

Moreover, MIR techniques, such as MM-CRDS, are capable of detecting explosive concentrations at low ppb levels without preconcentration. These LODs are suitable for detecting high-volatility explosives (e.g., TATP, EGDN, DNT, DNB, and TNT) at room temperature; however, considerable improvements are required for effective detection of low-volatility explosives such as RDX and, especially, HMX. By combining preconcentration with MM-CRDS, we anticipate detecting explosives at the parts-per-trillion level, sufficient for many field applications.

Recent years have seen significant advances in the application of MIR CEAS to difficult problems, largely due to improvements in QCLs operating in the MIR. High-power, commercially available DFB and EC lasers now exist at 4–14 μm, making rugged, fieldable explosives and CWA detectors possible. Laser power, stability, and spectral properties continue to improve, and new spectral regions (such as the THz regime) are an active area of research.[146–148] Additionally, the underlying maturity of CEAS is leading to combination methods such as DiTOF-CRDS, thermal decomposition CRDS, and preconcentration, many of which push the detectable concentrations down by several orders of magnitude. With these gains, CEAS holds great potential for the quantitative detection of CWAs and explosives in the vapor phase, filling critical gaps in our ability to protect civilians and military forces from attack.

ACKNOWLEDGMENTS

We would like to thank the US Department of Defense, National Institutes of Health and the US Environmental Protection Agency for supporting this work under the auspices of the Small Business Innovation Research program.

REFERENCES

1. Moore, D. S. Instrumentation for trace detection of high explosives. *Rev. Sci. Instrum.* **2004**, *75*, 2499.
2. Demtröder, W. *Laser Spectrsoscopy: Basic Concepts and Instrumentation*, 2nd edn, Springer-Verlag: New York, 1998.
3. Struve, W. S. *Fundamentals of Molecular Spectroscopy*, John Wiley & Sons: New York, 1989.
4. Berden, G., Peeters, R., Meijer, G. Cavity ring-down spectroscopy: Experimental schemes and applications. *Int. Rev. Phys. Chem.* **2000**, *19*, 566–602.
5. O'Keefe, A., Deacon, D. G. Cavity ring-down optical spectrometer for absorption measurements using pulsed laser sources. *Rev. Sci. Instrum.* **1988**, *59*, 2544.
6. Scherer, J. J., Paul, J. B., O'Keefe, A., Saykally, R. J. Cavity ringdown laser absorption spectroscopy: History, development, and application to pulsed molecular beams. *Chem. Rev.* **1997**, *97*, 25–52.
7. Atkinson, D. B. Solving chemical problems of environmental importance using cavity ring-down spectroscopy. *Analyst* **2003**, *128*, 117–125.

8. Linnartz, H. Cavity ring-down spectroscopy of molecular transients of astrophysical interest. In Berden, G., Engeln, R., (eds), *Cavity Ring-Down Spectroscopy: Techniques and Applications*, Blackwell: Washington, DC, 2009.

9. Hancock, G., Orr-ewing, A. J. Applications of cavity ring-down spectroscopy in atmospheric chemistry. In Berden, G., Engeln, R., (eds), *Cavity Ring-Down Spectroscopy: Techniques and Applications*, Blackwell: Washington, DC, 2009, pp. 181–211.

10. Leen, J. B., Yu, X.-Y., Gupta, M., Baer, D. S., Hubbe, J. M., Kluzek, C. D., Tomlinson, J. M., Hubbell, M. R. Fast *in situ* airborne measurement of ammonia using a mid-infrared off-axis ICOS spectrometer. *Environ. Sci. Technol.* **2013**, *47*, 10446–10453.

11. Brent, L. C., Thorn, W. J., Gupta, M., Leen, B., Stehr, J. W., He, H., Arkinson, H. L., Weinheimer, A., Garland, C., Pusede, S. E., et al. Evaluation of the use of a commercially available cavity ringdown absorption spectrometer for measuring NO_2 in flight, and observations over the Mid-Atlantic States, during DISCOVER-AQ. *J. Atmos. Chem.* **2013**.

12. Berman, E. S. F. F., Fladeland, M., Liem, J., Kolyer, R., Gupta, M. Greenhouse gas analyzer for measurements of carbon dioxide, methane, and water vapor aboard an unmanned aerial vehicle. *Sensors Actuators B Chem.* **2012**, *169*, 128–135.

13. Leen, J. B., Berman, E. S. F., Liebson, L., Gupta, M. Spectral contaminant identifier for off-axis integrated cavity output spectroscopy measurements of liquid water isotopes. *Rev. Sci. Instrum.* **2012**, *83*, 044305.

14. Wankel, S. D., Huang, Y., Gupta, M., Provencal, R., Leen, J. B., Fahrland, A., Vidoudez, C., Girguis, P. R. Characterizing the distribution of methane sources and cycling in the deep sea via *in situ* stable isotope analysis. *Environ. Sci. Technol.* **2013**, *47*, 1478–1486.

15. Vogler, D. E., Sigrist, M. W. Near-infrared laser based cavity ringdown spectroscopy for applications in petrochemical industry. *Appl. Phys. B Lasers Opt.* **2006**, *85*, 349–354.

16. Ciaffoni, L., Hancock, G., Harrison, J. J., Helden, J. H. Van, Langley, C. E., Peverall, R., Ritchie, G. A. D., Wood, S. Demonstration of a mid-infrared cavity enhanced absorption spectrometer for breath acetone detection. *Anal. Chem.* **2013**, *85*, 846–850.

17. Cao, W., Duan, Y. Current status of methods and techniques for breath analysis. *Crit. Rev. Anal. Chem.* **2007**, *37*, 3–13.

18. Silva, M. L., Sonnenfroh, D. M., Rosen, D. I., Allen, M. G., O'Keefe, A. Integrated cavity output spectroscopy measurements of NO levels in breath with a pulsed room-temperature QCL. *Appl. Phys. B Lasers Opt.* **2005**, *81*, 705–710.

19. Moeskops, B. W. M., Naus, H., Cristescu, S. M., Harren, F. J. M. Quantum cascade laser-based carbon monoxide detection on a second time scale from human breath. *Appl. Phys. B* **2006**, *82*, 649–654.

20. Todd, M. W., Provencal, R. A., Owano, T. G., Paldus, B. A., Kachanov, A., Vodopyanov, K. L., Hunter, M., Coy, S. L., Steinfeld, J. I., Arnold, J. T. Application of mid-infrared cavity-ringdown spectroscopy to trace explosives vapor detection using a broadly tunable (6–8 μm) optical parametric oscillator. *Appl. Phys. B Lasers Opt.* **2002**, *75*, 367–376.

21. Ramos, C., Dagdigian, P. J. Detection of vapors of explosives and explosive-related compounds by ultraviolet cavity ringdown spectroscopy. *Appl. Opt.* **2007**, *46*, 620–627.

22. Snels, M., Venezia, T., Belfiore, L. Detection and identification of TNT, 2,4-DNT and 2,6-DNT by near-infrared cavity ringdown spectroscopy. *Chem. Phys. Lett.* **2010**, *489*, 134–140.

23. Usachev, A. D., Miller, T. S., Singh, J. P., Yueh, F.-Y., Jang, P.-R., Monts, D. L. Optical properties of gaseous 2,4,6-trinitrotoluene in the ultraviolet region. *Appl. Spectrosc.* **2001**, *55*, 125–129.

24. Taha, Y. M., Odame-Ankrah, C. A., Osthoff, H. D. Real-time vapor detection of nitroaromatic explosives by catalytic thermal dissociation blue diode laser cavity ring-down spectroscopy. *Chem. Phys. Lett.* **2013**, *582*, 15–20.

25. Qu, Z., Gao, C., Han, Y., Du, X., Li, B. Detection of chemical warfare agents based on quantum cascade laser cavity ring-down spectroscopy. *Chinese Opt. Lett.* **2012**, *10*, 1–3.

26. Fiedler, S. E., Hese, A., Ruth, A. A. Incoherent broad-band cavity-enhanced absorption spectroscopy of liquids. *Rev. Sci. Instrum.* **2005**, *76*, 023107.

27. Ball, S. M., Langridge, J. M., Jones, R. L. Broadband cavity enhanced absorption spectroscopy using light emitting diodes. *Chem. Phys. Lett.* **2009**, *398*, 68–74.

28. Langridge, J. M., Laurila, T., Watt, R. S., Jones, R. L., Kaminski, C. F., Hult, J. Cavity enhanced absorption spectroscopy of multiple trace gas species using a supercontinuum radiation source. *Opt. Express* **2008**, *16*, 10178–10188.

29. Galli, I., Bartalini, S., Borri, S., Cancio, P., Mazzotti, D., De Natale, P., Giusfredi, G. Molecular gas sensing below parts per trillion: radiocarbon-dioxide optical detection. *Phys. Rev. Lett.* **2011**, *107*, 270802.

30. Du, X., Farinas, A. D., Crosson, E. R., Balslev-Clausen, D., Blunier, T. Development of a field-deployable isotopic N_2O analyzer based on mid-infrared cavity ring-down spectroscopy. *Proc. SPIE* **2011**, 8032, 803207.

31. Werle, P., Slemr, F., Maurer, K., Kormann, R., Mücke, R., Jänker, B. Near- and mid-infrared laser-optical sensors for gas analysis. *Opt. Lasers Eng.* **2002**, *37*, 101–114.

32. Novelli, P. C. CO in the atmosphere: measurement techniques and related issues. *Chemosph. Glob. Chang. Sci.* **1999**, *1*, 115–126.

33. Lackner, M. Tunable diode laser absorption spectroscopy (TDLAS) in the process industries: A review. *Rev. Chem. Eng.* **2007**, *23*, 65–147.

34. Röpcke, J., Lombardi, G., Rousseau, A., Davies, P. B. Application of mid-infrared tuneable diode laser absorption spectroscopy to plasma diagnostics: a review. *Plasma Sources Sci. Technol.* **2006**, *15*, S148–S168.

35. Eng, R. S., Butler, J. F., Linden, K. J. Tunable diode laser spectroscopy: An invited review. *Opt. Eng.* **1980**, *19*, 945–960.

36. Frish, M. B., Laderer, M. C., Wainner, R. T., Wright, A. O., Patel, A. H., Stafford-Evans, J., Morency, J. R., Allen, M. G., Green, B. D. The next generation of TDLAS analyzers. *Proc. SPIE* **2007**, 6765, 676506.

37. Frish, M. B., Wainner, R. T., Laderer, M. C., Green, B. D., Allen, M. G. Standoff and miniature chemical vapor detectors based on tunable diode laser absorption spectroscopy. *IEEE Sens. J.* **2010**, *10*, 639–646.

38. Scott, D. C., Herman, R. L., Webster, C. R., May, R. D., Flesch, G. J., Moyer, E. J. Airborne laser infrared absorption spectrometer (ALIAS-II) for *in situ* atmospheric measurements of N_2O, CH_4, CO, HCl, and NO_2 from balloon or remotely piloted aircraft platforms. *Appl. Opt.* **1999**, *38*, 4609.

39. Duffin, K., McGettrick, A. J., Johnstone, W., Stewart, G., Moodie, D. G. Tunable diode-laser spectroscopy with wavelength modulation: a calibration-free approach to the recovery of absolute gas absorption line shapes. *J. Light. Technol.* **2007**, *25*, 3114–3125.

40. Harig, R., Rusch, P., Dyer, C., Jones, A., Moseley, R., Truscott, B. Remote measurement of highly toxic vapors by scanning imaging fourier-transform spectrometry. In Jensen, J. O., Thériault, J.-M., (eds), *Optics East 2005*, International Society for Optics and Photonics, 2005, pp. 599510–599512.

41. Walter, W. T. Sensitive detection of chemical agents and toxic industrial chemicals using active open-path FTIRs. Proc. SPIE 2004, 5270. In Vo-Dinh, T., Gauglitz, G., Lieberman, R. A., Schaefer, K. P., Killinger, D. K., (eds), *Optical Technologies for Industrial, Environmental, and Biological Sensing*, International Society for Optics and Photonics, 2004, pp. 144–150.

42. Manolakis, D., Model, J., Rossacci, M., Zhang, D., Ontiveros, E., Pieper, M., Seeley, J., Weitz, D. Software algorithms for false alarm reduction in LWIR hyperspectral chemical agent detection. *Proc. SPIE* **2008**, 6966, 69661U.

43. Banas, A., Banas, K., Bahou, M., Moser, H. O., Wen, L., Yang, P., Li, Z. J., Cholewa, M., Lim, S. K., Lim, C. H. Post-blast detection of traces of explosives by means of Fourier transform infrared spectroscopy. *Vib. Spectrosc.* **2009**, *51*, 168–176.

44. Heise, H. M., Mu, U., Ga, A. G., Ho, N. Improved chemometric strategies for quantitative FTIR spectral analysis and applications in atmospheric open-path monitoring. *F. Anal. Chem. Technol.* **2001**, *5*, 13–28.

45. Minnich, T. R., Scotto, R. L. Use of open-path ftir spectroscopy to address air monitoring needs during site remediations. *Remediation* **1999**, *9*, 1–16.

46. Farhat, S. K., Todd, L. A. Evaluation of open-path FTIR spectrometers for monitoring multiple chemicals in air. *Appl. Occup. Environ. Hyg.* **2000**, *15*, 911–923.

47. Marshall, T. L., Chaffin, C. T., Hammaker, R. M., Fateley, W. G. An introduction to open-path FT-IR atmospheric monitoring. *Environ. Sci. Technol.* **1994**, *28*, 224A–232A.

48. Griffiths, P. R., Shao, L., Leytem, A. B. Completely automated open-path FT-IR spectrometry. *Anal. Bioanal. Chem.* **2009**, *393*, 45–50.

49. Shao, L., Roske, C. W., Griffiths, P. R. Detection of chemical agents in the atmosphere by open-path FT-IR spectroscopy under conditions of background interference: II. Fog and rain. *Anal. Bioanal. Chem.* **2010**, *397*, 1521–1528.

50. Shao, L., Roske, C. W., Griffiths, P. R. Detection of chemical agents in the atmosphere by open-path FT-IR spectroscopy under conditions of background interference: I. High-frequency flashes. *Anal. Bioanal. Chem.* **2010**, *397*, 1511–1519.

51. Werle, P., Mücke, R., Slemr, F. The limits of signal averaging in atmospheric trace-gas monitoring by tunable diode-laser absorption spectroscopy (TDLAS). *Appl. Phys. B* **1993**, *57*, 131–139.

52. Herriott, D. R., Schulte, H. J. Folded optical delay lines. *Appl. Opt.* **1965**, *4*, 883.

53. White, J. U. Long optical paths of large aperture. *J. Opt. Soc. Am.* **1942**, *32*, 285–288.

54. McManus, J. B., Kebabian, P. L., Zahniser, M. S. Astigmatic mirror multipass absorption cells for long-path-length spectroscopy. *Appl. Opt.* **1995**, *34*, 3336.

55. McManus, J. B., Zahniser, M. S., Nelson, D. D. Dual quantum cascade laser trace gas instrument with astigmatic herriott cell at high pass number. *Appl. Opt.* **2011**, *50*, A74–85.

56. Tuzson, B., Mangold, M., Looser, H., Manninen, A., Emmenegger, L. Compact multipass optical cell for laser spectroscopy. *Opt. Lett.* **2013**, *38*, 257–259.

57. Silver, J. A. Simple dense-pattern optical multipass cells. *Appl. Opt.* **2005**, *44*, 6545–6556.

58. Phillips, M. C., Taubman, M. S., Bernacki, B. E., Cannon, B. D., Stahl, R. D., Schiffern, J. T., Myers, T. L. Real-time trace gas sensing of fluorocarbons using a swept-wavelength external cavity quantum cascade laser. *Analyst* **2014**, *139*, 2047-2056.

59. Paldus, B. A., Zare, R. N. Absorption spectroscopies: from early beginnings to cavity-ringdown spectroscopy. In Busch, K. W., Busch, M. A., (eds), *Cavity Ringdown Spectroscopy*, American Chemical Society: Washington, DC, 1999, pp. 49–70.

60. Hildenbrand, J., Herbst, J., Wöllenstein, J., Lambrecht, A. Explosive detection using infrared laser spectroscopy. *Proc. SPIE* **2009**, 7222, 72220B.

61. Curl, R. F., Tittel, F. K. Tunable infrared laser spectroscopy. *Annu. Rep. Prog. Chem. Sect. C* **2002**, *98*, 1–56.

62. Ingle, J. D., Crouch, S. R. *Spectrochemical Analysis*, Prentice Hall: New York, 1988.

63. Jongma, R. T., Boogaarts, M. G. H., Holleman, I., Meijer, G. Trace gas detection with cavity ring down spectroscopy. *Rev. Sci. Instrum.* **1995**, *66*, 2821.

64. Zalicki, P., Zare, R. N. Cavity ring-down spectroscopy for quantitative absorption measurements. *J. Chem. Phys.* **1995**, *102*, 2708.

65. Paldus, B. A., Kachanov, A. A. An historical overview of cavity-enhanced methods. *Can. J. Phys.* **2005**, *83*, 975–999.

66. Ball, S. M., Jones, R. L. Broad-band cavity ring-down spectroscopy. *Chem. Rev.* **2003**, *103*, 5239–5262.

67. Berden, G., Peeters, R., Meijer, G. Cavity ring-down spectroscopy: Experimental schemes and applications. *Int. Rev. Phys. Chem.* **2000**, *19*, 565–607.

68. Wheeler, M. D., Newman, S. M., Orr-Ewing, A. J., Ashfold, M. N. R. Cavity ring-down spectroscopy. *J. Chem. Soc. Faraday Trans.* **1998**, *94*, 337–351.

69. Stacewicz, T., Wojtas, J., Bielecki, Z., Nowakowski, M., Mikołajczyk, J. Cavity ring down spectroscopy: Detection of trace amounts of substance. *Opto-Electronics Rev.* **2012**, *20*, 53–60.

70. Busch, K. W., Busch, M. A. (eds). *Cavity-Ringdown Spectroscopy*, Washington, DC: American Chemical Society, 1999.

71. Berden, G., Engeln, R. *Cavity Ring-Down Spectroscopy Techniques and Applications*, Blackwell: West Sussex, UK, 2009.

72. Herbelin, J. M., McKay, J. A. Development of laser mirrors of very high reflectivity using the cavity-attenuated phase-shift method. *Appl. Opt.* **1981**, *20*, 3341–3344.

73. Anderson, D. Z., Frisch, J. C., Masser, C. S. Mirror reflectometer based on optical cavity decay time. *Appl. Opt.* **1984**, *23*, 1238.

74. O'Keefe, A., Scherer, J. J., Paul, J. B., Saykally, R. J. Cavity-ringdown laser spectroscopy history, development, and applications. In Busch, K. W., Busch, M. A., (eds), *Cavity Ringdown Spectroscopy*, Washington, DC: American Chemical Society, 1999, pp. 71–92.

75. Siegman, A. E. *Lasers*, Mill Valley, CA: University Science Books, 1986.

76. Bilger, H. R., Wells, P. V, Stedman, G. E. Origins of fundamental limits for reflection losses at multilayer dielectric mirrors. *Appl. Opt.* **1994**, *33*, 7390–7396.

77. Marquardt, D. W. An algorithm for least-squares estimation of nonlinear parameters. *J. Soc. Ind. Appl. Math.* **1963**, *11*, 431–441.

78. Lehmann, K. K., Huang, H. optimal signal processing in cavity ring-down spectroscopy. In Laane, J., (ed.), *Frontiers of Molecular Spectroscopy*, Amsterdam: Elsevier, 2009, pp. 623–658.

79. Lehmann, K. K., Berden, G., Engeln, R. An introduction to cavity ring-down spectroscopy. In Berden, G., Engeln, R., (eds), *Cavity Ring-Down Spectroscopy: Techniques and Applications*, Washington, DC: Blackwell, 2009.

80. Foltynowicz, A., Schmidt, F. M., Ma, W., Axner, O. Noise-immune cavity-enhanced optical heterodyne molecular spectroscopy: Current status and future potential. *Appl. Phys. B* **2008**, *92*, 313–326.

81. Curl, R. F., Capasso, F., Gmachl, C., Kosterev, A. a., McManus, B., Lewicki, R., Pusharsky, M., Wysocki, G., Tittel, F. K. Quantum cascade lasers in chemical physics. *Chem. Phys. Lett.* **2010**, *487*, 1–18.

82. Wysocki, G., Lewicki, R., Curl, R. F., Tittel, F. K., Diehl, L., Capasso, F., Troccoli, M., Hofler, G., Bour, D., Corzine, S., et al. Widely tunable mode-hop free external cavity quantum cascade lasers for high resolution spectroscopy and chemical sensing. *Appl. Phys. B* **2008**, *92*, 305–311.

83. Rao, G. N., Karpf, A. External cavity tunable quantum cascade lasers and their applications to trace gas monitoring. *Appl. Opt.* **2011**, *50*, A100–A115.

84. Tittel, F. K., Richter, D., Fried, A. Mid-infrared laser applications in spectroscopy. In Sorokina, I. T., Vodopyanov, K. L., (eds), *Topics Appl. Phys.*, Berlin: Springer-Verlag , 2003, Vol. 89, pp. 445–516.

85. Lehmann, K. K., Romanini, D. The superposition principle and cavity ring-down spectroscopy. *J. Chem. Phys.* **1996**, *105*, 10263.

86. Hodges, J. T., Looney, J. P., van Zee, R. D. Response of a ring-down cavity to an arbitrary excitation. *J. Chem. Phys.* **1996**, *105*, 10278.

87. Paul, J. B., Lapson, L., Anderson, J. G. Ultrasensitive absorption spectroscopy with a high-finesse optical cavity and off-axis alignment. *Appl. Opt.* **2001**, *40*, 4904–4910.
88. Baer, D. S., Paul, J. B., Gupta, M., O'Keefe, A. Sensitive absorption measurements in the near-infrared region using off-axis integrated cavity output spectroscopy. *Appl. Phys. B.* **2002**, 75, 261–265.
89. Hodges, J. T., Looney, J. P., van Zee, R. D. Response of a ring-down cavity to an arbitrary excitation. *J. Chem. Phys.* **1996**, *105*, 10278.
90. Romanini D., Kachanov A.A., Sadeghi N., Stoeckel, F. CW cavity ring down spectroscopy. *Chem. Phys. Lett.* **1997**, *264*, 316–322.
91. Spence, T. G., Harb, C. C., Paldus, B. a., Zare, R. N., Willke, B., Byer, R. L. A laser-locked cavity ring-down spectrometer employing an analog detection scheme. *Rev. Sci. Instrum.* **2000**, *71*, 347.
92. Engeln, R., Berden, G., Meijer, G. Fourier transform and polarization dependent cavity-ringdown spectroscopy. In Busch, K. W., Busch, M. A., (eds), *Cavity Ringdown Spectroscopy*, Washington, DC: American Chemical Society, 1999, pp. 146–161.
93. Ball, S. M., Jones, R. L. Broad-band cavity ring-down spectroscopy. In Berden, G., Engeln, Ri., (eds), *Cavity Ring-Down Spectroscopy: Techniques and Applications*, West Sussex, UK: Blackwell, 2009, Vol. 103, pp. 5239–5262.
94. Ball, S. M., Langridge, J. M., Jones, R. L. Broadband cavity enhanced absorption spectroscopy using light emitting diodes. *Chem. Phys. Lett.* **2004**, *398*, 68–74.
95. Laurila, T., Burns, I. S., Hult, J., Miller, J. H., Kaminski, C. F. A calibration method for broad-bandwidth cavity enhanced absorption spectroscopy performed with supercontinuum radiation. *Appl. Phys. B* **2011**, *102*, 271–278.
96. Johnston, P. S., Lehmann, K. K. Cavity enhanced absorption spectroscopy using a broadband prism cavity and a supercontinuum source. *Opt. Express* **2008**, *16*, 15013–15023.
97. Meshkov, A. I., De Lucia, F. C. Broadband absolute absorption measurements of atmospheric continua with millimeter wave cavity ringdown spectroscopy. *Rev. Sci. Instrum.* **2005**, *76*, 083103.
98. Kebabian, P. L., Wood, E. C., Herndon, S. C., Freedman, A. A practical alternative to chemiluminescence-based detection of nitrogen dioxide: Cavity attenuated phase shift spectroscopy. *Environ. Sci. Technol.* **2008**, *42*, 6040–6045.
99. Paul, J. B., Scherer, J. J., Keefe, A. O., Lapson, L., Anderson, J. G., Gmachl, C., Capasso, F., Cho, A. Y. Infrared cavity ringdown and integrated cavity output spectroscopy for trace species monitoring. *Proc. SPIE. Spectrosc. Sens. Syst.* **2002**, *4577*, 1–11.
100. Moyer, E. J., Sayres, D. S., Engel, G. S., St. Clair, J. M., Keutsch, F. N., Allen, N. T., Kroll, J. H., Anderson, J. G. Design considerations in high-sensitivity off-axis integrated cavity output spectroscopy. *Appl. Phys. B* **2008**, *92*, 467–474.
101. Decker, J. A., Rogers, H. W. Revised airborne exposure limits for chemical wargare agents. In Kolodkin, V. M., Ruck, W., (eds), *Ecological Risks Associated with the Destruction of Chemical Weapons*, Dordrecht: Springer, 2006, pp. 279–287.
102. Brown, M. A., Brix, K. A. Review of health consequences from high-, intermediate- and low-level exposure to organophosphorus nerve agents. *J. Appl. Toxicol.* **1998**, *18*, 393–408.
103. Webber, M. E., Pushkarsky, M., Patel, C. K. N. Optical detection of chemical warfare agents and toxic industrial chemicals: Simulation. *J. Appl. Phys.* **2005**, *97*, 113101.
104. Sharpe, S. W., Johnson, T. J., Chu, P. M., Kleimeyer, J., Rowland, B., Northwest, P., Ground, D. P., Desert, W. Quantitative, infrared spectra of vapor phase chemical agents. *Proc. SPIE Chem. Biol. Sens. IV* **2003**, *5085*, 19–27.
105. Pushkarsky, M. B., Webber, M. E., Macdonald, T., Patel, C. K. N. High-sensitivity, high-selectivity detection of chemical warfare agents. *Appl. Phys. Lett.* **2006**, *88*, 044103.

106. Mukherjee, A., Dunayevskiy, I., Prasanna, M., Go, R., Tsekoun, A., Wang, X., Fan, J., Patel, C. K. N. Sub-parts-per-billion level detection of dimethyl methyl phosphonate (DMMP) by quantum cascade laser photoacoustic spectroscopy. *Appl. Opt.* **2008**, *47*, 1543–1548.

107. Patel, C. K. N. Laser based *in situ* and standoff detection of chemical warfare agents and explosives. *Proc. SPIE Opt. Based Biol. Chem. Detect. Def. V* **2009**, *7484*, 748402–748414.

108. Takeuchi, E. B., Rayner, T., Weida, M., Crivello, S., Day, T. Standoff detection of explosives and chemical agents using broadly tuned external-cavity quantum cascade lasers (EC-QCLs). *Proc. SPIE Opt. Photonics Counterterrorism Crime Fight. III* **2007**, *6741*, 674104–674109.

109. Haibach, F., Erlich, A., Deutsch, E. Mid-infrared absorption spectroscopy using quantum cascade lasers. *Proc. SPIE* **2011**, 8032, 803208.

110. Chen, X., Choa, F.-S., Holthoff, E., Pellegrino, P., Fan, J. Standoff chemical detection with parts per million level calibrated detection sensitivity. *Proc. SPIE, Secur. Sens.* **2013**, *8710*, 871007.

111. Poznich, C. R., Thomas, D. W., Industries, E. Minimization of IR absorption by germanium at elevated temperatures. *Proc. SPIE* **1990**, 1326, 106–119.

112. O'Keefe, A., Gupta, M., Owano, T. G., Baer, D. S. Absorption spectroscopy instrument with increased optical cavity power without resonant frequency build-up. US Patent 7,468,797, 2008.

113. Leen, J. B., O'Keefe, A. Optical re-injection in cavity-enhanced absorption spectroscopy. *Rev. Sci. Instrum.* **2014**, *85*, 093101.

114. Bartelt-Hunt, S. L., Knappe, D. R. U., Barlaz, M. A. A review of chemical warfare agent simulants for the study of environmental behavior. *Crit. Rev. Environ. Sci. Technol.* **2008**, *38*, 112–136.

115. Butrow, A. B., Buchanan, J. H., Tevault, D. E. Vapor pressure of organophosphorus nerve agent simulant compounds. *J. Chem. Eng. Data* **2009**, *54*, 1876–1883.

116. Chu, P. M., Guenther, F. R., Rhoderick, G. C., Lafferty, W. J. The NIST quantitative infrared database. *J. Res. Natl. Inst. Stand. Technol.* **1999**, *104*, 59.

117. Sharpe, S. W., Johnson, T. J., Sams, R. L., Chu, P. M., Rhoderick, G. C., Johnson, P. A. Gas-phase databases for quantitative infrared spectroscopy. *Appl. Spectrosc.* **2004**, *58*, 1452–1461.

118. Johnson, T. J., Sams, R. L., Sharpe, S. W. The PNNL quantitative infrared database for gas-phase sensing: A spectral library for environmental, hazmat and public safety standoff detection. *Proc. SPIE Chem. Biol. Point Sensors Homel. Def.* **2004**, *5269*, 159–167.

119. Sharpe, S. W., Johnson, T. J., Sams, R. L. A quantitative infrared spectral library of vapor phase chemicals: applications to environmental monitoring and homeland defense. *Proc. SPIE* 2004, 5584, 77–84.

120. Beebe, K. R., Pell, R. J., Seasholtz, M. B. *Chemometrics: A Practical Guide*, John Wiley & Sons: New York, 1998.

121. Karoui, R., Downey, G., Blecker, C. Mid-infrared spectroscopy coupled with chemometrics: a tool for the analysis of intact food systems and the exploration of their molecular structure-quality relationships: A review. *Chem. Rev.* **2010**, *110*, 6144–6168.

122. The California Air Resources Board. California reformulated gasoline, California Air Resources Board, 2012.

123. Yinon, J. Field detection and monitoring of explosives. *TrAC Trends Anal. Chem.* **2002**, *21*, 292–301.

124. Ewing, R. G., Waltman, M. J., Atkinson, D. a., Grate, J. W., Hotchkiss, P. J. The vapor pressures of explosives. *TrAC Trends Anal. Chem.* **2013**, *42*, 35–48.

125. Hummel, R., Dubroca, T. Laser- and optical-based techniques for the detection of explosives. *Encyclopedia of Analytical Chemistry*, John Wiley & Sons: Hoboken, NJ, 2013.

126. Munson, C. A., Gottfried, J. L., Lucia, F. C. De, Mcnesby, K. L., Miziolek, A. W. *Laser-Based Detection Methods for Explosives ARL-TR-4279*, Aberdeen Proving Ground: MD, 2007.

127. Caygill, J. S., Davis, F., Higson, S. P. J. Current trends in explosive detection techniques. *Talanta* **2012**, *88*, 14–29.

128. Steinfeld, J. I., Wormhoudt, J. Explosives detection: A challenge for physical chemistry. *Annu. Rev. Phys. Chem.* **1998**, *49*, 203–232.

129. Burks, R. M., Hage, D. S. Current trends in the detection of peroxide-based explosives. *Anal. Bioanal. Chem.* **2009**, *395*, 301–313.

130. Kolla, P. The Application of analytical methods to the detection of hidden explosives and explosive devices. *Angew. Chem. Int. Ed. Engl.* **1997**, *36*, 800–811.

131. Stott, W. A., Davidson, W. A., Rd, G. C. High throughput real time chemical ntraband detection. *Proc. SPIE* **1993**, 2092, 53–63.

132. Nambayah, M., Quickenden, T. I. A quantitative assessment of chemical techniques for detecting traces of explosives at counter-terrorist portals. *Talanta* **2004**, *63*, 461–467.

133. Sharpe, S. W., Johnson, T. J., Sheen, D. M., Atkinson, D. a. Relative Infrared (IR) and Terahertz (THz) Signatures of Common explosives. *Proc. SPIE* 2006, 6378, 63780A1-63780A5.

134. Johnson, T. J., Valentine, N. B., Gassman, P. L., Atkinson, D. a., Sharpe, S. W., Williams, S. D. On the Relative Utility of Infrared (IR) versus Terahertz (THz) for Optical Sensors. In *Proceedings of SPIE Vol. 6756 Chemical and Biological Sensors for Industrial and Environmental Monitoring III*, Ewing, K. J., Gillespie, J. B., Chu, P. M., Marinelli, W. J., Eds., 2007, Vol. 6756, pp. 675604–1 – 675604–675607.

135. Janni, J., Gilbert, B. D., Field, R. W., Steinfeld, J. I. Infrared absorption of explosive molecule vapors. *Spectrochim. Acta A* **1997**, *53*, 1375–1381.

136. Bauer, C., Sharma, a. K., Willer, U., Burgmeier, J., Braunschweig, B., Schade, W., Blaser, S., Hvozdara, L., Müller, a., Holl, G. Potentials and limits of mid-infrared laser spectroscopy for the detection of explosives. *Appl. Phys. B* **2008**, *92*, 327–333.

137. Martin, M., Crain, M., Walsh, K., McGill, R. A., Houser, E., Stepnowski, J., Stepnowski, S., Wu, H.-D., Ross, S. Microfabricated vapor preconcentrator for portable ion mobility spectroscopy. *Sensors Actuators B Chem.* **2007**, *126*, 447–454.

138. Voiculescu, I., Mcgill, R. a., Zaghloul, M. E., Mott, D., Stepnowski, J., Stepnowski, S., Summers, H., Nguyen, V., Ross, S., Walsh, K., et al. Micropreconcentrator for enhanced trace detection of explosives and chemical agents. *IEEE Sens. J.* **2006**, *6*, 1094–1104.

139. Hallowell, S. F. Screening people for illicit substances: A survey of current portal technology. *Talanta* **2001**, *54*, 447–458.

140. Gresham, G. L., Davies, J. P., Goodrich, L. D., Blackwood, L. G., Liu, B. Y., Thimsen, D., Yoo, S. H., Hallowell, S. F. Development of particle standards for testing detection systems: mass of RDX and particle size distribution of composition 4 residues. In *Proceedings of SPIE Vol. 2276, Cargo Inspection Technologies*, Lawrence, A. H., Ed., 1994, Vol. 2276, pp. 34–44.

141. Fuller, A. M., Hummel, R. E., Schöllhorn, C., Holloway, P. H. Stand off detection of explosive materials by differential reflection spectroscopy. In Christesen, S. D., Sedlacek III, A. J., Gillespie, J. B., Ewing, K. J., (eds), *Proceedings of SPIE Vol 6378 Chemical and Biological Sensors for Industrial and Environmental Monitoring II,*, 2006, Vol. 6378, pp. 637819–637819–11.

142. Wojtas, J., Stacewicz, T., Bielecki, Z., Rutecka, B., Medrzycki, R. Towards optoelectronic detection of explosives. *Opto-Electronics Rev.* **2013**, *21*, 210–219.

143. Wojtas, J., Mikolajczyk, J., Bielecki, Z. Aspects of the application of cavity enhanced spectroscopy to nitrogen oxides detection. *Sensors* **2013**, *13*, 7570–7598.

144. Miller, A., Richman, B., Viteri, C. R., McKeever, J. Selective cavity-enhanced trace gas detection via diffusion time-of-fight. In *Proceedings of SPIE Vol. 8358 Chemical, Biological, Radiological, Nuclear, and Explosives (CBRNE) Sensing XIII*, Fountain, A. W., Ed., 2012, Vol. 8358, pp. 83581C–1–83581C–9.
145. Stockton, J. K., Tuchman, A. K. Bayesian estimation for selective trace gas detection. *Appl. Phys.* **2008**, *96*, 567–570.
146. Gerecht, E., Douglass, K. O., Plusquellic, D. F. Chirped-pulse terahertz spectroscopy for broadband trace gas sensing. *Opt. Express* **2011**, *19*, 8973–8984.
147. Tonouchi, M. Cutting-edge terahertz technology. *Nat. Photonics* **2007**, *1*, 97–105.
148. Scalari, G., Walther, C., Fischer, M., Terazzi, R., Beere, H., Ritchie, D., Faist, J. THz and sub-THz quantum cascade lasers. *Laser Photonics Rev.* **2009**, *3*, 45–66.

12 Detection and Recognition of Explosives Using Terahertz-Frequency Spectroscopic Techniques

Henry O. Everitt and Frank C. De Lucia

CONTENTS

12.1 INTRODUCTION

This chapter considers the challenge of using terahertz (THz) techniques for the detection and recognition of explosives. There are two general approaches to this problem: (1) imaging, which uses the penetration capability of THz radiation to "see" the image of the explosive, and (2) spectroscopy, which allows the chemical signature to be recognized. The former is beyond the scope of this volume and has been well addressed previously.[1] The latter can be divided into two subcategories: the detection and recognition of the explosive itself and the detection and recognition of its precursors, additives, or decay products. Most reports claiming detection of these materials, hereafter collectively called "explosives," have only considered laboratory measurements of them in pure form and have not considered the greater challenge of recognizing these materials among naturally occurring compounds or mixed with other compounds and shielded in containers. When these challenges are considered seriously, the exquisite recognition specificity afforded by gas-phase THz spectroscopy may only be used to detect and recognize explosives in very limited circumstances.

This chapter focuses on the application of spectroscopic techniques to the detection and recognition of explosives. After a discussion of the basic physics and technology of the spectral region, we will consider THz approaches for molecular detection and recognition. Because an excellent article on the widely discussed prospects for direct THz spectroscopic detection of explosives has recently appeared,[2] we will address the THz detection of solid explosives primarily by reviewing the highlights and conclusions of this article. Here we will focus on innovative methods for atmospheric point and remote sensing and recognition in cases where the explosives contain a measurable vapor pressure.

12.1.1 BACKGROUND

The THz spectral region is of great promise and some disappointment. It has perhaps been best described by Graham Jordan in the opening plenary talk of the SPIE Symposium on Optics and Photonics in Bruges, as an area of "whispered excitement."[3] This promise and excitement arise not only because the THz is virgin territory for many, but also because radiation at these wavelengths gives a unique and often useful combination of penetrability and imaging resolution, combined with the prospect of using spectral resonances for materials recognition.

However, for a potential THz application to be successful, these qualitative features must stand up to quantitative analysis and comparison with alternative solutions. Here we will discuss the detection of explosives in the context of technology choices and the physics that links applications and technology.

12.1.2 NOMENCLATURE AND COMMUNITIES

There are two distinct communities that have interests in the spectral region between roughly 0.1 THz (3 mm) and 10 THz (0.03 mm). The community that has arisen primarily from the microwave electronics tradition typically refers to this as the millimeter and submillimeter (mm/submm) spectral region.[4-7] The community that uses gas-phase lasers, ultrashort laser pulses, difference frequency laser mixing, or other photonic approaches typically refer to this as the THz spectral region.[8-11] Over time, these designations are becoming blurred in some areas of application, so, for the purposes of this chapter, we will refer to this spectral region as the *SMM/THz*.

Additionally, a third community, commonly referred to as the microwave community and associated with wireless communications, is rapidly approaching this spectral region, with the potential to alter it fundamentally by developing mass-market, low-cost technology. While different in market size and history, its technology closely resembles that of the electronic mm/submm community. Because the size of the communications community increases very rapidly as frequency goes below 100 GHz, we will not attempt here to do more than note its potential to impact the THz region dramatically in the near future. Interested readers are referred to a recent review that considers the expansion of the communications applications into the SMM/THz spectral region.[12]

There is remarkably little cross-referencing and awareness between these communities, to some extent because the technologies that have emerged often have very different characteristics and applications. One consequence of this is that web searches for key words such as "THz" can give a rather distorted view of activity in the SMM/THz spectral region as a function of time. Another more serious consequence is that much effort has been devoted to the development of applications based on inappropriate technology. Since most of the readers of this chapter come from the optics tradition, and there have been many tutorials, reviews, and press releases about laser-based THz systems,[13–18] this chapter will focus on electronic technologies. Because of issues involving power, spectral purity, and frequency calibration, virtually the entire mm/submm spectroscopic scientific database derives from electronic techniques.[19–21] Our purpose is not to compare the various technologies but to discuss the salient molecular physics.

12.1.3 BANDWIDTHS, TECHNOLOGY, PHENOMENA, AND NAMES

It will emerge in the following that the bandwidths of the several available SMM/THz technologies and phenomena under investigation play a key role in the technical, scientific, and social organization of these communities and their interactions. This is because a very large fraction of applications involve either rather narrow bandwidths (low-pressure gases, radar, communications, etc.) of <1 MHz or very broad bandwidths (solids and liquids, atmospheric-pressure gases) of ~100,000 MHz. The former have often been referred to as "SMM" techniques and the latter as "THz" techniques. It should also be remarked that the thermal approaches, whose practitioners refer to this same spectral region as the far infrared (FIR), also provide powerful and often simple methods for the study of broadband phenomena.

12.1.4 HISTORY

While it is not the purpose of this chapter to provide a history of the development of this spectral region, a number of interesting articles have been written that provide useful context. Wiltse has included in his history of millimeter and submillimeter waves a particularly interesting account of the early years, tracing the development back to the experiments of Hertz and including a discussion of both early thermal and early electronic approaches.[22] By the late 1950s, much of the fundamental physics had been defined, and the technology to explore it had been developed.

Gordy provides an overview.[4] More recently, Siegel provided an interesting history including discussions of current technology and their applications.[18] An interesting set of bookends for this spectral region is the April 1966 issue of *The Proceedings of the IEEE (Millimeter Waves and Beyond)*[5] and the inaugural issue of the *IEEE Transactions on Terahertz Science and Technology.*[23]

This development of the SMM/THz has been an extraordinarily successful undertaking, having attracted interest and contributions from a much broader community. In addition to the closely connected physical chemists, the communities of astrophysicists and atmospheric scientists have emerged as important participants, witnessed by substantial systems analysis reviews that led to the construction of major THz astronomical systems.[24–26]

However, it is noteworthy that, while there are a number of extremely successful applications, with large user communities and substantial investment, none of these are mass applications used beyond the bounds of specialized communities. Of all the regions of the electromagnetic spectrum, and in spite of substantial attention in the media to "public" applications in SMM/THz, our spectral region is still uniquely unexploited for widespread applications such as the detection and recognition of explosives. The situation is improving, however, as technological advances from both the electronics mm/submm community and the photonics THz community are producing commercially viable systems that will inevitably lead to practical, fieldable solutions.

12.2 THE PHYSICS

In this section, we will discuss some of the basic physics of the SMM/THz spectral region. This will include both the fundamental physics of system noise as well as the physics of interactions between SMM/THz radiation and physical systems of interest.

12.2.1 BLACKBODY RADIATION AND NOISE

In much of the optical spectral regions, the background blackbody radiation and the noise associated with it play only a minor role, and other forms of noise, such as shot noise, are dominant. However, the SMM/THz is immersed in the long-wavelength tail of the 300 K blackbody radiation. Because there are many misconceptions about the role of noise in SMM/THz systems, we will discuss this noise first.

Many years ago, Lewis considered this issue in some detail.[27] The energy density $\rho(\nu)$ between frequency ν and $\nu + d\nu$ is given by

$$\rho(\nu) = \frac{8\pi h \nu^3}{c^3} \left(\frac{1}{e^{h\nu/kT} - 1} \right) d\nu \qquad (12.1)$$

where T is the blackbody temperature, h is Planck's constant, c is the speed of light, and k is Boltzmann's constant. In the long-wavelength limit typical of the SMM/THz, $h\nu \ll kT$, and Equation 12.1 can be simplified to [28]

$$\rho(\nu) = \frac{8\pi k T \nu^2}{c^3} d\nu \qquad (12.2)$$

In this limit, simple manipulations lead to the familiar relation for the power P in a single mode, as given in Equation 12.3:

$$P = kTd\nu \qquad (12.3)$$

At 300 K, the power within a typical 1 MHz SMM spectroscopic linewidth is $P = 4.14 \times 10^{-15}$ W.

In the long-wavelength limit of the SMM/THz, source temperature is a useful concept. Solving Equation 12.3 for T yields the temperature required of a blackbody source to produce a power P in a single mode in bandwidth $\Delta\nu$:

$$T = \frac{P}{k\Delta\nu} \qquad (12.4)$$

This leads directly to the concept of source brightness, in units of watts per hertz. Source brightness is often a better figure of merit than source power. In terms of interactions with physical systems, what matters is not the total available power but rather the power within the system bandwidth, because only this power interacts with the physical system. Moreover, the power outside this linewidth is often detrimental because it can add to overall system noise.

The noise limit in a SMM/THz system originates in the fluctuation in the number of photons in the thermal background.[27] Because the photons within each mode are governed by Bose–Einstein statistics, it is easy to show that[28,29]

$$P_N \sim kT(B\nu_{max})^{1/2} = kT\nu_{max}\left(\frac{B}{\nu_{max}}\right)^{1/2} \qquad (12.5)$$

for broadband detectors in the long-wavelength limit, where ν_{max} is the maximum frequency response of the detector, and B is the bandwidth of the integrating amplifier. The noise-equivalent power (NEP) is defined by normalizing this to $B=1$ Hz,

$$NEP \equiv kT\nu_{max}^{1/2}. \qquad (12.6)$$

With $\nu_{max} = 1$ THz, $B = 1$ Hz, and $T = 300$ K, $P_N = 4.14 \times 10^{-15}$ W/Hz$^{1/2}$.

In contrast, heterodyne detectors can sample a small bandwidth, $\Delta\nu$, at any point in the long-wavelength tail of the blackbody distribution, and the noise power, P_N, in this bandwidth is

$$P_N = kT\Delta\nu, \qquad (12.7)$$

from which we may define $NEP' = kT$ with $\Delta\nu = 1$ Hz. This leads to $P_N = 4 \times 10^{-21}$ W. To place this on the same basis as Equation 12.6, one must assume an integrating bandwidth of $B=1$ Hz as well, leading to the same numerical value. However, it should be noted that whereas $\nu_{max} = 1$ THz is typical of broadband detectors, a bandwidth of $\Delta\nu = 1$ Hz is *not* typical for heterodyne systems. Since this bandwidth is set by the (adjustable) bandwidth of the intermediate frequency (IF) amplifier, we set this bandwidth to be $\Delta\nu = B$.

Figure 12.1 shows noise and power as a function of the frequency cutoff ν_{max}, with $B = 1$ Hz for a broadband detector. For a heterodyne system, it would be less. The noise power given by Equation 12.7 is many orders of magnitude below the thermal power. Confusion of the two, along with neglect of the single-mode antenna theorem, has led to the widespread misconception that continuous-wave SMM systems are "plagued by noise."[28] To the contrary, it is the quiet background that has made much of the activity in the SMM possible, even with modest power levels. Expressed another way, Equation 12.6 shows that even the modest power of 1 mW corresponds to $T \sim 10^{14}$ K for a bandwidth of ~1 MHz, very bright indeed in comparison to 300 K noise.

12.2.2 TOWNES NOISE

Because this volume is primarily aimed at laser methods, it is important to note a significant difference between fundamental noise limits of optical spectroscopy (shot noise) and SMM/THz spectroscopy (Townes noise).

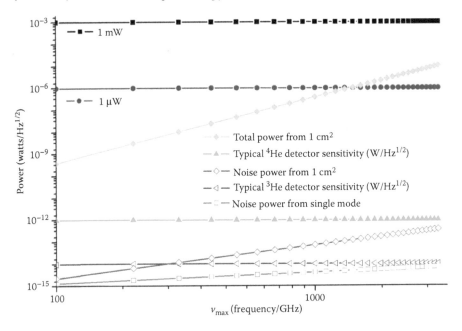

FIGURE 12.1 Noise and power in the SMM spectral region as a function of the maximum frequency response ν_{max} of a broadband detector. The upper black and red traces show 1 mW and 1 μW of source power, respectively, illustrating how much larger electronically generated power is than thermal noise (purple or dark green traces). Similarly, the blackbody thermal power radiated from 1 cm² at 300 K (light blue trace) is orders of magnitude greater than the thermal noise by any of the noise measures shown in the remaining trace, including the light green and dark blue traces, which represent the sensitivity of an InSb or Si bolometer at ⁴He or ³He temperatures, respectively. (From De Lucia, F. C., *J. Opt. Soc. Am.*, B 21, 1273–1297, **2004**.)

Because SMM/THz frequencies are 10–1000 times lower than optical frequencies, emission lifetimes (which scale as v^{-3}) are much longer, and experiments and sensors ordinarily operate in absorption rather than emission. Additionally, except at cryogenic or interstellar temperatures, experiments are done in the long-wavelength tail of the blackbody radiation. As a result, there is often an intimate interaction between the noise in sources, detectors, and backgrounds.

Absorption spectrometers operate in a very different noise regime from imagers, radars, and communications systems. They detect a small fractional change in a relatively large amount of power rather than a small amount of power against background noise. Accordingly, we need to consider an additional, and often dominant, source of noise, which we shall refer to as Townes noise.[30] This noise arises because in the waveguide mode containing the probe power and the thermal noise power, it is the linear superposition of their electric fields, not of their powers, that determines the noise characteristics.[6] The voltage V of the probe power P in a waveguide of impedance Z is

$$V^2 = 2ZP \tag{12.8}$$

and the noise fluctuation voltage ΔV in a bandwidth Δv is

$$\left(\Delta V\right)^2 = 4ZkT\Delta v \tag{12.9}$$

The total power in the waveguide is then

$$\frac{\left(V \pm \Delta V\right)^2}{2Z} = P \pm 2\left(2kTP\Delta v\right)^{1/2} + 2kT\Delta v \tag{12.10}$$

As a numerical example, with a 1 s integration time (which also sets the bandwidth over which noise is measured) and 1 mW of probe power, the cross term is of the order 10^{-12} W. This noise contribution is larger than the last term (and larger than Equation 12.7) by approximately

$$\left(\frac{P}{kT\Delta v}\right)^{1/2} \tag{12.11}$$

This factor is $>10^8$ for our numerical example.

12.2.3 INTERACTIONS WITH MATTER

One of the most widely touted characteristics of SMM/THz radiation—and one of significance to the detection of explosives—is its purported ability to "see through walls." While it is true that SMM/THz radiation may penetrate many materials that are opaque to shorter wavelength radiation, this is a limited capability confined mostly to sub-terahertz frequencies in a manner that depends strongly on frequency and water content. Of course, SMM/THz radiation cannot penetrate metallic containers, so we need only discuss the attenuation by dielectric barriers and containers.

To assess this potential, the frequency dependence of a material's attenuation must be known. Figure 12.2 illustrates the one-way attenuation of common building materials at sub-THz frequencies. Note the exponential growth in attenuation with increasing frequency, especially of wooden materials, whose attenuation grows to ~40–50 db at 0.5 THz and is much worse when wet. A similar exponential growth of attenuation with frequency has been observed with clothing materials, but with lower losses varying between 2 and 10 db at 1 THz.[31] Results from Gatesman et al.[32] show a similar rapid rise in attenuation with increasing frequency, with most common building materials (including glass) showing no transmission near 1 THz using a system with >70 db dynamic range. This would seem to preclude widely discussed applications such as seeing through concrete walls[15,33] and *in situ* detection of cancer. Foam is an exception to the strong attenuation of most building materials and made possible the now obsolete application of THz imaging techniques to locate defects buried in space shuttle insulation.[35]

It has been noted that this rapid rise of attenuation with frequency has often been missed or omitted in the analysis of potential SMM/THz applications.[2] The important point here is that imaging through materials is *very* different at 100 GHz than 1 THz. This has particular importance for the interpretation of results of imaging systems for which broadband radiation is used, especially those for which there is considerably more power at lower frequency with commensurately lower attenuation.

12.2.3.1 Interactions with Gases

The vast majority of scientific results used by the wider scientific and technical communities are based on SMM/THz interactions with gas-phase molecules. An

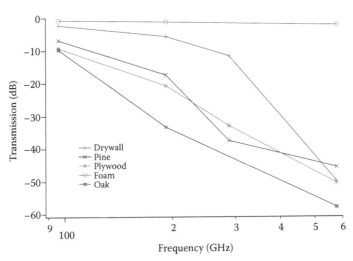

FIGURE 12.2 One-way attenuation of common materials as a function of frequency from 90 to 600 GHz. All of these building materials were ~2 cm thick.

extensive literature and a number of databases exist.[19–21,36] These range from fundamental studies of molecular structure and collision dynamics to studies done to provide a basis for the large astrophysical and atmospheric science communities who base many of their remote sensing experiments on the strong and specific rotational interactions of gases with SMM/THz radiation.

It is well known that different regions of the electromagnetic spectrum probe different types of molecular transitions: the electronic degree of freedom corresponds to optical and ultraviolet radiation, the vibrational degree of freedom to the infrared (IR), the rotational degree of freedom to the MM/SMM/THz, and various fine, hyperfine, and magnetic interactions to the radio region. The rotational signature is much less familiar than the others; as evidence, note that large-scale applications for all of the other signatures have been developed for the commercial sector, ranging from alcohol breathalyzers to magnetic resonance imagers.

While the details of rotational spectroscopy[6,7,37] can bring joy to an aficionado and despair to much of the rest of the scientific community, we would like to mention a few basic principles:

1. The fingerprint that arises is linked, in first order, to the moment of inertia tensor of the molecule. This means that, unlike with vibrational spectroscopy in the IR, the signature depends upon the *whole* molecule, not functional groups (e.g., vibrational energies associated with specific bond).
2. Molecules with small moments of inertia, such as water (H_2O) and hydrochloric acid (HCl), have sparse spectra whose intensities peak in the low THz, whereas heavier molecules (such as chlorine nitrate) have much denser spectra that peak at a few hundred gigahertz or lower.[38]
3. At room temperature, the thermal quantum, kT, is larger than the rotational energy level spacing. As a result, many rotational levels are populated. The number of levels with significant population is referred to as the rotational partition function, Q_r, and ranges from ~100 to 10^5 for molecules of interest. This has two important consequences:
   a. Because many levels are populated, if an instrument can resolve the rotational structure, the fingerprint is so specific that it is reasonable to speak of "absolute" recognition specificity.[39]
   b. Because the population is divided among many levels, this reduces the strength of any one transition. However, the basic strength of the electric dipole moment interaction, the quietness of the background, the ability to build systems whose limits approach the background limit, and the low optimum pressure produce a very sensitive technique.
4. In many circumstances, it is advantageous to take advantage of the small Doppler broadening in the SMM/THz (~1 MHz) by operating at low pressure (~10 mTorr), thus reducing sample requirements. A negative consequence is that the collisional broadening produced by the 10^5 times higher atmospheric pressure hinders most applications at or near atmospheric pressure.

Figure 12.3 shows a series of spectral expansions of an important astrophysical molecule containing a complex asymmetric rotor, ethyl cyanide (CH_3CH_2CN), in

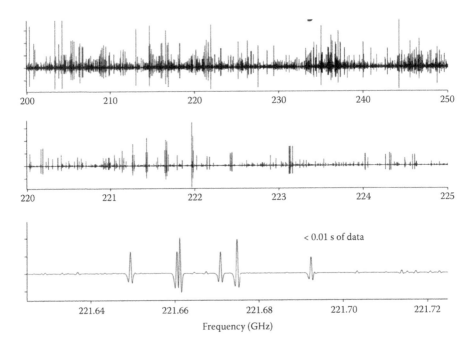

FIGURE 12.3 Rotational fingerprint of 1 mTorr of ethyl cyanide (CH_3CH_2CN) in the region near 600 GHz as a series of spectral expansions. (From Fortman, S. M., Medvedev, I. R., Neese, C. F., and De Lucia, F. C., *Astrophys. J.*, 714, 476–486, **2010**.)

the region around 600 GHz.[40] For comparison, water has only about 3 lines in this spectral region, while ethyl cyanide has ~10,000, showing the very rapid change of spectral density as a function of molecular size and mass.

12.2.3.2 Atmospheric Propagation

It is well known that the atmosphere significantly affects the performance of systems in the SMM/THz because of its strong and highly variable attenuation of radiation.[1,41,42] Figure 12.4 shows the attenuation for several atmospheric conditions from the microwave into the ultraviolet. Considering that the plot is of a logarithmic property on a log scale, note the rapid rise of the absorptions due to molecular rotation as the frequency is increased out of the microwave into the THz (100–10,000 GHz), peaking at several THz, before falling exponentially into the IR. In the IR (10,000–300,000 GHz), strong vibrational bands due to water and carbon dioxide become important, before the frequency becomes higher than the fundamental of molecular motion and the transparency of the visible is reached (>300,000 GHz). Above the visible into the ultraviolet, electronic transitions rapidly make the atmosphere opaque, leading to the "vacuum ultraviolet." Note the sensitivity of attenuation to fog and rain at MM/SMM/THz frequencies, particularly that fog is more transparent in this region than in the IR. In a heavy fog, it is the humidity in the highly saturated air between the water droplets that leads to

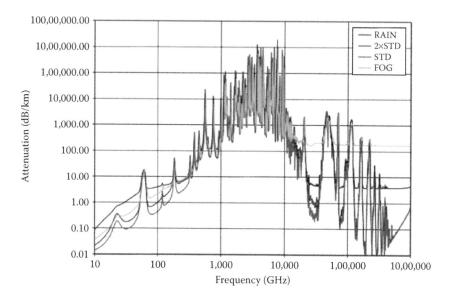

FIGURE 12.4 Atmospheric propagation as a function of frequency, showing in red the atmospheric attenuation due to water vapor and oxygen for a standard atmosphere, in blue the attenuation for a atmosphere with twice as much water, in black with the addition of rain, and in green with the addition of fog. (Courtesy of B. Wallace.)

the absorption, while the droplets themselves contribute more to scattering than to absorption.

12.2.3.3 Interactions with Solids and Liquids

Solids, liquids, and atmospheric-pressure gases do not have the sharp resonances associated with low-pressure gases. Nevertheless, these are exceptionally important analytes of interest, and considerable attention has been paid to them. In solids, molecules are not free to rotate, and there is no rotational spectrum. As a result, signatures must come either from low-lying vibrational modes of large molecules or from "classical" resonances associated with the texture and the structure of the target.[2] The latter may either serve as target signatures or clutter interference, often depending upon their reproducibility and sensitivity to their environment.

The observation of resonances in solids is challenging. They are often broad and difficult to separate from standing wave effects (1 mm corresponds to 150/n GHz for a material with refractive index n), especially when the target is embedded in lossy material. Nevertheless, there have been many reproducible reports of resonances in solids.[15,35,43–45] Figure 12.5 shows perhaps the most well-known spectra of solids in the SMM/THz, the spectra of several explosives between 0.5 and 4 THz.[46,47] These spectra are "well behaved" in that they are not sensitive to their environment and satisfy Beer's law.[2] While narrower resonances at lower frequencies in explosives have been reported,[48] these have not been confirmed.[2] Resonances of similar widths (10 cm^{-1}) have been reported in crystalline pharmaceuticals.[49] Most, if not all, of the distinct, reproducible resonances reported are of materials in crystalline form.

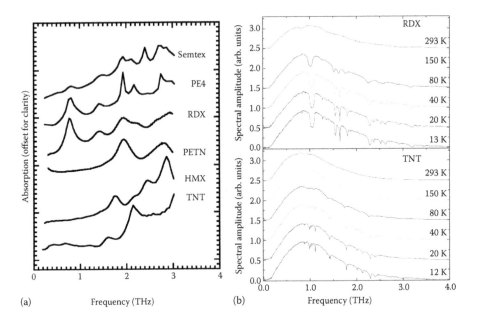

FIGURE 12.5 (a) THz-time domain spectra of several explosives. (b) Spectra of thin films of two explosives (RDX and TNT) as a function of temperature. (From (a) Kemp, M. C., Taday, P. F., Cole, B. E., Cluff, J. A., Fitzgerald, A. J., Tribe, W. R., Security applications of terahertz technology. In Hwu, R. J., Woolard, D. L. (eds), *Proceedings of SPIE v5070*, pp. 44–52. Orlando, FL, **2003**; (b) Melinger, J. S., Laman, N., Grischkowsky, D., *Appl. Phys. Lett.*, 93, 011102–011102-3, **2008**.)

An example of the impact of environment is sucrose, whose strong spectral features disappear in its amorphous form.[50]

Melinger et al. have shown an interesting line of attack on many of these challenges.[45] Briefly, in Figure 12.5b they have shown that by using a waveguide surface to provide orientation and order to a sample and by cooling the sample to cryogenic temperatures, much narrower and well-defined resonances can be observed.

12.3 APPLICATIONS

We will focus our attention on two potential methods for the detection of explosives in the SMM/THz and some of the science and technology that enables them. They are (1) direct spectroscopic detection of the explosive and (2) indirect spectroscopic detection via detection of precursors or decay products.

12.3.1 GAS-PHASE POINT SENSING

It can be argued that gas-phase spectroscopy is the central application of the SMM/THz spectral region—most of the applications that are in use are based on it. Because one of the authors has written a review of this field,[36] we will be very brief here. In

Section 12.2.3.1, the fundamental interaction between gases and radiation has been laid out. From a system point of view, point sensing is more attractive because gas-phase spectroscopy and related applications involve collecting the sample itself or its vapors, then pumping away the atmospheric gases and reducing the pressure of the remaining vapors to reveal the narrow Doppler broadened lines of low-pressure gases of interest. For this point-sensing application, a premium is placed on source brightness, frequency agility, and absolute frequency calibration. As a result, electronic sources based on frequency synthesis in the microwave frequency regime followed by frequency multiplication are very advantageous.

In general, explosives have very low equilibrium vapor pressures. This is especially true of many of the most important explosives, such as trinitrotoluene (TNT, 9.15 ppb), pentaerythritol tetranitrate (PETN, 0.0107 ppb), and cyclotrimethylene-trinitramine (RDX, 0.00485 ppb). Partial pressures in field situations are often orders of magnitude smaller, and the explosives are often held in containers that further limit out-gassing. However, if it is possible to sample, for example, shipping containers for the more volatile co- and by-products, this may be an advantageous strategy.

One point-sensing approach is to collect and heat solid particulate samples, because the vapor pressures can be raised considerably. However, it is unclear how crowded and resolvable the spectra of these relatively large molecules will be. Another approach is to decompose the solid particulates, resulting in smaller decay by-product molecules with much higher vapor pressures. A significant disadvantage of this approach is that the "absolute" specificity associated with the detection of a high-resolution SMM/THz rotational spectrum is lost. Instead, while the detection of the by-products may be absolutely specific, the specificity of the detection of the parent explosive depends upon how well the decomposition process is understood. In complex field mixtures, this can be very challenging. However, it should be noted that many widely used analytical techniques, including mass spectroscopy, ion mobility spectroscopy, and gas chromatography, make similar inferences. Since the SMM/THz fragment detection is both quantitative and highly specific, it is possible that an SMM/THz approach could be advantageous. However, to the best of the authors' knowledge, this approach has not been explored. This stands in stark contrast to the many person-years that have been devoted to similar decomposition approaches based on other fragment detection techniques.

Even in the infancy of microwave spectroscopy, its potential as an analytical method was recognized. In their 1955 book, Townes and Schawlow devoted an entire chapter to this subject and expressed surprise that this application had not yet emerged.[6] In the mid 1970s, Hewlett-Packard developed a system that turned out to be of more interest to structural chemists than the analytical community. Its lack of analytical success can be traced to the facts that it was large, expensive, relatively slow, and not too sensitive, and that it had limited spectral range.

Technological developments and new system approaches now make practical gas-phase analytical point sensors based on the SMM/THz techniques.[38,51,52] These include advances in solid-state technology,[53] microcomputing for the large information content of rotational spectra, and simplified system designs.[51,54] In the few cases for which a sufficient quantity of explosive precursor or decomposition fragments can be detected, it is necessary to consider how a sensitive SMM/THz sensor system performs.

Low-pressure gas-phase point sensors in the SMM/THz are based on the highly specific and strong rotational fingerprint discussed in Section 12.2.3.1. An example of a mixture of 20 gases is shown in Figure 12.6, along with an expansion of a small region that can be used to identify 8 of the gases. Many similar regions exist. Since the frequency of each spectral line is known to $\sim 1/10^7$ and intensities are known quantitatively as well, any identification is highly specific. In fact, in this mixture, a simple statistical analysis showed that the probability of false alarm for the least favorable of the 20 gases was 10^{-107}, close enough to zero to claim "absolute" specificity *in a complex mixture*.

Figure 12.7 shows a compact (1 cu. ft.), self-contained, room-temperature, low-pressure gas-phase point sensor used to take spectra such as those in Figures 12.6 and 12.8.[39] In this system, frequency agile electronic synthesis methods in the region around 10 GHz are used to drive frequency multiplier chains to provide both the spectroscopic probe and local oscillator power for a heterodyne receiver in the 210–270 GHz region. The detector is a heterodyne receiver linked to the source frequency. This system approaches the fundamental Townes noise limit of Section 12.2.2.

Because even in a complex mixture the resolution of this system is so high that most of the spectrum is white space, it is advantageous to use the frequency agility of the system so that the system spends most of its time at the location of strong fingerprint lines. For demonstrations such as the one illustrated in Figure 12.8, a false alarm probability $\ll 10^{-10}$ was desired, and it was found that inspection of only six

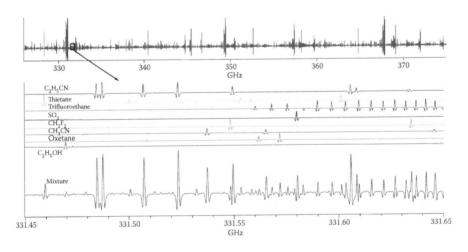

FIGURE 12.6 Spectrum (50 GHz) of a mixture of 20 gases on a highly compressed scale (*top*), of which 0.4% of this spectrum is expanded to show individual lines (*bottom*). Library spectra of eight gases indicating their presence in the gas mixture are shown in the middle. The scan time for the 50 GHz segment was ~3 s, and the scan for the expanded region used for the analysis ~10 ms. The full 3 s spectrum expanded to the scale of the lower trace and to show noise would occupy a graph about 50×10 m. (From Medvedev, I. R., Neese, C. F., Plummer, G. M., and De Lucia, F. C., *Opt. Lett.*, 35 (10), 1533–1535, **2010**.)

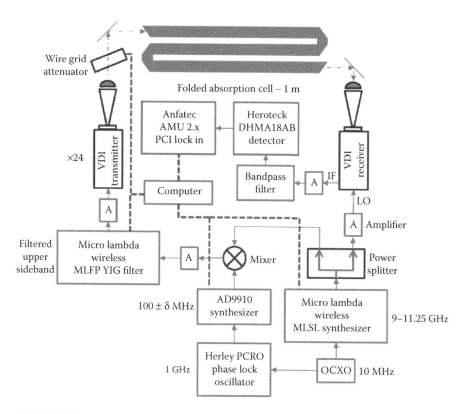

FIGURE 12.7 A compact gas sensor based on solid-state frequency multiplication and heterodyne detection. (From Medvedev, I. R., Neese, C. F., Plummer, G. M., and De Lucia, F. C., *Opt. Lett.,* 35 (10), 1533–1535, **2010.**)

lines from each of the 32 potential gases in the mixture would easily satisfy this requirement.

Figure 12.8 shows the results for one such mixture. Along the central axis, 5 MHz "snippets" are displayed successively around each of 182 line centers, sorted according to decreasing strength. If a gas is present in the mixture, a line (of correct intensity and frequency) must appear in each of the six snippets assigned to that chemical. Automated recognition and quantification are obtained by least-squares methods that calculate the amount of each gas by comparison with intensity-calibrated library spectra. The results of this analysis are shown in the table at the right of this figure. In red, the standard deviations for the gases that were placed in the mixture are shown. In blue are much smaller concentrations due to gases not placed in the mixture, but produced by chemical reaction in the test mixture. The absent gases are in black. This automated process takes much less than one second for analysis. The importance and accuracy of the intensity calibration are shown in the several inserts. For example, there are spectral resonances in some of the six snippets that "belong" to cyanogen chloride (ClCN). However, the intensity-calibrated fit in black correctly attributes these resonances to other gases (4, 8, 17, and 20) that

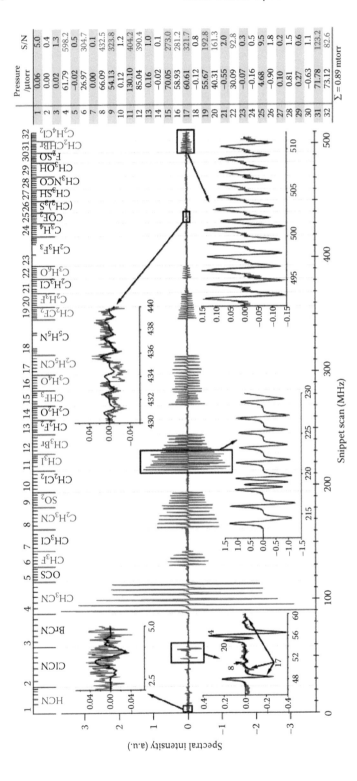

FIGURE 12.8 A comparison between the observed spectra (*red*) and the fitted spectra (*black*) for a mixture of gases. The results of an automated analysis are shown in the table at the right, indicating the mixture contained 14 intended and 2 unintended molecules of 32 in the library. (From Medvedev, I. R., Neese, C. F., Plummer, G. M., De Lucia, F. C., *Opt. Lett.*, 35 (10), 1533–1535, **2010.**)

are in the mixture. It can be seen in the table that the least-squares analysis correctly deduces that no ClCN is present.

12.3.2 GAS-PHASE REMOTE SENSING

The remote sensing of gases with high specificity is much more challenging. At atmospheric pressure, spectral lines in all spectral regions are pressure broadened to about 5 GHz, eliminating the incredible recognition specificity afforded by low-pressure point sensing of gas-phase molecules. If this were not enough to deter applications of SMM/THz spectroscopy to remote sensing, the severe, humidity-dependent attenuation by atmospheric water vapor not only limits the range to <1 km but also introduces a huge fluctuating baseline from atmospheric turbulence, above which these weak lines must be measured. Not surprisingly, applications of THz spectroscopy to remote sensing have been thoroughly disappointing for fundamental reasons.

Nevertheless, there is hope in the form of a newly proposed double-resonance spectroscopic approach that may recover recognition specificity and avoid the effects of atmospheric turbulence.[55–57] The technique takes advantage of the comparative rareness of a coincidence between the frequency of any IR rotational–vibrational transition of a molecule and any of the 40–60 lines from a carbon dioxide (CO_2) laser. When such a rare coincidence occurs in a given gas, molecules are photoexcited from a heavily populated rotational state in the ground vibrational level to a sparsely populated rotational state in an excited vibrational level (Figure 12.9). Consequently, the SMM/THz absorption coefficients of the rotational transitions directly connected to these pumped rotational states are strongly modulated by the IR pump. This is not new physics: if the pump is sufficiently strong, rotational population inversions may be created, and the resulting "optically pumped far infrared" (OPFIR) lasers were the first powerful THz sources ever developed.[8]

What is new is the recognition that the combination of IR transition/CO_2 laser line and the associated SMM/THz rotational transition, which are unique to a given molecule, may be used for remotely detecting and recognizing the constituents of a trace gas cloud. Specifically, THz/IR double-resonance spectroscopy in which the IR pump is tuned to coincide with the rovibrational transition (long arrow) and the SMM/THz probe is tuned to coincide with any of the four rotational transitions (short arrows) connected to the two pump levels may be used to discriminate target gases with even isotopic specificity. Basic requirements for this technique to work are that the CO_2 laser be operated in pulsed mode and that the pulse widths be shorter than the atmospheric collision time (~200 ps) so that the pump-induced modulation of the SMM/THz rotational absorption is not destroyed by collisional relaxation. Such short pulses have the added benefit of "freezing" atmospheric fluctuations, which occur on millisecond timescales. A continuous-wave SMM/THz probe is then frequency-tuned to monitor the pump-induced modulation of the target gas.

For this technique to work in a practical setting, the THz probe will be colocated with the IR pump, requiring that the THz beam be reflected back to a heterodyne receiver after passing through the target trace gas. Both natural and man-made reflectors may be used, and perhaps the most favorable configuration is a "trip-wire" sensor that interrogates the space between a fixed SMM/THz transceiver and a fixed

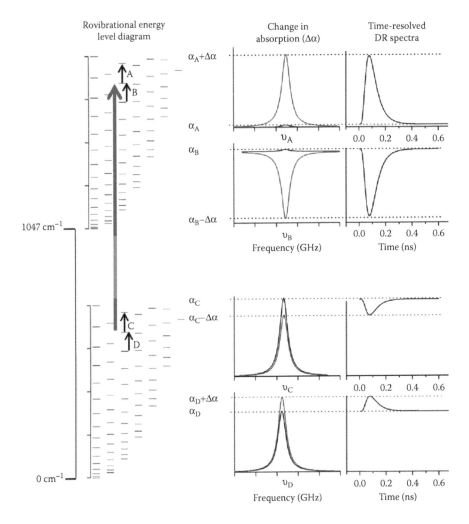

FIGURE 12.9 Schematic diagram of the double-resonance approach, illustrating the rare pump coincidence between a CO_2 laser line (*long red arrow*) and a rovibrational transition in the molecule. The pump-induced changes in rotational absorption A, B, C, D as a function of frequency and time are illustrated in the panels to the right for the case of pulsed excitation.

retroreflector. Since this probe is still limited by atmospheric attenuation, the maximum range is on the order of 1 km in a manner that depends strongly on the atmospheric dew point.

An example of THz/IR double-resonance (DR) spectra is presented in Figure 12.10 for a trace gas cloud of the recently banned pesticide bromomethane. Each trace on the left represents the modulated SMM/THz absorption strength induced by a given CO_2 pump line (labeled on the right-hand side of the figure) when multiplied by the concentration of the cloud (in ppm-m) and the energy of the IR pump (in J). Note that each trace induces changes at many frequencies, not just the four indicated. This occurs because at atmospheric pressure the pump coincidence requirement is relaxed

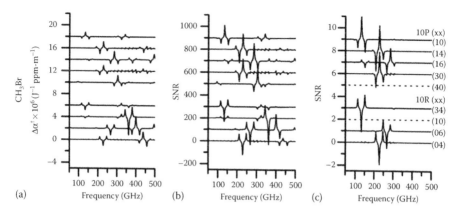

FIGURE 12.10 Example double-resonance spectra (a) and estimated SNRs for a standard scenario (b) and worst-case scenario (c) for detection of a 100 ppm-m trace gas cloud of CH_3Br.

through pressure broadening of the rovibrational transition; nevertheless, a given molecule's DR signatures remain unique. Indeed, this multiplicity of DR signatures is advantageous as it allows the selection of transitions in SMM/THz atmospheric transparency windows to improve the signal-to-noise ratio (SNR) for detection. The detection methodology then requires that an optimal DR signature be selected for monitoring.

Figure 12.10 also shows estimates of the SNR for a 100 ppm-m cloud located between the SMM/THz transceiver and a retroreflector using the optimal DR signature in bromomethane. Two scenarios are considered in the center and right panels: a standard scenario with the retroreflector 100 m away and a 50°F dewpoint, and a worst-case scenario with the retroreflector 1 km away and a 75°F dewpoint. Both assumed the same 1 W SMM/THz probe power, the same 1 J/pulse IR pump energy, and signal averaging over 100 pulses. Even in the worst-case scenario, bromomethane is detectable and recognizable with an SNR>2. Other analytes have stronger or weaker SNR values, depending on the type of rovibrational transition involved, the rotational partition function (i.e., the size and complexity) of the molecule, and the degree of atmospheric attenuation at the probe frequency.

12.3.3 DETECTION OF SOLID EXPLOSIVES

Probably no other application has received more attention than the detection of hidden explosives.[58] The concept is attractive: use the penetration powers of the THz to see through the covering to reveal the specific spectroscopic signatures of explosive compounds.

After some false starts, the spectra of the more important species have been recorded (see Figure 12.5)[46] and reproduced.[59] However, the challenges for converting these spectra into a reliable system for detecting and recognizing explosives are substantial and perhaps insurmountable. While there have been reports of sharp resonances below 500 GHz,[48] despite their potential importance, they have not been reproduced. The resonances that have been observed are broad (several hundred GHz) and at frequencies above 700 GHz, where the transmission of obstructing material is much lower (see Section 2.3.3). These resonances derive primarily from low-frequency phonon and vibronic modes of the solid, and, like all such "structural" modes, they depend quite sensitively on preparation conditions and interactions with hosts. Additionally, remote sensing is only practical below 500 GHz (Figure 12.4), meaning there are fewer than 100 spectral resolution elements available for spectral recognition, so there can be little recognition specificity when judged against the many other sources of broad spectra in this region,[60] including most famously Hershey Bars. Indeed, simply showing a sample has a spectral feature does not mean that it can be unambiguously recognized. It is noteworthy that almost all the literature reporting spectral features of explosives have involved placing a known explosive in a spectrometer, while there are almost no reports of blind investigations to assess how well these spectrometers can discriminate explosive signatures from the copious environmental interferents common to most detection scenarios of interest.

In an unusually perceptive article demonstrating the difference between an attractive concept and a quantitative analysis, Kemp concludes

> the relative weakness of spectral features in reflection, the limited frequency range available due to absorption by clothing, fluctuations and distortions in the spectra due to scattering from rough surfaces, attenuation by water vapor in the atmosphere, and other effects, all combine to make it very difficult for terahertz spectroscopy to be developed as a practical technique for this application. Whilst it [is] impossible to say that it will never be done, potential system developers should develop performance models to explore the capabilities and limits of their proposed approach, taking into account all of the factors identified in this paper.[2]

Indeed, systems engineering of a working spectroscopic recognition instrument requires that each of the requirements for success be met individually, collectively, and simultaneously. The most challenging combination of this in the THz is often that requirements for penetration and atmospheric propagation (which are best in the 0.1–0.3 THz range) be satisfied at the same time as spectroscopic requirements (which are often best >1.0 THz).

12.4   CHEAP AND PRACTICAL

For SMM/THz applications, the best is yet to come. For most of the history of the SMM/THz, we have been limited by technology and cost. At great expense in time and money, progress has been made, but we are still a cottage industry. However, the wireless community, driven largely by an insatiable appetite for bandwidth, is rapidly driving the mass market to our doorsteps. Chip-level technology already exists at costs of tens of dollars to 100 GHz. Not only is the mass-market technology approaching us, the physics for smaller devices and lower power is favorable. The wavelengths are short, but still macroscopic, so small systems can be fabricated. Moreover, because the photons are small and the noise low in the SMM/THz, very low powers satisfy many requirements. Soon, we will be in the extreme opposite of where we have been for most of the history of the SMM/THz; the technology will be almost free. It is up to us to figure out how to take advantage of the opportunities that will emerge as "our" spectral region becomes mainstream.

12.5 SUMMARY

While the THz spectroscopic detection of hidden solid explosives has been widely discussed, in the authors' opinion (and [2] as well), this application is extremely challenging. While it is possible that it is viable in specialized scenarios, even in these limited cases it is unlikely to be the technological solution of choice. Similarly, because of limited vapor pressure and the expected dense rotational spectra of most explosives, the direct detection of gas-phase rotational spectra in point sensors is also very challenging.

There are two much less considered approaches. The first is the detection of smaller gas-phase molecules that are either co- or by-products of the explosives themselves. While this would sacrifice the "absolute" specificity associated with the direct detection of the spectra of explosives, this is a widely used analytical approach and one that is commonly used in many explosives detectors. The second is to use an IR-THz double-resonance approach for the detection of the co- or by-products that provides means for remote detection and obtaining specificity via the 2-D spectroscopic signature of the double resonance.

While beyond the scope of this chapter, the nonspectroscopic detection of explosives via THz imaging is viable (and if one wants to extend the THz to 0.03 THz, already proven in the form of the ubiquitous airport scanner) and will become of increasing importance as the costs decrease and the performance of electronic devices expands into the THz.

12.6 ACKNOWLEDGMENTS

The authors would like to thank the Army Research Office, DTRA, DARPA, NASA, NSF, and the Semiconductor Research Corporation for their support of this work. The authors would also like to thank Doug Petkie, Dane Phillips, and Elizabeth Tanner for their help in developing the double-resonance spectroscopic technique.

REFERENCES

1. Appleby, R., Anderton, R. N., Millimeter-wave and submillimeter-wave imaging for security and surveillance. *Proc. IEEE* **2007**, *95*, 1683–1690.
2. Kemp, M. C., Explosive detection by terahertz spectroscopy—A bridge too far? *IEEE Trans. THz Sci. Tech.* **2011**, *1*, 282–292.
3. Jordan, G., Whispered excitement about the THz. In *SPIE Symposium: Optics/Photonics in Security and Defense*, Bruges, Belgium, **2005**.
4. Gordy, W., Millimeter and submillimer waves in physics. *Polytechnic Press* **1960**, *9*, 1–23.
5. King, D. D., Millimeter waves and beyond. *Proc. IEEE* **1966**, *54* (4).
6. Townes, C. H., Schawlow, A. L., *Microwave Spectroscopy*. McGraw-Hill: New York, **1955**.
7. Gordy, W., Cook, R. L., *Microwave Molecular Spectra*. 3rd edn., John Wiley & Sons: New York, **1984**; Vol. 18, p. 929.

8. Chang, T. Y., Bridges, T. J., Laser action at 452, 496, and 541 µm in optically pumped CH₃F. *Opt. Commun.* **1970**, *1* (9), 423–426.
9. Auston, D. H., Smith, P. R., Generation and detection of millimeter waves by picosecond photoconductivity. *Appl. Phys. Lett.* **1983**, *43*, 631–633.
10. Brown, E. R., McIntosh, K. A., Nichols, K. B., Dennis, C. L., Photomixing up to 3.8 THz in low-temperature-grown GaAs. *Appl. Phys. Lett.* **1995**, *66*, 285–287.
11. Rochat, M., Ajili, L., Willenberg, H., Faist, J., Beere, H., Davies, G., Linfield, E., Ritchie, D., Low-threshold terahertz quantum-cascade lasers. *Appl. Phys. Lett.* **2002**, *81* (8), 1381–1383.
12. Song, H.-J., Nagatsuma, T., Present and future of terahertz communications. *IEEE Trans. THz Sci. Tech.* **2011**, *1*, 256–263.
13. Ferguson, B., Zhang, X.-C., Materials for terahertz science and technology. *Nat. Mater.* **2002**, *1*, 26–33.
14. Mittleman, D. M., *Sensing with Terahertz Radiation.* Springer-Verlag, Berlin, **2003**.
15. Tonouchi, M., Cutting-edge terahertz technology. *Nature Photonics* **2007**, *1*, 97–106.
16. Baxter, J. B., Guglietta, G. W., Terahertz spectroscopy. *Anal. Chem.* **2011**, *83*, 4342–4368.
17. Gallerano, G. P., Overview of terahertz radiation sources. *Proceedings of the 2004 FEL Conference* **2004**, 216–221.
18. Siegel, P. H., Terahertz technology. *IEEE Trans. Microwave Theory Tech.* **2002**, *50*, 910–928.
19. Pickett, H. M., Poynter, R. L., Cohen, E. A., Delitsky, M. L., Pearson, J. C., Muller, H. S. P., Submillimeter, millimeter, and microwave spectral line catalog. *J Quant. Spectrosc. Radiat. Transf.* **1998**, *60*, 883–890.
20. Muller, H. S. P., Schloder, F., Stutzki, J., Winnewisser, G., The cologne database for molecular spectroscopy. *J. Mol. Struct.* **2005**, *742*, 215–227.
21. Rothman, L. S., Gordon, I. E., Barbe, A., Benne, D. C., Bernath, P. F., Birk, M., Boudon, V., et al., The HITRAN 2008 molecular spectroscopic database. *J Quant. Spectrosc. Radiat. Transf.* **2009**, *110*, 533–572.
22. Wiltse, J. C., History of millimeter and submillimeter waves. *IEEE Trans. Microwave Theory Tech.* **1984**, *MTT-32*, 1118–1127.
23. Siegel, P. H., Inaugural editorial. *IEEE Trans THz Sci. Tech.* **2011**, *1*, 1–4.
24. Turner, J. L., Wooten, H. A., Atacama large millimeter/submillimeter array. *Highl. Astron.* **2006**, *14*, 521–522.
25. Clery, D., Herschel will open a new vista on infant stars and galaxies. *Science* **2009**, *324*, 584–586.
26. Becklin, E. E., Stratospheric observatory for infrared astronomy (SOFIA). *Adv. Space Res.* **2005**, *36*, 1087–1090.
27. Lewis, W. B., Fluctuations in streams of thermal radiation. *Proc. Phys. Soc.* **1947**, *59*, 34–40.
28. De Lucia, F. C., Noise, detectors, and submillimeter-terahertz system performance in nonambient environments. *J Opt. Soc. Am.* **2004**, *B 21*, 1273–1297.
29. Putley, E. H., The ultimate sensitivity of sub-mm detectors. *Infr. Phys.* **1964**, *4*, 1–8.
30. Townes, C. H., Geschwind, S., Limiting sensitivity of a microwave spectrometer. *J Appl. Phys.* **1948**, *19*, 795L.
31. Bjarnason, J. E., Chan, T. L. J., Lee, A. W. M., Celis, M. A., Brown, E. R., Millimeter-wave, terahertz, and mid-infrared transmission through common clothing. *Appl. Phys. Lett.* **2004**, *85*, 519–521.
32. Gatesman, A. J., Danylov, A., Goyette, T. M., Dickinson, J. C., Giles, R. H., Goodhue, W., Waldman, J., Nixon, W. E., Hoen, W., Terahertz behavior of optical components and common materials. *Proc. SPIE* **2006**, *6212*, 62120E.
33. Kallmeyer, M., T-ray technology not just science fiction anymore. *Columbus Dispatch* **2010**.

34. The Ohio State University, Ohio State students win coveted international technology business competition. Materials Research Society Spring Meeting. San Francisco. 10 April **2007**. http://fisher.osu.edu/centers/entrepreneurship/press-releases/.

35. Zandonella, C., T-ray specs. *Nature* **2003**, *424*, 721–722.

36. De Lucia, F. C., The submillimeter: A spectroscopist's view. *J Mol. Spectrosc.* **2010**, *261* (1), 1–17.

37. Gordy, W., Smith, W. V., Trambarulo, R. F., *Microwave Spectroscopy*. John Wiley & Sons: New York, **1953**.

38. Albert, S., Petkie, D. T., Bettens, R. P. A., Belov, S. P., De Lucia, F. C., FASSST: A new gas-phase analytical tool. *Anal. Chem.* **1998**, *70*, 719A–727A.

39. Medvedev, I. R., Neese, C. F., Plummer, G. M., De Lucia, F. C., Submillimeter spectroscopy for chemical analysis with absolute specificity. *Opt. Lett.* **2010**, *35* (10), 1533–1535.

40. Fortman, S. M., Medvedev, I. R., Neese, C. F., De Lucia, F. C., A new approach to astrophysical spectra: The complete experimental spectrum of ethyl cyanide (CH_3CH_2CN) between 570 and 645 GHz. *Astrophys. J.* **2010**, *714*, 476–486.

41. Liebe, H. J., *Atmospheric Water Vapor: Nemesis for Millimeter Wave Propagation*. Academic Press: New York, **1980**.

42. Liebe, H. J., An updated model for millimeter wave propagation in moist air. *Radio Sci.* **1985**, *20*, 1069–1089.

43. Kemp, M. Terahertz applications: Work at Teraview (lecture). In Chamberlain, M. (ed.), *Terahertz Sources and Systems for Security Use*, at Her Majesty's Government Communications Centre: Grey College, Durham University, 28–30 September, **2005**.

44. Khan, U. A., Al-Moayed, N., Nguyen, N., Korolev, K. A., Afsar, M. N., Broadband dielectric characterization of tumorous and nontumorous breast tissues. *IEEE Trans. Microw. Theory Tech.* **2007**, *55*, 2887–2893.

45. Melinger, J. S., Laman, N., Grischkowsky, D., The underlying terahertz vibrational spectrum of explosives solids. *Appl. Phys. Lett.* **2008**, *93*, 011102–011102-3.

46. Kemp, M. C., Taday, P. F., Cole, B. E., Cluff, J. A., Fitzgerald, A. J., Tribe, W. R., Security applications of terahertz technology. In Hwu, R. J., Woolard, D. L. (eds), *Proceedings of SPIE v5070*, pp. 44–52. Orlando, FL, **2003**.

47. Kemp, M. C., Baker, C., Gregory, I., Stand-off explosives detection using terahertz technology. In Schubert, H., Rimski-Korsakov, A. (eds), *Stand-off Detection of Suicide Bombers and Mobile Subjects*, pp. 151–165. Springer, Dordrecht, The Netherlands, **2006**.

48. Choi, M. K., Bettermann, A., van der Weide, D. W., Potential for detection of explosive and biological hazards with electronic terahertz systems. *Phil. Trans. R. Soc. Lond. A* **2004**, *362*, 337–349.

49. Ajito, K., Ueno, Y., THz chemical imaging for biological applications. *IEEE Trans. Terahertz Sci. Technol.* **2011**, *1*, 293–300.

50. Walthers, M., Fischer, B., Jepsen, P. U., Noncovalent intermolecular forces in polycrystalline and amorphous saccharides in the far infrared. *Chem. Phys.* **2003**, *288*, 261–268.

51. Medvedev, I., Behnke, M., De Lucia, F. C., Fast analysis of gases in the submillimeter/terahertz with "absolute" specificity. *Appl. Phys. Lett.* **2005**, *86*, 154105.

52. Medvedev, I., Behnke, M., De Lucia, F. C., Chemical analysis in the submillimeter spectral region with a compact solid state system. *Analyst* **2006**, *131*, 1299–1307.

53. Crowe, T. W., Bishop, W. L., Porterfield, D. W., Hesler, J. L., Weikle, R. M., Opening the terahertz window with integrated diode circuits. *IEEE J. Solid-State Circuits* **2005**, *40*, 2104–2110.

54. Medvedev, I. R., Winnewisser, B. P., Winnewisser, M., Behnke, M., De Lucia, F. C., Petkie, D. T., Bettens, R. P. A., Kisiel, Z., Fast scan submillimeter spectroscopy technique. In *OSU International Symposium on Molecular Spectroscopy*, Columbus, OH, **2006**.

55. De Lucia, F. C., Petkie, D. T., Everitt, H. O., A double resonance approach to submillimeter/terahertz remote sensing at atmospheric pressure. *IEEE J. Quan. Elec.* **2009**, *45* (1–2), 163–170.

56. Phillips, D. J., Tanner, E. A., De Lucia, F. C., Everitt, H. O., Infrared-terahertz double-resonance spectroscopy of CH_3F and CH_3Cl at atmospheric pressure. *Phys. Rev. A* **2012**, *85*, 052507/1–13.

57. Phillips, D. J., Tanner, E. A., Everitt, H. O., Medvedev, I. R., Neese, C. F., Holt, J., De Lucia, F. C. In Anwar, M. D., N.K.; Crowe, T.W. (eds), *Infrared/Terahertz Double Resonance for Chemical Remote Sensing: Signatures and Performance Predictions*, *SPIE Conference on Terahertz Physics, Devices, and Systems IV: Advanced Applications in Industry and Defense*, Orlando, FL, 5–6 April, **2010**, p. 76710F.

58. Liu, H.-B., Zhong, H., Karpowicz, N., Chen, Y., Zhang, X.-C., Terahertz spectroscopy and imaging for defense and security applications. *Proc. IEEE* **2007**, *95*, 1514–1527.

59. Cook, D. J., Decker, B. K., Maislin, G., Allen, M. G., Through container THz sensing: Applications for explosives screening. *Proc. SPIE* **2004**, *5354*, 55.

60. Tribe, W. R., Newnham, D. A., Taday, P. F., Kemp, M. C., Hidden object detection: Security applications of terahertz technology. *Proc. SPIE* **2004**, *5354* (1), 168–76.

13 Detection of Explosives and Energetic Components via Short-Pulse Laser-Based Techniques

John J. Brady and Robert J. Levis

CONTENTS

13.1 INTRODUCTION

The detection of explosives and energetic formulations is an extraordinary challenge for analytical science given the wide range of molecules to be detected, the complexity of their formulations, and the low vapor pressure of several species of interest. For instance, a typical black powder formulation can have up to 20 components that can form classification signatures. The rapid analysis of such complex mixtures is non-trivial. In addition, the vapor pressure of 2,4,6-trinitrotoluene (TNT) is on the order of 10^{-4} torr, while that of 1,3,5-trinitroperhydro-1,3,5-triazine (RDX) is three orders of magnitude lower at room temperature. To detect vapor-phase signatures for either of these molecules remains a significant problem for modern analytical methods given the sub-part-per-billion sensitivity required. Typical detectors for explosives and energetics operate either as a point detector, where the detector is in immediate proximity to the sample, as a remote detector, where the system is at some distance from the operator (robot deployment), or as a standoff detector, where both the system and operator are at some distance from the sample. Each mode of operation has advantages and limitations. Point detectors are often the most sensitive, with the greatest capability for correct classification of materials. In the field, such systems expose the operator and equipment to hazardous conditions. Standoff detectors, at present, have the lowest probability of accurate classification, but have the advantage that neither operator nor equipment is in the vicinity of the object under inspection (although this does not rule out the danger from prior false-negative inspections). Remote detectors represent a compromise where the equipment, but not the operator, is deployed for the inspection.

The gold standard for the detection of explosive and energetic compounds is ion mobility spectrometry. Ion mobility's capability to detect energetic compounds was first explored by F. W. Karesek in his experiments with nitroaromatics in air [1,2]. The method relies on the charging of energetics, commonly through interaction with a radioactive source, followed by extraction into a flight tube for atmospheric pressure time-of-flight (TOF) detection. While extraordinarily sensitive and versatile, the system remains a point detector that is subject to interference, in part because of the relatively low resolution of the detection system. Laser-based methods hold considerable promise for providing complementary detection technologies in multiple implementations, including as a release method for low vapor pressure materials for subsequent high-resolution mass spectrometry, as a standoff detection system using Raman spectroscopy, and as a remote detection system using coherent anti-Stokes Raman spectroscopy (CARS). In each of these detection modalities, the molecular nature of the energetic material is the key to the detection strategy. In the laser release–mass spectrometry method, it is the molecular weight of the parent, and associated fragments of the molecules, that allows high-fidelity classification of the formulation of interest. In the remote Raman experiment, it is the rotational and vibrational modes of the gas-phase signature molecules that enable classification. Since each molecule in the formulation has specific modes of motion, a unique spectrum will result.

This chapter represents an up-to-date compilation (2013) of laser-based analytical probes of energetic materials, with a focus on ultrafast laser systems. The femtosecond (fs) laser is based on two technological innovations that have enabled the production of turnkey systems: Kerr lens mode-locking and chirped pulse amplification. The first technology is the capability to mode-lock the fluorescence of a titanium-doped sapphire (Ti:sapphire) crystal over a bandwidth of approximately 750–850 nm to produce a few optical cycle pulses, as developed by Sibbett in 1988 [3]. This mode-locking occurs when a multitude of longitudinal modes interfere constructively to provide a pulse duration that scales inversely with the bandwidth (i.e., a measure of the colors contained in the pulse) of the laser pulse. Mode-locking is commonly accomplished using Kerr lensing through the intensity-dependent index of refraction, $n(t) = n_0 + I(t) \, n_2$, where n_0 is the linear refractive index and n_2 is the second-order refractive index ($n_2 = 3 \times 10^{-19}$ W/cm^2 for air). In this approach, the intensity of the ultrashort laser pulse creates a lens in the Ti:sapphire lasing medium through the Gaussian intensity distribution. The intensity-dependent lens is incorporated into the design of the laser cavity to enhance the formation of the shortest possible pulse. Here, the continuous-wave (CW) mode of the cavity is damped through a low quality factor (Q), while the short-pulse mode is amplified through a high Q factor. The fluorescence from the pumped Ti:sapphire crystal is collimated using a pair of reflective lenses: one pointing to the output mirror and the second pointing toward a prism pair that compensates for the group velocity dispersion accumulated by the pulse passing through the Ti:sapphire crystal. A broadband mirror then reflects the pulse back through the prism pair to the gain medium. A typical oscillator cavity generates a 12 fs pulse at 100 MHz with a few nanojoules of energy per pulse. This energy is not sufficient to perform the experiments described in this chapter, so further amplification is required.

The second technology required for intense fs pulses is chirped pulse amplification, as developed by Strickland and Mourou in 1985 [4]. In this method, the ultrashort pulse emanating from the oscillator is chirped from a pulse width of 20 fs to a pulse width of 100 ps using a pair of gratings. Such a pulse can then be amplified without causing optical damage in a second Ti:sapphire crystal pumped using ~20 mJ of pulsed 531.2 nm light from an ytterbium (Yb) laser. Pulses are directed into the amplification cavity using a Q switch at 1 kHz and the pulses are amplified by a factor of 10^6. The amplified pulses are then directed into a compressor where the initial chirp is reversed to produce the ultrashort, ultraintense pulse. The shortest pulse arises when the relative phase between each frequency component within the bandwidth is set to zero at one point in space–time, leading to constructive interference between the colors at that point and destructive interference at all other points in the laser cavity. This phase profile is termed "flat phase" and forms the shortest possible duration in space–time through the Fourier limit. Commercial systems now routinely produce 5 mJ pulses with pulse duration of 35 fs.

Fiber laser technology offers a complementary approach to the generation of ultrashort pulses. The approach uses different mode locking and amplification methods based entirely on optical fiber technology [5]. While the method has the potential to be more robust, the energy per pulse is limited to less than 100 μJ at the current time, which is an order of magnitude too low for many of the standoff experiments described in this chapter.

These ultrashort pulses, when amplified, lead to many opportunities to explore new analytical methods. When focused, the pulse reaches intensities of 10^{13} W/cm^2. This in turn leads to coupling to all molecules through multiphoton excitation causing translational, rotational, vibrational, and electronic excitation. Each of these modes of excitation will be explored in this chapter. In addition, the large bandwidth leads to the opportunity to perform various nonlinear four-wave-mixing spectroscopies including impulsive vibrational excitation and CARS, which will also be reviewed.

13.2 POINT DETECTION OF ENERGETICS USING ULTRAFAST LASERS

The point detection of energetics can be enabled using a variety of separation-based or optical-based techniques. Separation-based methods, such as ion mobility spectrometry [6–9], gas chromatography [10–12], and mass spectrometry [13–17], are commonly used to safeguard personnel and high-risk targets in a variety of settings (e.g., transportation, military, ports, borders, etc.) and are often used in security applications because of their high-throughput, inherent sensitivity [6,18], and capability to distinguish hazards from benign background contaminants (e.g., perfume, lotions, hand sanitizers, etc.). To enable analysis, gas-phase and condensed-phase explosive samples are typically collected via solid-phase microextraction (SPME) [6] or via specialized swabs [19], respectively. The inherent use of swabs and their composition [20] results in an inconsistent amount of material transferred to the detection system [21]. The *in situ*, direct analysis of energetic materials and suspect surfaces has the potential to decrease complications in the detection process caused by incomplete transfer to either the SPME fiber or the swab.

Recently, an *in situ*, laser-based ambient analysis method has been introduced for the detection of hazardous materials in complex backgrounds. This technique, known as laser electrospray mass spectrometry (LEMS) [22–29], focuses an ultrafast laser pulse (~70 fs, ~400 μJ–1 mJ, centered at 800 nm) to a spot size of approximately 200 μm diameter on an adsorbed sample. The high intensity of the laser pulse (~10^{13} W/cm^2) enables nonresonant, multiphoton absorption of the laser pulse energy, eliminating the need for the application of a matrix, and induces vaporization of a condensed-phase sample through several possible mechanisms. In one possible vaporization mechanism [25], the excitation of the condensed-phase system into a repulsive adsorbate–substrate or a repulsive adsorbate–adsorbate electronic state occurs via the nonresonant multiphoton absorption of the fs laser pulse or the hot electrons subsequently produced (1–10 eV) [30–32]. Once in the repulsive state, the stored potential energy from electronic repulsion is converted into kinetic energy, enabling vaporization of the adsorbed species before thermalization or dissociation can occur. Another potential vaporization mechanism involves the direct vibrational excitation of intramolecular bonds within the condensed-phase system and may result from collisions with the hot electrons created at the surface [25,33]. The excitation of the intramolecular bonds leads to expansion of the sample against neighboring molecules and the surface. This expansion can create a pressure pulse, ejecting molecules and clusters from the surface with a temperature equal to that of the local surface [34]. Once ejected, the molecules and clusters will undergo expansion into the ambient air, where collisions with the background nitrogen and oxygen decrease their internal energy. As a result, the temperature of the sample is sufficiently lowered to prevent molecular dissociation. The vaporized material is subsequently captured, solvated, and ionized by an electrospray plume (typically composed of 7:3 methanol:water, containing 50 mmol Na$^+$ and K$^+$ for energetic material analysis). The electrospray plume is directed into the inlet of the mass spectrometer and travels in a direction perpendicular to the laser-vaporized material. This allows for rapid, universal analysis of the sample via TOF mass spectrometry, as shown in Figure 13.1a.

The nonresonant, fs laser vaporization should, in principle, couple into and vaporize all molecules, implying that sample preparation is not necessary. In addition, the nonthermal nature of the vaporization process has enabled the analysis of thermally labile samples (e.g., taggants for explosive materials [24], smokeless powders [22], high explosives [24,35], and simulated improvised explosive devices (IEDs) [26])

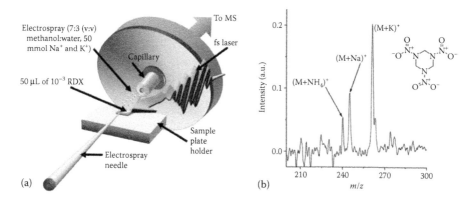

FIGURE 13.1 (a) A schematic of the non-resonant laser vaporization and electrospray ionization apparatus. The analyte is vaporized from a sample plate holder mounted onto a three-dimensional translation stage. (b) The mass spectrum resulting from analysis of a 50 μL aliquot of 10^{-3} M RDX dried on a metal surface. The ammoniated, sodiated, and potassiated adducts of RDX are detected in the spectrum at m/z 240, 245, and 261, respectively. The chemical structure of RDX is shown in the inset of (b).

without any observable fragmentation and without any preprocessing procedures [22,24,26]. For example, LEMS has demonstrated the capability to detect taggants, such as 2,3-dimethyl-2,3-dinitrobutane, but also the capability to detect the sodiated and potassiated parent ions for military-grade and homemade explosives, such as RDX (Figure 13.1b) and 3,4,8,9,12,13-hexaoxa-1,6-diazabicyclo[4.4.4]tetradecane (HMTD), respectively [24]. In addition, it has also been demonstrated that LEMS is capable of detecting RDX from various surfaces, such as stainless steel and sand, as well as of remote detection at a distance of 2 m via a Venturi air-jet pump for sample transfer [35]. Despite these successes, a key component of any explosives detection system is adaptability to new targets, such as the inorganic salts (e.g., ammonium nitrate) found in some IED components. The adaptability of the LEMS technique has been demonstrated through the detection of IED formulations including chlorate, perchlorate, sugar, ammonium nitrate, black powder, and postblast black powder. Each of these samples contains a complex mixture of cations, anions, and small molecules. To enable the classification of the formulations, complexation agents including 1-monooleoyl-rac-glycerol (monoolein), 1,9-nonanediyl-bis(3-methylimidazolium) difluoride, and 1,3-bis-[6-(3-benzyl-1-imidazolio)-hexyl]imidazolium trifluoride were incorporated into the electrospray plume [26]. The addition of the complexation agents enabled the simultaneous detection of both the cationic and anionic components from the inorganic salts in a single experiment, possibly reducing misidentification or false positives. To detect the ammonium ions, the lipid monoolein was added to the electrospray solution to provide a series of four distinct signature ions. As a result of this approach, nearly perfect classification of the formulations was achieved. Note that the detection of species is limited to those that are ionized by the electrospray plume. Therefore, species that are not captured or solvated will be unable to undergo adduct formation within the electrospray plume, preventing detection. Therefore, additional research should be performed to develop an electrospray solution containing the appropriate adducts to ionize any energetic material.

Recently, the components in smokeless powders, including ethyl central-ite, methyl centralite, dibutyl phthalate, and dimethyl phthalate, were detected under atmospheric conditions without additional sample preparation using LEMS (Figure 13.2a–e) [22]. The LEMS analysis of the smokeless powders also revealed several new mass spectral features that had not been identified previously. As mentioned previously, the obtained mass spectral signatures can be used in offline principal component analysis and discrimination (Figure 13.2f). Principal component analysis in conjunction with a K-nearest neighbor or linear discriminant analysis has enabled perfect classification of the obtained mass spectra [22]. The rapid means to classify could provide a triage method for samples given the short analysis time, providing forensic scientists with a reduced sample set for generating adjudicative data. We anticipate that the method's capability to detect all molecules in a sample in a single experiment would be valuable for forensic analysis.

The sensitivity of the LEMS technique has been reported to be approximately 50–100 ng per laser shot, as determined for simulated IEDs [26]. This sensitivity has been determined to be two orders of magnitude higher than that for traditional electrospray ionization mass spectrometry [26,29]. The sensitivity reported for the LEMS technique should not be taken as the absolute limit of detection (LOD) for

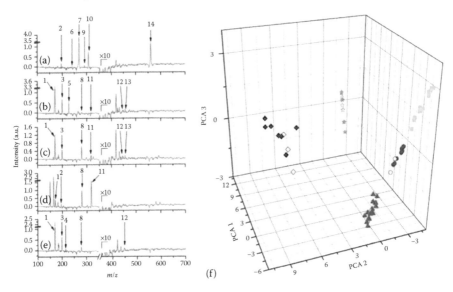

FIGURE 13.2 Laser electrospray mass spectra showing the identified component peaks for five smokeless powders: (a) Federal pistol, (b) Remington pistol, (c) Hornady pistol, (d) Remington rifle, and (e) high grade. (f) Principal component analysis of five different commercially available powders projected into three dimensions for high grade (*red*, •), Remington rifle (*blue*, ◆), Remington pistol (*cyan*, ■), Hornady pistol (*magenta*, ▲), and Federal pistol (*orange*, ★), which shows separation between manufacturers' powders. The open symbols (○, ◇, □, △, and ☆) and the filled, colored symbols (•, ◆, ■, ▲, and ★) represent the training and testing sets for each smokeless powder, respectively.

the nonresonant, fs laser vaporization method. The LOD ultimately depends on laser parameters (wavelength, pulse duration, intensity, etc.), mass spectral sensitivity (linear TOF vs. reflectron TOF vs. orbitrap), amount of material deposited, explosive film morphology, and chemical noise from interferents. Nevertheless, the LEMS technique has proved to be extremely versatile in its point detection capabilities. However, the technique does suffer from the drawback that true standoff could never be achieved because of the use of mass spectrometry as an analysis/detection method. In addition, the use of an fs laser and mass spectrometer greatly increases the cost, space, and power requirements of the technique. The development of portable fiber-based fs lasers and miniature mass spectrometers could lower the cost, space, and power requirements, possibly enabling a portable, ruggedized detection scheme with remote detection capabilities. In fact, preliminary experiments with fiber-based laser vaporization in the Center for Advanced Photonics Research at Temple University show extremely encouraging results in this regard.

13.3 STANDOFF DETECTION OF ENERGETICS USING ULTRAFAST LASERS

Optical-based techniques offer a unique advantage over separation-based techniques with respect to the capability to detect explosives and related materials at point, remote, and standoff distances. Commonly, these laser-based optical approaches focus and irradiate a gas-phase or condensed-phase sample with either CW or pulsed nanosecond lasers. The light that is emitted, reflected, or scattered is collected and directed into a detector for analysis. The collected light allows for the determination of the electronic structure, vibrational structure, or atomic constituents of the target species with high spectral resolution. The use of short-pulsed lasers opens new avenues of research because of the differences in laser–material interaction, which may allow advances toward more robust and sensitive techniques in comparison with CW and nanosecond systems. The techniques that are discussed below are typically configured to be used for the standoff detection of energetics. In this configuration, the operator and the equipment are not in the vicinity of the explosive device, reducing the risk to both the operator and the equipment. Despite this arrangement, these techniques may be adapted for the "remote" detection of material if necessary.

13.3.1 Laser-Induced Breakdown Spectroscopy

Laser-induced breakdown spectroscopy (LIBS) [36–39] is an optical detection technique that allows qualitative and quantitative analysis of material based on atomic emission from the constituents contained in a laser-generated plasma [40] (refer to Chapter 7). The majority of LIBS research has been performed with high-energy, nonresonant nanosecond lasers, but the technique can be performed with any pulsed laser that is capable of producing a plasma, including picosecond [41,42] and fs lasers [43–46]. Figure 13.3a displays a representative schematic of the experimental set-ups for both long- and short-pulsed LIBS. For long-pulse-duration (i.e., nanosecond) LIBS experiments, a laser with sufficient energy is focused onto or into a medium, producing free electrons via one of two mechanisms: by multiphoton ionization or

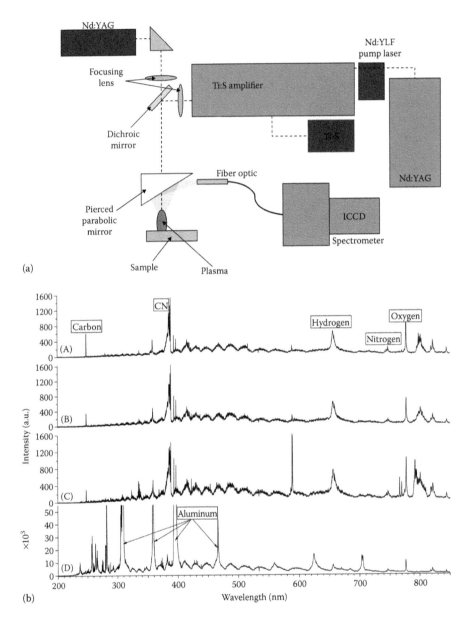

FIGURE 13.3 (a) A schematic displaying the experimental setups for both fs and nanosecond laser-induced breakdown spectroscopy for comparative measurements. (b) Femtosecond LIBS spectra of bulk explosives collected using an echelle spectrometer. Note: the intensity axis is different for aluminum. The laser line at 800 nm is observed in the explosive spectra. (A) CompB, (B) C-4, (C) RDX, (D) aluminum. (From De Lucia Jr., F.C., Gottfried, J.L., and Miziolek, A.W., *Opt. Exp.*, 17, 419–425, 2009. With permission.)

by a thermionic process [47]. The free electrons in the medium allow for further absorption of the laser pulse's energy, enabling additional heating of the sample and avalanche ionization. These processes enable the subsequent breakdown of the material [47–49] into its atomic constituents, producing a dense plasma (10^{17}–10^{19} cm^{-3}) at high temperatures (6,000–20,000 K) [50,51]. The high-temperature plasma continues to interact with the nanosecond laser pulse through plasma shielding. This allows further absorption of the laser pulse energy via inverse bremsstrahlung. As a result, the plasma can reach higher temperatures and higher electron densities. Time-resolved measurements in metallic systems reveal that after the initial electronic excitation, the internal modes of the condensed-phase system thermalize on the picosecond to nanosecond timescale [52]. Therefore, when nanosecond lasers are used to perform the LIBS technique, there is sufficient time for the system to come to thermal equilibrium with the plasma temperature. This heating leads to increased atomic emissions but also a higher-intensity background with prolonged continuum emission (i.e., bremsstrahlung and recombination radiation from the plasma as the electrons and ions recombine, thus cooling the plasma [47]). In addition, the supplementary heating causes additional melting and ablation [41], leaving large craters at the target's surface.

Recent work with fs laser pulses suggests that the laser–matter interaction is fundamentally different with ultrashort lasers. For example, fs laser pulses have the capability to exceed the ionization threshold of the target molecule without requiring a large amount of laser pulse energy. Femtosecond pulses have sufficient intensity ($>10^{13}$ W/cm^2) to form the free electrons necessary for plasma formation via multiphoton excitation. Note that the fs pulse is gone by the time the plasma is formed, eliminating the interaction between the laser pulse and the resultant plasma. Plasma shielding in the sample is thus prevented, allowing all of the laser energy to be deposited prior to material removal. In addition, the lower pulse energy thresholds [53] and the fast pulse energy deposition may prevent thermal damage to the sample from occurring. Reducing damage to the sample prevents layer mixing and results in cleaner ablation profiles [53–55], thus enabling depth profiling [47]. Additionally, in traditional nanosecond LIBS the continuum emission is electronically filtered out using a gated intensified charge-coupled device (CCD) array, allowing the atomic emission spectrum from the excited atoms and ions to be observed. This electronic filtering is not required when an fs laser is used for plasma generation. The plasmas formed are temporally shorter [41] and spatially smaller, enabling gate-free analysis of the sample [45,47,54]. Such gate-free analysis of samples could reduce the cost of the optical system and enable higher repetition-rate lasers. These advantages suggest that fs lasers may offer unique benefits over traditional nanosecond lasers in LIBS studies, and researchers have begun to explore the use of fs lasers for LIBS analysis [38,56–61].

Femtosecond laser pulses have been used to investigate explosives materials such as RDX, CompB, and C-4 (Figure 13.3b) [59]. Short laser pulses have been found to yield more molecular fragments than elemental constituents when compared to nanosecond lasers operated at lower fluence [38,59,62]. The increased yield of molecular components could give signatures for energetics and possibly eliminate background matrix effects from the substrate and the atmosphere. In addition, researchers have demonstrated that dual-pulse LIBS using fs lasers provides an enhancement in atomic emission similar to dual-pulse nanosecond LIBS [63]. This research [63] and work performed by others [59] have also shown that as the energy of the fs laser pulse was decreased, the capability to discriminate the material of interest increased. The advantages that fs pulses provided (i.e., more molecular fragments and less atomic constituents) were negated for explosives sampling when the fluence of the laser was increased [59]. At higher fluences, continuum background emission and atmospheric nitrogen and oxygen emission lines are observed, limiting the use of gate-free analysis [59]. This suggests that the use of low-fluence, fs pulses could provide unique signatures for hazardous materials identification, aiding in the discrimination process.

Although there are a number of advantages that one can observe when using an fs laser to produce plasma for LIBS, there are a number of challenges as well. One drawback to fs lasers is the complexity of the lasers themselves. Fs laser pulses require mode-locking within the cavity of an oscillator. Ruggedization of Ti:sapphire oscillators and amplifiers still has to be performed. Laser manufacturers are working to ruggedize fs lasers, with fiber-based fs lasers displaying optimal stability at the cost of lower laser power per pulse: 40 μJ versus 4 mJ for commercially available systems.

Dispersion is another challenge for fs systems that arises because of the ultrashort pulse duration of the laser. Any given material will have dispersive properties due to the frequency-dependent index of refraction, including air. Therefore, as the ultrashort laser pulse propagates through the air, the pulse will become chirped (with the redder colors leading the bluer colors) and the degree of dispersion is dependent on the amount of material through which the pulse propagates. This dispersion can be precompensated for, to a certain extent, but there may be some variability of the laser's pulse duration at the target. Such dispersion effects can be controlled in fs lasers by altering the phase of the pulse via a spatial light modulator [64] prior to propagation to the target. Precompensation does not have to be performed for nanosecond laser pulses, because the dispersion of the pulse is negligible due to their limited bandwidth.

Another issue with fs lasers arises from the inherent high energy and the ultrashort pulse duration. The irradiance or intensity of the beam is many orders of magnitude higher than that of nanosecond lasers because of the short pulse duration. Therefore, nonlinear effects such as Kerr lensing, self-phase modulation [65], and filamentation [65,66] can occur in air, and these effects are uncommon with lower-intensity nanosecond laser pulses. When a fs laser beam of sufficient power propagates through the air toward the target, Kerr lensing can begin to exceed the diffraction effects. Kerr lensing leads to a self-focusing of the laser beam because of the intensity-dependent refractive index of the material: $n(I) = n_o + n_2 I$. Once the intensity of the pulse reaches approximately 10^{13} W/cm^2, ionization of the air will occur, causing a negative lens effect on the laser pulse [65], essentially limiting the intensity of the pulse to 5×10^{13} W/cm^2 [67,68]. When the processes of Kerr lensing and plasma defocusing are in balance, the plasma channel (or filament) typically propagates for a meter and multiple cycles of filamentation may propagate for several hundred meters [69] to kilometers [70] for pulses having sufficient intensity.

Therefore, the issue of filamentation could be advantageous, allowing for the delivery of ultrashort pulses to a target at virtually any distance by altering the phase [71], power, or divergence angle of the pulse [72]. The filament can be initiated anywhere close to the desired target without the need for focusing optics. Nanosecond laser pulses are limited in standoff distance for LIBS analysis by the available numerical aperture of the focusing optic and the laser power [73]. For example, to change from 50 to 100 m would require alteration of the optical components to enable LIBS analysis using a nanosecond laser. In the case of fs remote filament-induced breakdown spectroscopy (R-FIBS) [74–76], no alteration of the optical components would be necessary to change the standoff distance. The use of R-FIBS could allow for kilometer-range detection, and, in theory [75,77], eliminate the effect of the lens-to-target distance [75]. R-FIBS may produce higher signal-to-noise measurements because of the intensity clamping effect of the filament [78]. Using this method of R-FIBS, preliminary investigations have begun allowing for the remote detection of DNT with infrared (IR) and ultraviolet (UV) filaments [74]. In addition, since the ultrafast fs pulses result in an increase of molecular fragments relative to elemental constituents, matrix effects should be minimized in a similar manner to that seen with traditionally focused, low-fluence, fs laser pulses.

13.3.2 Raman Scattering

Vibrational spectroscopy allows for the measurement of the molecular composition of a given sample. Each molecule in the detection volume produces a unique and distinctive vibrational "fingerprint" in the optical spectrum, which can be used for high-specificity chemical detection. Spontaneous, incoherent Raman scattering is a vibrational spectroscopy technique that relies on the inelastic scattering of light (refer to Chapter 5). In spontaneous Raman scattering, a photon of frequency v_o, from a narrowband light source, excites a molecule into a virtual state. The molecule then spontaneously returns to a different vibrational state and emits a photon of frequency $v_o \pm v_{vib}$, where v_{vib} is the vibrational frequency of the new state [79]. The Stokes ($v_o - v_{vib}$) and anti-Stokes ($v_o + v_{vib}$) frequency sidebands observed in the recorded spectra are unique to the molecule and have enabled the detection of explosive material [80] at a distance of approximately 50 m. The target molecule's small Raman cross section [81] ($\sim 10^{-30}$ cm^{-2}/molecule) reduces the sensitivity of incoherent Raman scattering. For example, spontaneous Raman scattering has an estimated LOD in the microgram range [82,83], as compared to LIBS with a LOD in the nanogram range [82]. The sensitivity of the technique can be augmented by increasing the scattering efficiency, by decreasing the wavelength through the v^4 scattering dependence [84] and/or by raising the incident laser power. These approaches can introduce a variety of complications, including interference from solar emissions lines, fluorescence, and ablation/decomposition of the target material.

The sensitivity of Raman spectroscopy can be increased when molecular vibrations are coherently driven, leading to an N^2 response (where N is the number of excited sample molecules) rather than the N scaling for spontaneous Raman. One method, known as CARS [79,85], excites molecular oscillations by both temporally and spatially overlapping a pump beam of frequency v_o and a Stokes beam of frequency v_s in the medium. A Raman-active vibrational mode, v_{vib}, is coherently driven when the oscillation frequency matches the beat frequency $v_o - v_s$ of the pump and Stokes laser pulses. The coherently driven vibrational mode is subsequently probed via a third beam of frequency v_{pr}, producing an anti-Stokes beam [86,87] of frequency v_{as}, where $v_{as} = v_{pr} + v_o - v_s$. This four-wave-mixing process facilitates the emission of coherent light, increasing the Raman signal intensity by several orders of magnitude as compared to the scattered signal obtained when using spontaneous Raman [88–90]. One drawback to traditional CARS is that the Stokes frequency has to be scanned to obtain a spectrum.

The use of fs-duration laser pulses may allow for the rapid acquisition of the Raman spectrum without the need to scan the Stokes frequency. In this scenario, a fs laser pulse having a sufficiently wide distribution of frequency components will contain multiple Stokes components, enabling multiple Raman-active vibrational modes to be driven simultaneously. The simultaneous excitation of multiple modes leads to the acquisition of a scattering spectrum, typically 1200 cm^{-1} wide for a standard fs pulse duration. Despite this advantage, the Stokes or pump frequency would still have to be altered via an optical parametric amplifier, as the ground-electronic state vibrational spectrum may have resonances spanning from 0 to ~4200 cm^{-1} for a given material of interest. Researchers have begun to take advantage of the high intensity of fs lasers and the ensuing nonlinear effects such as Kerr lensing, self-phase modulation [65], and filamentation [65,66] to perform single-shot measurements over the entire vibrational spectrum. The process of laser filamentation leads to substantial reshaping of the temporal, spatial, and spectral properties of the laser pulse [65,91,92]. The reshaping of the laser pulse leads to a broadening in the spectral bandwidth and pulse shortening. The formation of an ultrashort fs laser pulse (<10 fs) through filamentation enables the acquisition of a complete Raman spectrum (i.e., 0 and ~4200 cm^{-1}) because of the large bandwidth generated [93]. In addition, the pulse shortening that occurs in the transparent medium and the production of a dispersion-compensated supercontinuum may lead to impulsive Raman excitation. Impulsive Raman excitation occurs when the laser pulse duration is shorter than the vibrational period(s) of the target molecule [94], and enables the ultrashort, ultrabroadband fs laser pulse to act as both the pump and the Stokes pulse [95]. The impulsively driven modes can be subsequently probed via a narrowband pulse, allowing for acquisition of a complete Raman scattering spectrum with high spectral resolution without scanning the Stokes frequency. This approach has recently allowed for the remote detection of IED components such as nitro-based compounds and fuel at distances ~2 m using a 2.5 mJ filament and a 12 µJ narrowband probe pulse [96] in a crossing geometry (Figure 13.4). Such a configuration has enabled the detection of the carbon–hydrogen (C–H) stretch in gasoline at a distance of ~15 m.

This crossing geometry is typically employed for microscopy purposes because of the higher spatial resolution and spatial filtering. This crossing geometry would not be applicable for ranged detection methods because both spatial and temporal overlap between the laser beams would have to be maintained. In addition, a collinear configuration would reduce the task of alignment and therefore maintain spatial overlap at the focus. Therefore, collinear or single-beam methods are very attractive

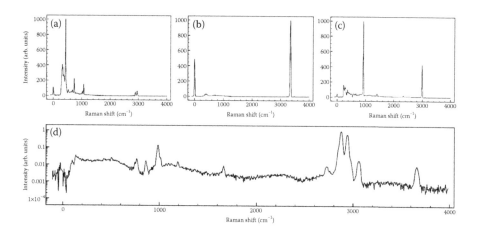

FIGURE 13.4 The measured filament-assisted impulsive Raman spectra of (a) triethylamine, (b) ammonia gas, (c) nitromethane, and (d) gasoline (log scale). The feature at ~3657 cm^{-1} in the gasoline spectrum is water.

for remote and standoff measurements. Recently, a number of researchers have begun to implement single-beam approaches using pulse shapers [97–105] or notch filters [106,107]. One method, single-pulse phase-contrast nonlinear Raman spectroscopy, uses a spatial light modulator (SLM) to apply a π phase step to a narrowband region of a spectrally filtered broadband pulse [97,98,103]. The applied phase gate induces a relative phase difference between the resonant (i.e., a signal generated by the Raman response of sample) and nonresonant (i.e., a frequency-independent background signal arising from the instantaneous electronic response) components of the generated CARS signal. As a result, the applied phase gate causes constructive interference for frequencies above the resonance, and destructive interference for frequencies below the resonance [97]. The use of this single-pulse phase-contrast Raman technique has enabled the detection of the Raman-active vibrational modes for various IED components such as potassium nitrate, sulfur, RDX, and urea at distances ≥ 5 m with an average power of 90–140 mW at the sample (intensity ~3×10^{11} W/cm^2). The π phase step can also be placed at the center wavelength of the fs laser pulse, causing stimulated Raman gain for the anti-Stokes sideband and stimulated Raman loss for the Stokes sideband [102]. Vibrational information is extracted by simply varying the spectral location of the π phase step and subtracting the corresponding spectra. As successful as these techniques have been with regard to the detection of hazardous material, spectra are typically limited to an approximate window of ~700 cm^{-1}. Once again, the bandwidth of the typical Ti:Sapphire, 30 fs transform-limited pulse used in the experiments prevents simultaneous acquisition of the entire fingerprint region.

Another technique, known as single-beam CARS [100,104,105], builds upon the work by Silberberg et al. [98] by shaping an ultrabroadband pulse generated in an argon-filled hollow waveguide by self-phase modulation. Using SLM technology, phase and amplitude shaping is performed on the ultrabroadband pulse for both compression and polarization shaping to create a narrowband pulse. The orthogonal polarization components of the ultrabroadband pump/Stokes pulse and the narrow-band probe pulse serve to reduce the nonresonant background associated with CARS while allowing a high spectral resolution spectrum to be obtained. The nonresonant background can be further reduced if the narrowband probe pulse is time-delayed, allowing for reduced complexity in retrieving the signature-specific resonant signal [108]. This scheme has enabled the detection of 2,4-dinitrotoluene at a concentration of ~10 µg/cm² embedded in a polystyrene film at a distance of 1 m using a 8 µJ laser pulse (pump, Stokes, and probe) [105].

A variant of the single-beam CARS method has been recently introduced and uses a resonant photonic crystal slab (RPCS) [106,107] instead of an SLM. The RPCS experiment employs a thin dielectric waveguide placed between two subwavelength gratings to create a tunable notch feature in the spectrum of the Ti:sapphire-generated fs pulse. The fs laser pulse is coupled into the RPCS at a particular angle transmitting most of the light. A notch is created in the spectrum when light diffracted from the grating is coupled into the waveguide layer causing destructive interference in a narrow spectral range of the transmitted pulse. The spectral position of the notch can be adjusted by altering the angle of the waveguide. This simple scheme, typically using a beam 1.5 mm in diameter and a peak power of ~10¹¹ W/cm², has enabled the detection of RDX (1 mg) at 24 m and sulfur (<500 µg), crystallized potassium nitrate (<1 mg), and crystallized urea (<4 mg) at 50 m (Figure 13.5) [107].

While some single-beam CARS methods have had success at detecting IED components at distances of up to 50 m [107], the strength of the CARS method, coherence, is also an inherent drawback. The CARS signal intensity or conversion efficiency is at

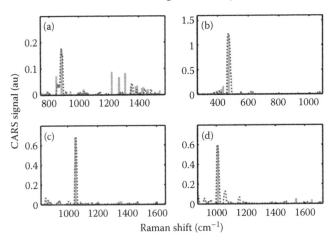

FIGURE 13.5 The single-beam CARS spectra of several samples (*dashed line*) obtained using a commercially available notch filter (i.e., the technique does not require the use of a pulse-shaper). The resolved CARS spectra for (a) cyclotrimethylene-trinitramine (RDX/T4) <1 mg, (b) sulfur powder <500 µg, (c) crystallized potassium nitrate (KNO₃) <1 mg, and (d) crystallized urea <4 mg. The spectra were obtained at a standoff distances of (a) 24 m and (b–d) 50 m, with integration times of: (a) 3 s, (b) 5 s, (c) 5 s, and (d) 1 s, respectively. The Raman lines (*solid line*) are plotted for reference. (From Natan, A., Levitt, J.M., Graham, L., Katz, O., and Silberberg, Y., *Appl. Phys. Lett.*, 100, 051111, 2012. With permission.)

its maximum when the phase matching condition is satisfied, $\Delta k = k_\text{o} + k_\text{pr} - k_\text{s} - k_\text{as} = 0$, where k_o, k_pr, k_s, and k_as are the momentum vectors for the pump, probe, Stokes, and anti-Stokes beams, respectively [79,85,88,89]. For collinear methods illustrated here, the anti-Stokes beam is produced in the same direction as the probe as a result of the phase matching [86,87]. The anti-Stokes signal is generated in the forward direction within the coherence length of the phase matching (up to several microns), and the signal is then scattered by the bulk sample in all directions, enabling the detection of trace solids in a standoff configuration [109]. Thus, the sample geometry may not have to be taken into account. In the case of thin films (i.e., <10 μm), the substrate can reflect the forward-scattered CARS beam in a direction normal to the surface, providing much higher signal in comparison to isotropic scattering. In addition, these methods typically require the use of the shortest possible pulse durations for the simultaneous excitation of all Raman-active vibrational modes in the molecule. Pulses as short as 8 fs are required, for instance to excite the hydrogen (i.e., H_2) stretch, and this requires several hundred nanometers of bandwidth in the visible region of the electromagnetic spectrum. Temporal dispersion of such a pulse will invariably occur, due to the frequency-dependent index of refraction during propagation through the air to the target. This dispersion can be precompensated for to a certain extent, but there will remain some variability in pulse duration as a function of propagation length, and this will affect the nonlinear excitation.

13.3.3 ROTATIONAL SPECTROSCOPY

Rotational spectroscopy is another method that could be used for the identification of hazardous material. Rotational spectroscopy measures the transition energies between quantized rotational states of molecules in the gas phase, and can be performed in either the frequency domain using microwave spectroscopy or in the time domain through impulsive excitation and nonlinear scattering. Frequency-resolved measurements in the microwave regime can be difficult to implement, due to the scanning that is required. Time-resolved measurements have the drawback of temporal scanning, but that limitation was recently overcome through a filament-based technique described in this section. Both approaches require isolated molecules in the gas phase.

Recently, a new rotational spectroscopy technique has been developed to provide the entire rotational spectrum in a single laser shot. The technique, known as spectral-to-temporal amplitude mapping polarization spectroscopy (STAMPS), measures the rotational period of a molecule rather than the energy of transitions between rotational states [110]. Briefly, after interaction with an intense, fs laser pulse, molecules with anisotripic polarizabilities along two axes experience a torque that aligns the molecules along the polarization direction of the laser field. After the pulse is gone, the excitation results in a periodic molecular alignment, often referred to as rotational revivals [111] or field-free molecular alignment [112]. This nonadiabatic alignment occurs when the pulse duration of the exciting laser pulse is much less than the rotational period ($T_r = 1/(2Bc)$) of the molecule, where $B = h/(8\pi^2 Ic)$ is the molecular rotational constant, I is the moment of inertia for the respective molecule, and c is the speed of light. The field-free molecular alignment that occurs is because of the preparation of a rotational wave packet through Raman transitions. The pre-pared rotational wave packet will subsequently undergo frequency-spread dephasing [113] prior to rephasing at fractional and integer multiples of the rotational period. As long as the rotational coherence of the molecules is preserved (i.e., collisional dephasing is negligible), alignment will recur. The degree of rotational alignment depends on the polarizability of the molecule, which in turn is a function of molecu-lar structure [110].

Typically, rotational revivals are measured by scanning the delay between pump and probe laser pulses. Such a configuration is time-consuming and it would there-fore be advantageous to acquire a spectrum in a single shot. Several methods have been developed thus far to perform single-shot rotational revival measurements [114–117]. STAMPS (Figure 13.6a) is a new method that is also capable of per-forming single-shot rotational revival measurements. STAMPS is unlike the pre-viously developed methods, in that it is based on chirping a coherent white-light continuum generated in a laser filament to map the temporal dynamics of a rotational wave packet through birefringence and cross-phase modulation [110]. The STAMPS method has allowed for the investigation of various gas-phase molecules such as oxygen and nitrogen (i.e., O_2 and N_2) but also gas-phase methanol (Figure 13.6b) over a time period of 65 ps in nonscanning measurements, affording direct access to the temporal evolution of a wave packet in relatively small molecules. The identification of methanol shows great promise for other IED components, but more work needs to be performed to extend the range, adaptability, and versatility of the technique prior to its implementation in security applications.

13.3.4 VIBRATIONAL SUM-FREQUENCY GENERATION

LIBS and Raman scattering may sample the underlying substrate at low material coverage. The additional signal from the substrate could complicate the analysis of

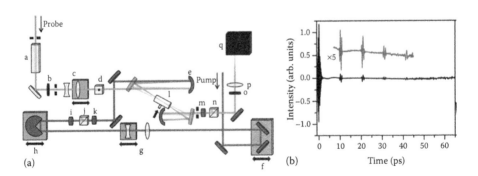

FIGURE 13.6 (A) A schematic of the STAMPS experimental scheme: (a) SF11 glass block; (b) short pass filter <750 nm; (c) beam expander; (d) and (j) polarizing cube; (e) 25 cm spheri-cal mirror; (f) mechanical delay stage; (g) beam reducer; (h) electronic delay stage; (i) and (k) half-wave plate; (l) sample tube; (m) quarter-wave plate; (n) polarizing cube (analyzer); (o) color glass filter; (p) 10 cm lens; (q) spectrometer. (B) The STAMPS signal of methanol with revivals shown at ×5 (inset).

the trace energetic material adsorbed onto the surface. Vibrational sum-frequency generation (VSFG) provides a means to probe only the surface of the sample. VSFG is a three-photon process that takes advantage of the second-order polarizability ($P^{(2)}$) of a given material to convert two photons to one (i.e., $\nu_1 + \nu_2 = \nu_{SFG}$). This conversion process can only occur in a material that does not display inversion symmetry, such as noncentrosymmetric crystals. Bulk material such as water, gases, and amorphous solids do possess inversion symmetry and will not undergo the nonlinear process of VSFG. At the interface, symmetry is broken, allowing the VSFG process and, as a result, VSFG allows the investigation of material at a particular net orientation only at the air/IED or IED/substrate interface. VSFG is typically performed by irradiating an interface with two laser pulses of frequency ν_1 and ν_2. At the interface where the symmetry is broken, a photon will be produced at frequency ν_{SFG}, and will propagate in a direction determined by the phase-matching conditions. The highly directional VSFG signal produced will be emitted collinear and perpendicular to the surface, so long as ν_1 and ν_2 are configured to be collinear and perpendicular to the surface. As a result, the VSFG signal could be used for the standoff detection of trace energetics deposited on surfaces. In addition, since the VSFG is intensity dependent, the use of picosecond/fs laser pulses will naturally increase the overall signal intensity and conversion efficiency.

Vibrational resonant enhancement is employed to obtain high molecular specificity in the VSFG process. The SFG signal undergoes an enhancement of several orders of magnitude when the energy of a photon matches the energy of an IR- and Raman-active vibrational mode. Scanning the frequency of the IR photon with respect to a photon in the visible region will produce peaks in the SFG spectrum, enabling identification of the adsorbed species in relation to background interferents. The Fourier transform infrared (FTIR) spectrum and the VSFG spectrum may not match exactly, because of surface effects causing the peaks to be blue-shifted or split in comparison to bulk measurements. Nevertheless, the use of VSFG has allowed for the point detection of a variety of materials including the high explosive octahydro-1,3,5,7-tetranitro-1,3,5,7-tetrazocine (HMX) [118,119], energetic oxidizers [120], and the binder Estane [118]. The theoretical LOD [118] for the high explosive HMX using VSFG was estimated to be approximately 20 pg or ~0.03 μg/cm². In addition, VSFG has also been used in the remote detection of explosive materials using picosecond laser pulses [121] with detection limits of ~0.51 μg/cm² for an explosive simulant, 1-amino-4-nitrobenzene, using a 10 μJ, 532 nm laser pulse and a 100 μJ, IR pulse.

Despite the success of VSFG, a spectrum can take several seconds to acquire because the frequency of the IR beam has to be scanned. One particular resonance could be used instead of acquiring a complete scan, increasing the overall speed of the technique. Another possible drawback to the technique is the propagation of an IR beam through the atmosphere. Water, along with other atmospheric contaminants, can affect the transmission of the IR beam, reducing the overall VSFG signal intensity. In addition, the VSFG signal will be produced in the same direction as the visible beam because of phase-matching conditions and its large momentum versus that of the IR beam. Therefore, in a manner similar to CARS, if the visible beam undergoes a specular reflection from a nonnormal reflective surface the highly coherent VSFG signal will propagate in the same direction. As a result, the VSFG signal will be lost as no light will be directed toward the detector. Analysis of non-normal reflective substrates can severely affect overall performance of VSFG for the detection of energetic threats.

13.3.5 RESONANCE-ENHANCED MULTIPHOTON IONIZATION

Laser photoionization mass spectrometry is another technique that has been used for the identification of gas-phase signatures. Laser photoionization mass spectrometry is performed by irradiating a sample with a nanosecond laser, causing subsequent ionization and fragmentation. The fragments (i.e., C^+, N^+, O^+, H^+, etc.) that are produced are typically monitored via a mass spectrometry technique such as TOF. Intact molecular ions are typically not observed because of the intensity and duration of the laser pulse, and the ensuing fragmentation makes identification of a given material arduous. A degree of selectivity can be regained by taking advantage of resonance enhancement through an excited electronic state. Resonance-enhanced multiphoton ionization (REMPI) [122–127] commonly uses a laser pulse that is resonant with a multiphoton ($N \geq 2$) transition to an electronically excited intermediate state in the molecule. The multiphoton resonant absorption process causes the absorption cross section of the molecule to increase by several orders of magnitude, leading to an enhancement in the ionization process after absorption of N photons. A chemical-specific spectrum is obtained by sweeping a tunable laser through several frequencies while monitoring the current between two electrodes or the ionization channels via a mass spectrometer.

REMPI uses narrowband nanosecond lasers to gain the selectivity/specificity required for identification. Using these long-pulse-duration lasers causes more fragmentation than using short-pulse lasers at comparable intensities [128]. The increased fragmentation results from energy being deposited into nuclear modes of the molecule during nanosecond ionization. The use of shorter-duration picosecond and fs laser pulses typically increases the amount of parent relative to fragment ions, possibly increasing the capability to perform molecular identification [128]. Therefore the use of fs laser pulses and pulse-shaping methods may lead to further selectivity and control using REMPI. Recently, an adaptive fs pulse-shaping method has been implemented to increase the ionization efficiency of nitromethane ~150% in a REMPI scheme [129].

Despite the demonstration of increased ionization efficiency for nitro-containing compounds, REMPI is a point detection method because of the requirement for a mass spectrometer or detection electrodes. A new approach recently developed known as radar REMPI enables remote detection of hazardous material [130] (Figure 13.7a). The radar REMPI technique works by focusing a laser pulse that is in resonance with an N photon transition. Subsequent absorption of M additional photons leads to ionization of the molecular species. For example, absorption of one photon at 226 nm is required to resonantly excite nitric oxide (NO), and one additional photon at 226 nm is required to ionize the molecule. Once ionized, a microwave signal is directed at the ionization region, causing scattering of the microwave signal. The microwave signal is backscattered by the electrons contained in the ionization region, enabling standoff detection of the photoionized sample. This

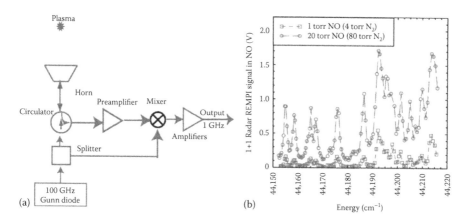

FIGURE 13.7 (a) Experimental setup for radar REMPI. (b) The 1 + 1 radar REMPI spectra in NO:N$_2$ mixture 1:4. (From Dogariu, A. and Miles, R.B., *Appl. Opt.*, 50, A68–A73, 2011. With permission.)

technique has allowed for the detection of NO at <1 ppm in air at a distance of 10 m using 1 mJ, 100 ps pulses at ~226 nm and a 20 mW, 100 GHz microwave source [130] (Figure 13.7b). The homodyne detection of the backscattered signal from the photo-ionized NO allowed the electric field of the microwave signal to be detected rather than the power. This allows for a higher signal-to-noise ratio and the observed signal strength falls inversely with distance rather than the distance squared.

The use of microwaves in this scheme is advantageous for several reasons. For example, the low spectral irradiance in the microwave region reduces the inherent background noise. The reduction in the background is unlike optical techniques based in the UV/visible region, such as Raman, LIBS, or fluorescence spectroscopy, where photons from the solar emission and other environmental sources could possibly interfere with detection of scattered or emitted photons. The long wavelength of the microwaves leads to point-coherent scattering from the photoionized sample, so phase matching is unimportant and scattering in the backward direction is strong. In addition, the microwave scattering that occurs falls into the Rayleigh scattering regime, since the microwave's wavelength is greater than the laser-induced plasma [131,132]. As a result, the scattered signal is proportional to the number of charges created and therefore proportional to the number of target species probed by the REMPI laser pulse [131].

Despite the success of radar REMPI, several seconds will be required to scan the excitation wavelength and acquire a complete fingerprint spectrum. The overall speed of the technique could be increased, at the cost of specificity, by using one particular resonance instead of acquiring a complete spectrum. Another drawback to the technique is that to date it requires the sample to be in the gas phase. It would be difficult to detect a range of explosive material with this technique, due to its current detection limits (~<1 ppm for NO) and the low vapor pressure of many explosives (e.g., TNT has a vapor pressure of ~10 ppb at 300 K and ammonium nitrate has a vapor pressure of ~12 ppb at 300 K). As a result, more work, such as the implementation of a stronger microwave signal, needs to be done to reduce the detection limits of the technique. Another possibility is to directly investigate adsorbed material. However, the use of microwaves could hinder such an approach if the energetic is deposited onto a metal substrate that may reflect the microwave "pulse" back at the detector, obscuring the backscattered signal from the sample.

13.3.6 BACKWARD LASING

All of the above optical-based techniques rely on the collection of light scattered or emitted from the sample in the backward direction. Coherent techniques, such as CARS, that produce a well-collimated beam can also be difficult to implement in the remote/standoff detection of hazardous material, because of the forward propagation of the scattered signal. Coherent methods for generating backward-propagating radiation are prohibited by momentum conservation and thus other solutions are sought. In one approach, a backward-propagating beam would be used as a means to generate a measurable signal from signature molecules at considerable standoff distances (100–1000 m). Thus, there has been a great deal of interest in generating backward lasing in a dilute gas-phase medium. The use of backward lasing would allow a spectroscopic fingerprint of a hazardous material to be sent directly back toward the detector.

Lasing in air can occur by creating a population inversion in either nitrogen or oxygen. Nitrogen lasers, in fact, create the population inversion via electron impact excitation in a fast plasma discharge. The electrons are rapidly heated in the plasma to kinetic energies of ~15 eV. An electron from the plasma can then collide with a nitrogen molecule in the ground state (X $^1\Sigma_g$), transferring the population to the upper C $^3\Pi_u$ state throughout the discharge channel. Radiative decay between the C $^3\Pi_u$ state and the lower B $^3\Pi_g$ state produces near-UV radiation. Since the plasma discharge maintains the hot and dense plasma, there is sufficient time to allow for population inversion to occur and therefore to produce lasing between the C $^3\Pi_u$ and B $^3\Pi_g$ states in the medium.

The goal of backward lasing is to create a coherent collimated beam at a remote or standoff distance that is sent back to the detector. A plasma discharge source is impractical for such measurements since it is a point source. To enable backward lasing in such measurement schemes, fs laser pulses could be used to generate a filament channel at the desired range. As noted earlier, when the intensity of the fs pulse exceeds the self-focusing threshold of air a filament will form. The filamentation process generates a plasma channel consisting of ionized oxygen and nitrogen, similar to a plasma discharge source. The density of ionized species in the laser-generated plasma within the filamentation channel is too low and the lifetime of the electrons generated is too short to create the necessary population inversion for lasing to occur. In addition, the use of higher-energy fs pulses will not simply increase the overall density of the ionized species within the plasma channel because of the intensity clamping effect in the filament. Despite these facts, backward-propagating stimulated emission has been reported using fs filamentation [133], yet true backward lasing in ambient air has been elusive using fs laser pulses. Nevertheless, in a proof-of-principle experiment, backward-propagating amplified spontaneous emission has been demonstrated using an argon/nitrogen (i.e., Ar/N$_2$) mixture contained in a gas cell. A filament driven by an intense, 80 fs, 3.9 μm mid-IR pulse with energies up to 8 mJ is used to excite Ar atoms that were present in tenfold excess relative to the nitrogen gas. After a three-body collision and dissociative recombination, the excited metastable Ar*(4 $^3P_2$) collides with nitrogen populating its C $^3\Pi_u$ state [134,135], forming the N_2 population inversion. Lasing was detected at 337 and 357 nm for input energies above 7 mJ/pulse. Unfortunately, nitrogen itself will serve to quench the C state population at pressures above 1.7 bar via the N_2 (C $^3\Pi_u$) + N_2 (X $^1\Sigma_g$) → N_4 → N_2 (B $^3\Pi_g$ vibrationally excited) + N_2 (X $^1\Sigma_g$) relaxation pathway.

Oxygen is an efficient quencher of excited N_2, suggesting that a free-space N_2 laser in atmospheric air will be challenging to create. Therefore, it may be advantageous to use the molecular oxygen in air as a lasing medium [136]. Recently, a 100 ps, 226 nm laser pulse was used to fragment diatomic oxygen (O_2) producing atomic oxygen (O) in the 2p ^3P state (Figure 13.8a). The laser radiation is subsequently absorbed by the atomic oxygen, enabling population transfer to the upper 3p ^3P state, thus creating a population inversion with the 3s ^3S state allowing for lasing to occur [136]. Amplified spontaneous emission at 845 nm (Figure 13.8b) from the 1 mm long cylinder of excitation led to a coherent backward-propagating pulse that in principle could be used as a seed for the optical-based detection methods listed

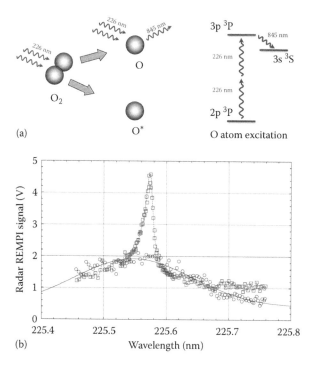

(a)

(b)

FIGURE 13.8 (a) Two-photon dissociation of the oxygen molecule and subsequent two-photon resonant excitation of the ground-state oxygen atom fragment result in emission at 845 nm. (b) Atomic oxygen 2 + 1 radar REMPI signals for preexisting oxygen atoms in a flame (*red squares*) and produced by photodissociation in air (*blue circles*). (From Dogariu, A., Michael, J.B., Scully, M.O., and Miles, R.B., *Science*, 331, 442–445, 2011. With permission.)

previously. Geometric focusing of a high-intensity beam would be required to create the backward-propagating pulse, preventing the extension of this technique to longer distances.

In addition, it was shown that a second pulse could be used to generate a denser plasma, leading to higher gain in the backward-propagating laser [136]. The use of a second, longer (>350 fs), higher-energy pulse (>1 J) accelerates the seed electrons generated by the first laser pulse. In the process, the additional seed electrons generated allow for a sufficiently dense and hot plasma to create the required population inversion leading to higher gain [137]. It has been suggested that this approach may also be used in conjunction with fs laser pulses to produce a sufficiently dense plasma from a filament channel to allow for backward lasing in ambient air [137].

The development of backward lasing in air has great promise for the detection of gas-phase trace contaminants. More work is required to enable spectroscopic investigations using this technique. For example, it has been suggested that the stimulated Raman scattering could be performed using backward lasing [138]. A possible drawback to the technique is that it would have difficulty in sampling condensed-phase material. This drawback may be overcome if the sensitivity of the technique is sufficient to sample the vapor produced by the target material.

13.3.7 FREQUENCY COMB SPECTROSCOPY

Ti:Sapphire lasers can emit pulses with a duration of 50 fs. These short pulses typically possess a bandwidth of approximately 30 nm, with colors ranging from 810 to 780 nm. In addition, the emitted pulses exhibit a periodic and regular separation determined by the length of the laser cavity. Despite their common bandwidth and regular separation, the fs pulses emitted are not identical due to changes in cavity length leading to a phase shift in the electric field underneath the carrier envelope, $\Delta\phi_{CE}$. When the carrier envelope phase is stabilized (i.e., the phase shift does not change), a Fourier transform of the emitted pulses will provide the frequency of every optical line m contained within the pulse, known as a frequency comb, $v_m = m f_{rep} + f_0$, where f_{rep} is the pulse repetition rate (i.e., $1/\tau$), and f_0 is the comb offset frequency (i.e., carrier envelope offset frequency given by $f_0 = (\Delta\phi_{CE} \cdot f_{rep})/2\pi)$) [139]. From this expression, all optical frequencies contained in the comb can be determined if f_{rep} and f_0 are also known [139,140]. The parameters f_{rep} and f_0 can be determined by using a fast photodiode and nonlinear interferometry [141,142]. The pulse repetition rate is determined by inputting a portion of the pulse into a fast photodiode and measuring the separation between pulses. Discernment of f_0 is more complicated, and commonly requires a self-referencing interferometric $f - 2f$ scheme. In this scheme, the low-frequency part of the spectrum is frequency doubled using a nonlinear optical crystal (e.g., beta-barium borate), generating a second frequency comb at twice the f_0 frequency. This frequency-doubled spectrum is interfered with the lower-frequency spectrum, creating a beat note between the harmonics of the laser light enabling the deduction of f_0.

Once f_{rep} and f_0 are known, any unknown frequency within the range of the comb can be determined by interfering the signal with the frequency comb, thus producing a beat note. An approximate frequency measurement can be determined from the lowest beat frequency observed. The lowest beat frequency is an indication of distance between the unknown frequency and the closest line in the frequency comb. This knowledge along with subsequent changes to the comb line frequency allows direct determination of the unknown frequency. Measuring a frequency outside the range of the frequency comb is challenging, and various techniques have been developed to help overcome this issue. For example, octave-spanning spectra can be produced from Ti:sapphire pulses typically by passing the laser pulse through a photonic crystal fiber (PCF) [141]. The octave-spanning spectrum would allow for a wider range of unknown frequencies to be determined, in addition to the possibility of self-referenced frequency combs [139]. The bandwidth can also be adjusted and shifted farther, making IR frequency measurement possible using Yb-doped fiber lasers or erbium (Er)-doped fiber lasers in conjunction with PCFs or highly nonlinear fibers (HNFs) [143]. Extension into the IR region has been enabled using optical parametric oscillators [144].

In the IR wavelength region, detection of gas-phase molecules may be improved due to strong vibrational transitions. The strong transitions observed for molecules in this spectral/frequency region lead to lower LODs, facilitating detection. Commonly a comb is coupled into a cavity in a similar manner to cavity ring-down spectroscopy (CRDS) (refer to Chapter 11). The pulse is injected into the cavity, where it will undergo multiple reflections due to the high-finesse mirrors. The overlap of the cavity and the comb only occurs in a limited spectral range, because of the available reflectivity bandwidth of the mirrors. The light exiting the cavity is detected via a virtually imaged phased array (VIPA) spectrometer, enabling determination of the trace gases contained within the cavity (Figure 13.9). This technique, commonly known as cavity-enhanced direct frequency comb spectroscopy (CE-DFCS), has enabled the analysis of trace gases such as iodine [145] and methane in human breath for the possible diagnosis of disease [146]. The technique shows promise for the detection of trace molecules in the gas phase, and may eventually enable the identification of explosives via their trace gas-phase signatures.

13.3.8 Terahertz Spectroscopy

The ultrafast optical techniques discussed so far rely on line of sight for detection of the hazardous material because of the use of IR and visible radiation. If an energetic material or IED were concealed, detection might be avoided if IR and visible radiation are not capable of penetrating the obscuring material. Thus, the exploration of millimeter and terahertz (THz) radiation (30 µm–3 mm, 10^{11}–10^{13} Hz) for the detection of hazardous material is of interest (refer to Chapter 12). Millimeter and THz radiation are capable of penetrating obscuring layers and are also nonintrusive and nonionizing. The use of radiation in these regions allows for the electromagnetic radiation to penetrate an optically opaque layer, perhaps enabling identification of unknowns contained inside or underneath an object. One limitation of using such radiation is that the spatial resolution of the measurement is reduced.

THz radiation can be produced by a variety of sources, such as backward-wave oscillators [147–150], gyrotrons, free-electron lasers, quantum cascade lasers [151–153], and so on. Recently, fs laser sources have begun to be implemented to enable

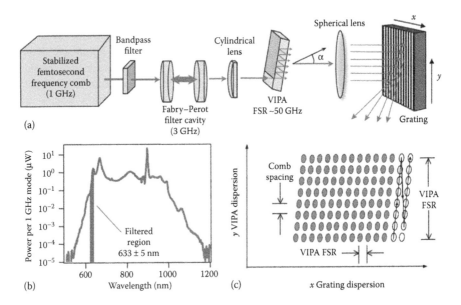

FIGURE 13.9 A high-resolution virtually imaged phased array (VIPA) disperser is used in combination with a diffraction grating to spatially resolve the stabilized frequency comb of a Ti:sapphire femtosecond laser. The full output spectrum of the laser and the 633 nm region isolated by the bandpass filter is shown in (b). The spectrometer output consists of a two-dimensional array of the frequency comb modes, where each "dot" represents an individual mode (c). Within a column (y), which is tilted by the grating dispersion, the dots are separated by the mode spacing (3 GHz in this case). Within each row (x), the dots are separated by the VIPA free spectral range (FSR, ~50 GHz in this case). The manner in which successive modes can be indexed and counted is indicated by the arrows in the rightmost two columns. For clarity, not all modes are shown in this diagram. (From Diddams, S.A., Hollberg, L., and Mbele, V., *Nature*, 445, 627–630, 2007. With permission.)

the generation of THz radiation via the use of organic and inorganic semiconductor sources such as 4-*N*,*N*-4-dimethylamino-4′-*N*′-methylstilbazolium tosylate (DAST) [154–156], gallium arsenide (GaAs) [157,158], and zinc telluride (ZnTe) [159]. These materials are commonly noncentrosymmetric materials and therefore possess a nonzero second-order polarizability. Upon irradiation with a fs laser pulse, a direct current second-order nonlinear polarization is produced, causing optical rectification [160–162]. Optical rectification in the medium generates polarization terms for both the sum and difference frequencies [79]. The sum-frequency components do not contribute to the generation of THz radiation and will not be discussed further. The production of THz radiation relies on the difference frequency generation between all the frequencies within the bandwidth of the fs laser pulse [161,163], typically producing a sub-picosecond THz radiation pulse [161] ranging from 0 to ~$1/\tau_d$ Hz, where τ_d is the duration of the fs pulse [160], so long as the phase-matching conditions are satisfied [164].

To recover the time profile of the THz radiation, the optical rectification process is reversed [163,164]. THz radiation is detected by focusing the THz pulse into an electro-optic crystal (e.g., ZnTe) along with a visible probe pulse. When the two beams are present in the crystal, birefringence is induced through the Pockels effect [160,165]. The probe beam will therefore undergo a phase modulation that can be measured by using a ¼ wave plate and a Wollaston prism [161]. The combination of this "balanced detection" scheme with time-delay scan of the probe with respect to the THz pulse allows for the retrieval of a time-domain trace of the electric field of the THz pulse [160,161]. The method relies on the use of materials with inversion symmetry to produce THz radiation.

Once the THz pulse is produced, the radiation has to be directed to a sample to collect an image in a reflective geometry. The irradiance may decrease, affecting the obtained data because of absorption by atmospheric constituents such as water. Therefore, it would be of value to create THz pulses in proximity to the target to be investigated. Recently, several groups have discovered that THz radiation could be produced through fs filamentation in air, possibly enabling the delivery of THz pulses at the target [166,167]. A fs laser pulse of sufficient intensity traveling through air will undergo Kerr lensing/self-focusing, leading to the formation of a plasma channel. Once the intensity of the pulse reaches approximately 10^{13} W/cm^2, ionization of the air occurs, which leads to plasma defocusing. The self-focusing and plasma defocusing balance and enable the plasma channel or filament to propagate through the air. The plasma formed in the filament channel is weakly ionized [168,169]: ~10^{16}/cm^3. The electron–atom collision frequency of the electrons contained within the plasma channel is relatively short when compared to the plasma frequency [166,169]. As a result, the electrons will oscillate in the plasma 2–3 times before being damped out [166]. These electron–atom collisions, along with spatial gradients associated with the laser pulse envelope, lead to ponderomotive forces, causing the electrons to oscillate [168]. This oscillation has the appearance of an electric dipole with a length approximately equal to the width of the plasma channel [162]. In addition, the plasma current moves along the propagation axis of the laser pulse emitting radially polarized broadband THz radiation because of a Cherenkov-like effect (i.e., a dipole-like electric charge oriented along the propagation axis of the filament and moving at the light velocity behind the self-guided laser pulse in the medium) [162,166,169,170]. Elliptically polarized THz radiation has recently been produced using fs filamentation via four-wave optical rectification or second-order optical rectification within the plasma channel where the centrosymmetry of the air is broken by the fs laser pulse [167,171]. The rapid development of these new sources may lead to breakthroughs in THz imaging.

13.4 CONCLUSIONS

The exploration of ultrafast lasers as a means for detecting explosive materials is a rapidly expanding area of research. The unique properties of ultrafast lasers, including broad bandwidth, phase-locked frequencies, and high laser intensity, have led to fundamentally new analytical techniques. Femtosecond laser pulses have been used to couple into molecular degrees of freedom including translational, rotational, vibrational, and electronic modes. Coupling into translational modes has produced the laser vaporization method, which is capable of universally releasing all molecular compounds from the solid state into the gas phase, where electrospray ionization can be performed on extremely low-vapor-pressure molecules. This combination of LEMS has resulted in a method capable of classifying many explosive and energetic formulations. The method is a point detector capable of rapidly identifying essentially any material. Coupling into rotational and vibrational modes of molecules has produced new laser-based detection methods capable of identifying signatures of energetics deposited onto surfaces and gas-phase molecules in a rapid manner that is amenable to standoff detection of dilute signatures. VSFG is a uniquely surface-sensitive probe, while filament-based impulsive Raman spectroscopy has enabled detection of gas-phase signatures at distances beyond 10 m. Coupling into electronic modes has enabled fs-LIBS and remote filament-based breakdown spectroscopy. In the latter case, standoff distances of beyond 100 m are enabled through the nonlinear mechanism of filament formation. Despite their preliminary success, each technique needs to be further refined to make them truly viable options to safeguard personnel and high-risk targets in a variety of settings (e.g., transportation, military, ports, borders, etc.) and security applications.

REFERENCES

1. Karasek, F. W. and Denney, D. W., Detection of 2,4,6-trinitrotoluene vapours in air by plasma chromatography. *J. Chromatogr. A* 1974, *93*(1), 141–147.
2. Karasek, F. W., Plasma chromatography. *Anal. Chem.* 1974, *46*(8), 710A–720A.
3. Spence, D. E., Kean, P. N., and Sibbett, W., 60-fsec pulse generation from a self-mode-locked Ti:sapphire laser. *Opt. Lett.* 1991, *16*(1), 42–44.
4. Strickland, D. and Mourou, G., Compression of amplified chirped optical pulses. *Opt. Commun.* 1985, *56*(3), 219–221.
5. Peng, X., Kim, K., Mielke, M., Jennings, S., Masor, G., Stohl, D., Chavez-Pirson, A., et al., High efficiency, monolithic fiber chirped pulse amplification system for high energy femtosecond pulse generation. *Opt. Express* 2013, *21*(21), 25440–25451.
6. Perr, J. M., Furton, K. G., and Almirall, J. R., Solid phase microextraction ion mobility spectrometer interface for explosive and taggant detection. *J. Sep. Sci.* 2005, *28*(2), 177–183.
7. Tam, M. and Hill, H. H., Secondary electrospray ionization-ion mobility spectrometry for explosive vapor detection. *Anal. Chem.* 2004, *76*(10), 2741–2747.
8. Buxton, T. L. and Harrington, P. D., Trace explosive detection in aqueous samples by solid-phase extraction ion mobility spectrometry (SPE-IMS). *Appl. Spectrosc.* 2003, *57*(2), 223–232.
9. Greg, G., Christine, M., Scott, W., and Richard, L., Characterization of high explosive particles using cluster secondary ion mass spectrometry. *Rapid Commun. Mass Spectrom.* 2006, *20*(12), 1949–1953.
10. Meurer, E. C., Chen, H., Riter, L., Cotte-Rodriguez, I., Eberlin, M. N., and Cooks, R. G., Gas-phase reactions for selective detection of the explosives TNT and RDX. *Chem. Commun.* 2004, *1*(X), 40–41.
11. Sigman, M. E. and Ma, C. Y., Detection limits for GC/MS analysis of organic explosives. *J. Forensic Sci.* 2001, *46*(1), 6–11.
12. Barshick, S. A. and Griest, W. H., Trace analysis of explosives in seawater using solid-phase microextraction and gas chromatography ion trap mass spectrometry. *Anal. Chem.* 1998, *70*(14), 3015–3020.
13. Wu, Z., Hendrickson, C. L., Rodgers, R. P., and Marshall, A. G., Composition of explosives by electrospray ionization Fourier transform ion cyclotron resonance mass spectrometry. *Anal. Chem.* 2002, *74*(8), 1879–1883.

14. Jehuda, Y., Joseph, E. M., and Richard, A. Y., Electrospray ionization tandem mass spectrometry collision-induced dissociation study of explosives in an ion trap mass spectrometer. *Rapid Commun. Mass Spectrom.* 1997, *11*(18), 1961–1970.

15. Martin, A. N., Farquar, G. R., Gard, E. E., Frank, M., and Fergenson, D. P., Identification of high explosives using single-particle aerosol mass spectrometry. *Anal. Chem.* 2007, *79*(5), 1918–1925.

16. Cotte-Rodriguez, I., Takats, Z., Talaty, N., Chen, H., and Cooks, R. G., Desorption electrospray ionization of explosives on surfaces: Sensitivity and selectivity enhancement by reactive desorption electrospray ionization. *Anal. Chem.* 2005, *77*(21), 6755–6764.

17. Cotte-Rodriguez, I., Hernandez-Soto, H., Chen, H. and Cooks, R. G., *In situ* trace detection of peroxide explosives by desorption electrospray ionization and desorption atmospheric pressure chemical ionization. *Anal. Chem.* 2008, *80*(5), 1512–1519.

18. Ewing, R. G., Atkinson, D. A., Eiceman, G. A., and Ewing, G. J., A critical review of ion mobility spectrometry for the detection of explosives and explosive related compounds. *Talanta* 2001, *54*(3), 515–529.

19. Waddell, R., Dale, D. E., Monagle, M., and Smith, S. A., Determination of nitroaromatic and nitramine explosives from a PTFE wipe using thermal desorption-gas chromatography with electron-capture detection. *J. Chromatogr. A* 2005, *1062*(1), 125–131.

20. Colon, Y., Ramos, C. M., Rosario, S. V., Castro, M. E., Hernandez, S. P., Mina, N., Chamberlain, R. T., and Lareau, R. T., Ion mobility spectrometry determination of smokeless powders on surfaces. *Int. J. Ion Mobil. Spec.* 2002, *5*(3), 127–131.

21. Strege, M. A., Total residue analysis of swabs by ion mobility spectrometry. *Anal. Chem.* 2009, *81*(11), 4576–4580.

22. Perez, J. J., Flanigan, P. M., Brady, J. J., and Levis, R. J., Classification of smokeless powders using laser electrospray mass spectrometry and offline multivariate statistical analysis. *Anal. Chem.* 2012, *85*(1), 296–302.

23. Brady, J. J., Judge, E. J., and Levis, R. J., Mass spectrometry of intact neutral macromolecules using intense nonresonant femtosecond laser vaporization with electrospray postionization. *Rapid Commun. Mass Spectrom.* 2009, *23*(19), 3151–3157.

24. Brady, J. J., Judge, E. J., and Levis, R. J., Identification of explosives and explosive formulations using laser electrospray mass spectrometry. *Rapid Commun. Mass Spectrom.* 2010, *24*(11), 1659–1664.

25. Brady, J. J., Judge, E. J., and Levis, R. J., Nonresonant femtosecond laser vaporization of aqueous protein preserves folded structure. *Proc. Natl. Acad. Sci. U S A* 2011, *108*(30), 12217–12222.

26. Flanigan, P. M., Brady, J. J., Judge, E. J., and Levis, R. J., Determination of inorganic improvised explosive device signatures using laser electrospray mass spectrometry detection with offline classification. *Anal. Chem.* 2011, *83*(18), 7115–7122.

27. Judge, E. J., Brady, J. J., Dalton, D., and Levis, R. J., Analysis of pharmaceutical compounds from glass, fabric, steel, and wood surfaces at atmospheric pressure using spatially resolved, nonresonant femtosecond laser vaporization electrospray mass spectrometry. *Anal. Chem.* 2010, *82*(8), 3231–3238.

28. Judge, E. J., Brady, J. J., and Levis, R. J., Mass analysis of biological macromolecules at atmospheric pressure using nonresonant femtosecond laser vaporization and electrospray ionization. *Anal. Chem.* 2010, *82*(24), 10203–10207.

29. Brady, J., Judge, E., and Levis, R., Analysis of amphiphilic lipids and hydrophobic proteins using nonresonant femtosecond laser vaporization with electrospray post-ionization. *J. Am. Soc. Mass Spectrom.* 2011, *22*(4), 762–772.

30. Arnolds, H., Levis, R. J., and King, D. A., Vibrationally assisted DIET through transient temperature rise: The case of benzene on Pt{111}. *Chem. Phys. Lett.* 2003, *380*(3–4), 444–450.

31. Arnolds, H., Rehbein, C., Roberts, G., Levis, R. J., and King, D. A., Femtosecond near-infrared laser desorption of multilayer benzene on Pt{111}: A molecular Newton's cradle? *J. Phys. Chem. B* 2000, *104*(14), 3375–3382.

32. Arnolds, H., Rehbein, C. E. M., Roberts, G., Levis, R. J., and King, D. A., Femtosecond near-infrared laser desorption of multilayer benzene on Pt{111}: Spatial origin of hyperthermal desorption. *Chem. Phys. Lett.* 1999, *314*(5–6), 389–395.

33. Williams, P. and Sundqvist, B., Mechanism of sputtering of large biomolecular ions by impact of highly ionizing particles. *Phys. Rev. Lett.* 1987, *58*(10), 1031–1034.

34. Johnson, R. E., Sundqvist, B. U. R., and Ens, W., Laser-pulse ejection of organic molecules from a matrix: Lessons from fast-ion-induced ejection. *Rapid Commun. Mass Spectrom.* 1991, *5*(11), 574–578.

35. Brady, J. J., Flanigan IV, P. M., Perez, J. J., Judge, E. J., and Levis, R. J., Multidimensional detection of explosives and explosive signatures via laser electrospray mass spectrometry. *Proc. SPIE* 2012, *8358*, 83580X.

36. De Lucia Jr., F. C., Gottfried, J. L., Munson, C. A., and Miziolek, A. W., Multivariate analysis of standoff laser-induced breakdown spectroscopy spectra for classification of explosive-containing residues. *Appl. Opt.* 2008, *47*(31), G112–G121.

37. De Lucia Jr., F. C., Harmon, R. S., McNesby, K. L., Winkel, R. J., and Miziolek, A. W., Laser-induced breakdown spectroscopy analysis of energetic materials. *Appl. Opt.* 2003, *42*(30), 6148–6152.

38. Dikmelik, Y., McEnnis, C., and Spicer, J. B., Femtosecond and nanosecond laser-induced breakdown spectroscopy of trinitrotoluene. *Opt. Express* 2008, *16*(8), 5332–5337.

39. Gottfried, J. L., De Lucia Jr., F. C., and Miziolek, A. W., Discrimination of explosive residues on organic and inorganic substrates using laser-induced breakdown spectroscopy. *J. Anal. Atom. Spectrom.* 2009, *24*(3), 288–296.

40. Hahn, D. W. and Omenetto, N., Laser-induced breakdown spectroscopy (LIBS), Part I: Review of basic diagnostics and plasma-particle interactions: Still-challenging issues within the analytical plasma community. *Appl. Spectrosc.* 2010, *64*(12), 335A–366A.

41. Eland, K. L., Stratis, D. N., Lai, T. S., Berg, M. A., Goode, S. R., and Angel, S. M., Some comparisons of LIBS measurements using nanosecond and picosecond laser pulses. *Appl. Spectrosc.* 2001, *55*(3), 279–285.

42. Stavropoulos, P., Palagas, C., Angelopoulos, G. N., Papamantellos, D. N., and Couris, S., Calibration measurements in laser-induced breakdown spectroscopy using nanosecond and picosecond lasers. *Spectrochim. Acta B* 2004, *59*(12), 1885–1892.

43. Rohwetter, P., Yu, J., Mejean, G., Stelmaszczyk, K., Salmon, E., Kasparian, J., Wolf, J. P., and Woste, L., Remote LIBS with ultrashort pulses: Characteristics in picosecond and femtosecond regimes. *J. Anal. Atom. Spectrom.* 2004, *19*(4), 437–444.

44. Santagata, A., Spera, D., Albano, G., Teghil, R., Parisi, G. P., De Bonis, A., and Villani, P., Orthogonal fs/ns double-pulse LIBS for copper-based-alloy analysis. *Appl. Phys. A* 2008, *93*(4), 929–934.

45. Schill, A. W., Heaps, D. A., Stratis-Cullum, D. N., Arnold, B. R., and Pellegrino, P. M., Characterization of near-infrared low energy ultra-short laser pulses for portable applications of laser induced breakdown spectroscopy. *Opt. Express* 2007, *15*(21), 14044–14056.

46. Baudelet, M., Guyon, L., Yu, J., Wolf, J. P., Amodeo, T., Frejafon, E., and Laloi, P., Femtosecond time-resolved laser-induced breakdown spectroscopy for detection and identification of bacteria: A comparison to the nanosecond regime. *J. Appl. Phys.* 2006, *99*(8), 5332–5337.

47. López-Moreno, C., Palanco, S., Javier Laserna, J., DeLucia Jr., F., Miziolek, A. W., Rose, J., Walters, R. A., and Whitehouse, A. I., Test of a stand-off laser-induced breakdown spectroscopy sensor for the detection of explosive residues on solid surfaces. *J. Anal. Atom. Spectrom.* 2006, *21*(1), 55–60.

48. Anisimov, S. I. and Khokhlov, V. A., *Instabilities in Laser-Matter Interaction.* CRC Press, Boca Raton, FL, 1995.

49. Cremers, D. A. and Radziemski, L. J., *Handbook of Laser-Induced Breakdown Spectroscopy.* Wiley, Chichester, 2006.

50. Capitelli, M., Casavola, A., Colonna, G., and De Giacomo, A., Laser-induced plasma expansion: Theoretical and experimental aspects. *Spectrochim. Acta B* 2004, *59*(3), 271–289.

51. Radziemski, L. J., Loree, T. R., Cremers, D. A., and Hoffman, N. M., Time-resolved laser-induced breakdown spectrometry of aerosols. *Anal. Chem.* 1983, *55*(8), 1246–1252.

52. Elsayed-Ali, H. E., Norris, T. B., Pessot, M. A., and Mourou, G. A., Time-resolved observation of electron-phonon relaxation in copper. *Phys. Rev. Lett.* 1987, *58*(12), 1212.

53. Semerok, A., Salle, B., Wagner, J. F., and Petite, G., Femtosecond, picosecond, and nanosecond laser microablation: Laser plasma and crater investigation. *Laser Part. Beams* 2002, *20*(1), 67–72.

54. Angel, S. M., Stratis, D. N., Eland, K. L., Lai, T. S., Berg, M. A., and Gold, D. M., LIBS using dual- and ultra-short laser pulses. *Fresen. J. Anal. Chem.* 2001, *369*(3–4), 320–327.

55. Chichkov, B. N., Momma, C., Nolte, S., von Alvensleben, F., and Tunnermann, A., Femtosecond, picosecond and nanosecond laser ablation of solids. *Appl. Phys. A* 1996, *63*(2), 109–115.

56. Dikmelik, Y. and Spicer, J. B., Femtosecond laser-induced breakdown spectroscopy of explosives and explosive-related compounds. *Proc. SPIE* 2005, *5794*, 757–761.

57. McEnnis, C., Dikmelik, Y., Spicer, J. B., and Dagdigian, P. J., Femtosecond laser-induced breakdown spectroscopy of trinitrotoluene. In *Proceedings of the IEEE Antennas and Propagation Society International Symposium*, pp. 4933–4936. 9–15 June, Honolulu, HI: IEEE.

58. Rao, S. V., Sreedhar, S., Kumar, M. A., Kiran, P. P., Tewari, S. P., and Kumar, G. M., Laser induced breakdown spectroscopy of high energy materials using nanosecond, picosecond, and femtosecond pulses: Challenges and opportunities. *Proc. SPIE* 2011, *8173*, 81731A.

59. De Lucia Jr., F. C., Gottfried, J. L., and Miziolek, A. W., Evaluation of femtosecond laser-induced breakdown spectroscopy for explosive residue detection. *Opt. Express* 2009, *17*(2), 419–425.

60. Sreedhar, S., Kumar, M. A., Kumar, G. M., Kiran, P. P., Tewari, S. P., and Rao, S. V. Laser-induced breakdown spectroscopy of RDX and HMX with nanosecond, picosecond, and femtosecond pulses, *Proc. SPIE* 2010, *7665*, 76650T.

61. Huan, H., Lih-Mei, Y., and Jian, L., Femtosecond fiber-laser-based, laser-induced breakdown spectroscopy. *Proc. SPIE* 2012, *8358*, 835817.

62. Boueri, M., Baudelet, M., Yu, J., Mao, X., Mao, S. S., and Russo, R., Early stage expansion and time-resolved spectral emission of laser-induced plasma from polymer. *Appl. Surf. Sci.* 2009, *255*(24), 9566–9571.

63. Piñon, V., Fotakis, C., Nicolas, G., and Anglos, D., Double pulse laser-induced breakdown spectroscopy with femtosecond laser pulses. *Spectrochim. Acta B* 2008, *63*(10), 1006–1010.

64. Xu, B., Coello, Y., Lozovoy, V. V., Harris, D. A., and Dantus, M., Pulse shaping of octave spanning femtosecond laser pulses. *Opt. Express* 2006, *14*(22), 10939–10944.

65. Couairon, A. and Mysyrowicz, A., Femtosecond filamentation in transparent media. *Phys. Rep.* 2007, *441*(2–4), 47–189.

66. Chin, S. L., Brodeur, A., Petit, S., Kosareva, O. G., and Kandidov, V. P., Filamentation and supercontinuum generation during the propagation of powerful ultrashort laser pulses in optical media (white light laser). *J. Nonlinear Opt. Phys.* 1999, *08*(01), 121–146.

67. Lange, H. R., Chiron, A., Ripoche, J. F., Mysyrowicz, A., Breger, P., and Agostini, P., High-order harmonic generation and quasiphase matching in xenon using self-guided femtosecond pulses. *Phys. Rev. Lett.* 1998, *81*(8), 1611–1613.

68. Becker, A., Aközbek, N., Vijayalakshmi, K., Oral, E., Bowden, C. M., and Chin, S. L., Intensity clamping and re-focusing of intense femtosecond laser pulses in nitrogen molecular gas. *Appl. Phys. B* 2001, *73*(3), 287–290.

69. La Fontaine, B., Vidal, F., Jiang, Z., Chien, C. Y., Comtois, D., Desparois, A., Johnston, T. W., Kieffer, J. C., Pepin, H., and Mercure, H. P., Filamentation of ultrashort pulse laser beams resulting from their propagation over long distances in air. *Phys. Plasmas* 1999, *6*(5), 1615–1621.

70. Kasparian, J., Rodriguez, M., Méjean, G., Yu, J., Salmon, E., Wille, H., Bourayou, R., et al., White-light filaments for atmospheric analysis. *Science* 2003, *301*(5629), 61–64.

71. Heck, G., Sloss, J., and Levis, R. J., Adaptive control of the spatial position of white light filaments in an aqueous solution. *Opt. Commun.* 2006, *259*(1), 216–222.

72. Jin, Z., Zhang, J., Xu, M. H., Lu, X., Li, Y. T., Wang, Z. H., Wei, Z. Y., Yuan, X. H., and Yu, W., Control of filamentation induced by femtosecond laser pulses propagating in air. *Opt. Express* 2005, *13*(25), 10424–10430.

73. Hecht, E., *Optics*. Addison-Wesley: Harlow, England, 2002.

74. Mirell, D., Chalus, O., Peterson, K., and Diels, J.-C., Remote sensing of explosives using infrared and ultraviolet filaments. *J. Opt. Soc. Am. B* 2008, *25*(7), B108–B111.

75. Stelmaszczyk, K., Rohwetter, P., Mejean, G., Yu, J., Salmon, E., Kasparian, J., Ackermann, R., Wolf, J.-P., and Woste, L., Long-distance remote laser-induced breakdown spectroscopy using filamentation in air. *Appl. Phys. Lett.* 2004, *85*(18), 3977–3979.

76. Tzortzakis, S., Anglos, D., and Gray, D., Ultraviolet laser filaments for remote laser-induced breakdown spectroscopy (LIBS) analysis: Applications in cultural heritage monitoring. *Opt. Lett.* 2006, *31*(8), 1139–1141.

77. Rohwetter, P., Stelmaszczyk, K., Wöste, L., Ackermann, R., Méjean, G., Salmon, E., Kasparian, J., Yu, J., and Wolf, J. P., Filament-induced remote surface ablation for long range laser-induced breakdown spectroscopy operation. *Spectrochim. Acta B* 2005, *60*(7–8), 1025–1033.

78. Judge, E. J., Heck, G., Cerkez, E. B., and Levis, R. J., Discrimination of composite graphite samples using remote filament-induced breakdown spectroscopy. *Anal. Chem.* 2009, *81*(7), 2658–2663.

79. Boyd, R. W., *Nonlinear Optics*, 3rd edn. Academic Press, Boston, MA, 2008.

80. Carter, J. C., Angel, S. M., Lawrence-Snyder, M., Scaffidi, J., Whipple, R. E., and Reynolds, J. G., Standoff detection of high explosive materials at 50 m in ambient light conditions using a small Raman instrument. *Appl. Spectrosc.* 2005, *59*(6), 769–775.

81. Kneipp, K., Kneipp, H., Itzkan, I., Dasari, R. R., and Feld, M. S., Ultrasensitive chemical analysis by Raman spectroscopy. *Chem. Rev.* 1999, *99*(10), 2957–2976.

82. Moros, J., Lorenzo, J. A., and Laserna, J. J., Standoff detection of explosives: Critical comparison for ensuing options on Raman spectroscopy-LIBS sensor fusion. *Anal. Bioanal. Chem.* 2011, *400*(10), 3353–3365.

83. Zachhuber, B., Ramer, G., Hobro, A., Chrysostom, E. H., and Lendl, B., Stand-off Raman spectroscopy: A powerful technique for qualitative and quantitative analysis of inorganic and organic compounds including explosives. *Anal. Bioanal. Chem.* 2011, *400*(8), 2439–2447.

84. Albrecht, A. C. and Hutley, M. C., On the dependence of vibrational Raman intensity on the wavelength of incident light. *J. Chem. Phys.* 1971, *55*(9), 4438–4443.

85. Cheng, J. X. and Xie, X. S., Coherent anti-Stokes Raman scattering microscopy: Instrumentation, theory, and applications. *J. Phys. Chem. B* 2004, *108*(3), 827–840.

86. Romanov, D., Filin, A., Compton, R., and Levis, R., Phase matching in femtosecond BOXCARS. *Opt. Lett.* 2007, *32*(21), 3161–3163.

87. Zumbusch, A., Holtom, G. R., and Xie, X. S., Three-dimensional vibrational imaging by coherent anti-Stokes Raman scattering. *Phys. Rev. Lett.* 1999, *82*(20), 4142–4145.

88. Evans, C. L. and Xie, X. S., Coherent anti-Stokes Raman scattering microscopy: Chemical imaging for biology and medicine. *Annu. Rev. Anal. Chem.* 2008, *1*, 883–909.

89. Tolles, W. M., Nibler, J. W., McDonald, J. R., and Harvey, A. B., A review of the theory and application of coherent anti-Stokes Raman spectroscopy (CARS). *Appl. Spectrosc.* 1977, *31*(4), 253–271.

90. Dogariu, A. and Pidwerbetsky, A., Coherent anti-stokes Raman spectroscopy for detecting explosives in real time. *Proc. SPIE* 2012, *8358*, 83580R–83589R.

91. Bergé, L., Skupin, S., Nuter, R., Kasparian, J., and Wolf, J.-P., Ultrashort filaments of light in weakly ionized, optically transparent media. *Rep. Prog. Phys.* 2007, *70*(10), 1633.

92. Odhner, J. H. and Levis, R. J., Direct phase and amplitude characterization of femtosecond laser pulses undergoing filamentation in air. *Opt. Lett.* 2012, *37*(10), 1775–1777.

93. Odhner, J. H., Romanov, D. A., and Levis, R. J., Self-shortening dynamics measured along a femtosecond laser filament in air. *Phys. Rev. Lett.* 2010, *105*(12), 125001.

94. Dhar, L., Rogers, J. A., and Nelson, K. A., Time-resolved vibrational spectroscopy in the impulsive limit. *Chem. Rev.* 1994, *94*(1), 157–193.

95. Lee, Y. J., Liu, Y., and Cicerone, M. T., Characterization of three-color CARS in a two-pulse broadband CARS spectrum. *Opt. Lett.* 2007, *32*(22), 3370–3372.

96. Odhner, J. H., McCole, E. T., and Levis, R. J., Filament-driven impulsive Raman spectroscopy. *J. Phys. Chem. A* 2011, *115*(46), 13407–13412.

97. Oron, D., Dudovich, N., and Silberberg, Y., Single-pulse phase-contrast nonlinear Raman spectroscopy. *Phys. Rev. Lett.* 2002, *89*(27), 273001.

98. Oron, D., Dudovich, N., and Silberberg, Y., Femtosecond phase-and-polarization control for background-free coherent anti-Stokes Raman spectroscopy. *Phys. Rev. Lett.* 2003, *90*(21), 213902.

99. Li, H. W., Harris, D. A., Xu, B., Wrzesinski, P. J., Lozovoy, V. V., and Dantus, M., Standoff and arms-length detection of chemicals with single-beam coherent anti-Stokes Raman scattering. *Appl. Opt.* 2009, *48*(4), B17–B22.

100. Wrzesinski, P. J., Pestov, D., Lozovoy, V. V., Xu, B. W., Roy, S., Gord, J. R., and Dantus, M., Binary phase shaping for selective single-beam CARS spectroscopy and imaging of gas-phase molecules. *J. Raman Spectrosc.* 2011, *42*(3), 393–398.

101. McGrane, S. D., Scharff, R. J., Greenfield, M., and Moore, D. S., Coherent control of multiple vibrational excitations for optimal detection. *New J. Phys.* 2009, *11*(10), 105047.

102. Frostig, H., Katz, O., Natan, A., and Silberberg, Y., Single-pulse stimulated Raman scattering spectroscopy. *Opt. Lett.* 2011, *36*(7), 1248–1250.

103. Katz, O., Natan, A., Silberberg, Y., and Rosenwaks, S., Standoff detection of trace amounts of solids by nonlinear Raman spectroscopy using shaped femtosecond pulses. *Appl. Phys. Lett.* 2008, *92*(17), 171116.

104. Bremer, M. T., Lozovoy, V. V., and Dantus, M., Nondestructive detection and imaging of trace chemicals with high-chemical specificity using single-beam coherent anti-stokes Raman scattering in a standoff configuration. *Proc. SPIE* 2012, *8358*, 835818.

105. Bremer, M. T., Wrzesinski, P. J., Butcher, N., Lozovoy, V. V., and Dantus, M., Highly selective standoff detection and imaging of trace chemicals in a complex background using single-beam coherent anti-Stokes Raman scattering. *Appl. Phys. Lett.* 2011, *99*(10), 101109.

106. Katz, O., Levitt, J. M., Grinvald, E., and Silberberg, Y., Single-beam coherent Raman spectroscopy and microscopy via spectral notch shaping. *Opt. Express* 2010, *18*(22), 22693–22701.

107. Natan, A., Levitt, J. M., Graham, L., Katz, O., and Silberberg, Y., Standoff detection via single-beam spectral notch filtered pulses. *Appl. Phys. Lett.* 2012, *100*(5), 051111.

108. Pestov, D., Murawski, R. K., Ariunbold, G. O., Wang, X., Zhi, M., Sokolov, A. V., Sautenkov, V. A., et al., Optimizing the laser-pulse configuration for coherent Raman spectroscopy. *Science* 2007, *316*(5822), 265–268.

109. Dogariu, A., Real-time standoff trace explosives detection and imaging with CARS. Final report to JHU/APL, 2013.

110. McCole, E. T., Odhner, J. H., Romanov, D. A., and Levis, R. J., Spectral-to-temporal amplitude mapping polarization spectroscopy of rotational transients. *J. Phys. Chem. A* 2013, *117*(29), 6354–6361.

111. Seideman, T., Revival structure of aligned rotational wave packets. *Phys. Rev. Lett.* 1999, *83*(24), 4971–4974.

112. Rouzée, A., Guérin, S., Faucher, O., and Lavorel, B., Field-free molecular alignment of asymmetric top molecules using elliptically polarized laser pulses. *Phys. Rev. A* 2008, *77*(4), 043412.

113. Lucht, R. P., Roy, S., Meyer, T. R., and Gord, J. R., Femtosecond coherent anti-Stokes Raman scattering measurement of gas temperatures from frequency-spread dephasing of the Raman coherence. *Appl. Phys. Lett.* 2006, *89*(25), 251112.

114. Chen, Y. -H., Varma, S., York, A., and Milchberg, H. M., Single-shot, space- and time-resolved measurement of rotational wavepacket revivals in H_2, D_2, N_2, O_2, and N_2O. *Opt. Express* 2007, *15*(18), 11341–11357.

115. Zamith, S., Ansari, Z., Lepine, F., and Vrakking, M. J. J., Single-shot measurement of revival structures in femtosecond laser-induced alignment of molecules. *Opt. Lett.* 2005, *30*(17), 2326–2328.

116. Hartinger, K. and Bartels, R. A., Single-shot measurement of ultrafast time-varying phase modulation induced by femtosecond laser pulses with arbitrary polarization. *Appl. Phys. Lett.* 2008, *92*(2), 021126.

117. Loriot, V., Tehini, R., Hertz, E., Lavorel, B., and Faucher, O., Snapshot imaging of post-pulse transient molecular alignment revivals. *Phys. Rev. A* 2008, *78*(1), 013412.

118. Kim, H., Lagutchev, A., and Dlott, D. D., Surface and interface spectroscopy of high explosives and binders: HMX and Estane. *Propell. Explos. Pyrot.* 2006, *31*(2), 116–123.

119. Surber, E., Lozano, A., Lagutchev, A., Kim, H., and Dlott, D. D., Surface nonlinear vibrational spectroscopy of energetic materials: HMX. *J. Phys. Chem. C* 2007, *111*(5), 2235–2241.

120. Rahm, M., Tyrode, E., Brinck, T., and Johnson, C. M., The molecular surface structure of ammonium and potassium dinitramide: A vibrational sum frequency spectroscopy and quantum chemical study. *J. Phys. Chem. C* 2011, *115*(21), 10588–10596.

121. Asher, W. E. and Willard-Schmoe, E., Vibrational sum-frequency spectroscopy for trace chemical detection on surfaces at stand-off distances. *Appl. Spectrosc.* 2013, *67*(3), 253–260.

122. Petty, G., Tai, C. and Dalby, F. W., Nonlinear resonant photoionization in molecular iodine. *Phys. Rev. Lett.* 1975, *34*(19), 1207–1209.

123. Johnson, P. M., Berman, M. R., and Zakheim, D., Nonresonant multiphoton ionization spectroscopy: The four-photon ionization spectrum of nitric oxide. *J. Chem. Phys.* 1975, *62*(6), 2500–2502.

124. Lehmann, K. K., Smolarek, J., and Goodman, L., Multiphoton resonance ionization bands in I_2. *J. Chem. Phys.* 1978, *69*(4), 1569–1573.

125. Parker, D. H., Sheng, S. J., and El-Sayed, M. A., Multiphoton ionization spectrum of trans-hexatriene in the 6.2 eV region. *J. Chem. Phys.* 1976, *65*(12), 5534–5535.

126. Berg, J. O., Parker, D. H., and El-Sayed, M. A., Symmetry assignment of two-photon states from polarization characteristics of multiphoton ionization spectra. *J. Chem. Phys.* 1978, *68*(12), 5661–5663.

127. Nieman, G. C. and Colson, S. D., A new electronic state of ammonia observed by multiphoton ionization. *J. Chem. Phys.* 1978, *68*(12), 5656–5657.

128. Weinkauf, R., Aicher, P., Wesley, G., Grotemeyer, J. and Schlag, E. W., Femtosecond versus nanosecond multiphoton ionization and dissociation of large molecules. *J. Phys. Chem.* 1994, *98*(34), 8381–8391.

129. Roslund, J., Shir, O. M., Dogariu, A., Miles, R., and Rabitz, H., Control of nitromethane photoionization efficiency with shaped femtosecond pulses. *J. Chem. Phys.* 2011, *134*(15), 154301–154310.

130. Dogariu, A. and Miles, R. B., Detecting localized trace species in air using radar resonance-enhanced multi-photon ionization. *Appl. Opt.* 2011, *50*(4), A68–A73.

131. Shneider, M. N. and Miles, R. B., Microwave diagnostics of small plasma objects. *J. Appl. Phys.* 2005, *98*(3), 033301–033303.

132. Miles, R. B., Zhang, Z. L., Zaidi, S. H., and Shneider, M. N., Microwave scattering from laser ionized molecules: A new approach to nonintrusive diagnostics. *AIAA J.* 2007, *45*(3), 513–515.

133. Luo, Q., Hosseini, A., Liu, W., and Chin, S. L., Lasing action in air induced by ultrafast laser filamentation. *Opt. Photon. News* 2004, *15*(9), 44–47.

134. Kartashov, D., Ališauskas, S., Andriukaitis, G., Pugžlys, A., Shneider, M., Zheltikov, A., Chin, S. L., and Baltuška, A., Free-space nitrogen gas laser driven by a femtosecond filament. *Phys. Rev. A* 2012, *86*(3), 033831.

135. Shneider, M. N., Baltuska, A. and Zheltikov, A. M., Population inversion of molecular nitrogen in an Ar: N_2 mixture by selective resonance-enhanced multiphoton ionization. *J. Appl. Phys.* 2011, *110*(8), 083112–083117.

136. Dogariu, A., Michael, J. B., Scully, M. O., and Miles, R. B., High-gain backward lasing in air. *Science* 2011, *331*(6016), 442–445.

137. Hemmer, P. R., Miles, R. B., Polynkin, P., Siebert, T., Sokolov, A. V., Sprangle, P., and Scully, M. O., Standoff spectroscopy via remote generation of a backward-propagating laser beam. *Proc. Nat. Acad. Sci.* 2011, *108*(8), 3130–3134.

138. Malevich, P. N., Kartashov, D., Pu, Z., Alisauskas, S., Pugzlys, A., Baltuska, A., Giniunas, L., et al., Ultrafast-laser-induced backward stimulated Raman scattering for tracing atmospheric gases. *Opt. Express* 2012, *20*(17), 18784–18794.

139. Adler, F., Thorpe, M. J., Cossel, K. C., and Ye, J., Cavity-enhanced direct frequency comb spectroscopy: Technology and applications. *Annu. Rev. Anal. Chem.* 2010, *3*(1), 175–205.

140. Telle, H. R., Steinmeyer, G., Dunlop, A. E., Stenger, J., Sutter, D. H., and Keller, U., Carrier-envelope offset phase control: A novel concept for absolute optical frequency measurement and ultrashort pulse generation. *Appl. Phys. B* 1999, *69*(4), 327–332.

141. Jones, D. J., Diddams, S. A., Ranka, J. K., Stentz, A., Windeler, R. S., Hall, J. L., and Cundiff, S. T., Carrier-envelope phase control of femtosecond mode-locked lasers and direct optical frequency synthesis. *Science* 2000, *288*(5466), 635–639.

142. Holzwarth, R., Udem, T., Hänsch, T. W., Knight, J. C., Wadsworth, W. J., and Russell, P. S. J., Optical frequency synthesizer for precision spectroscopy. *Phys. Rev. Lett.* 2000, *85*(11), 2264–2267.

143. Adler, F. and Diddams, S. A., High-power, hybrid Er:fiber/Tm:fiber frequency comb source in the 2 μm wavelength region. *Opt. Lett.* 2012, *37*(9), 1400–1402.

144. Adler, F., Cossel, K. C., Thorpe, M. J., Hartl, I., Fermann, M. E., and Ye, J., Phase-stabilized, 1.5 W frequency comb at 2.8–4.8 μm. *Opt. Lett.* 2009, *34*(9), 1330–1332.

145. Diddams, S. A., Hollberg, L., and Mbele, V., Molecular fingerprinting with the resolved modes of a femtosecond laser frequency comb. *Nature* 2007, *445*(7128), 627–630.

146. Thorpe, M. J., Balslev-Clausen, D., Kirchner, M. S., and Ye, J., Cavity-enhanced optical frequency combspectroscopy: Application to human breathanalysis. *Opt. Express* 2008, *16*(4), 2387–2397.

147. Mineo, M. and Paoloni, C., Corrugated rectangular waveguide tunable backward wave oscillator for terahertz applications. *IEEE Trans. Electron Devices* 2010, *57*(6), 1481–1484.

148. Gendriesch, R., Lewen, F., Winnewisser, G., and Hahn, J., Precision broadband spectroscopy near 2 THz: Frequency-stabilized laser sideband spectrometer with backward-wave oscillators. *J. Mol. Spectrosc.* 2000, *203*(1), 205–207.

149. Ge, X. H., Lv, M., Zhong, H., and Zhang, C. L., Terahertz wave reflection imaging system based on backward wave oscillator and its application. *J. Infrared Millim. Waves* 2010, *29*(1), 15–18.

150. Pronin, A. V., Goncharov, Y. G., Fischer, T., and Wosnitza, J., Phase-sensitive terahertz spectroscopy with backward-wave oscillators in reflection mode. *Rev. Sci. Instrum.* 2009, *80*(12), 123904.

151. Kohler, R., Tredicucci, A., Beltram, F., Beere, H. E., Linfield, E. H., Davies, A. G., Ritchie, D. A., Iotti, R. C., and Rossi, F., Terahertz semiconductor-heterostructure laser. *Nature* 2002, *417*(6885), 156–159.

152. Scalari, G., Walther, C., Fischer, M., Terazzi, R., Beere, H., Ritchie, D., and Faist, J., THz and sub-THz quantum cascade lasers. *Laser Photon. Rev.* 2009, *3*(1–2), 45–66.

153. Lee, A. W. M., Qin, Q., Kumar, S., Williams, B. S., and Hu, Q., Real-time terahertz imaging over a standoff distance (>25 m). *Appl. Phys. Lett.* 2006, *89*(14), 141125.

154. Kawase, K., Mizuno, M., Sohma, S., Takahashi, H., Taniuchi, T., Urata, Y., Wada, S., Tashiro, H., and Ito, H., Difference-frequency terahertz-wave generation from 4-dimethylamino-N-methyl-4-stilbazolium-tosylate by use of an electronically tuned Ti:sapphire laser. *Opt. Lett.* 1999, *24*(15), 1065–1067.

155. Schneider, A., Neis, M., Stillhart, M., Ruiz, B., Khan, R. U. A., and Gunter, P., Generation of terahertz pulses through optical rectification in organic DAST crystals: Theory and experiment. *J. Opt. Soc. Am. B* 2006, *23*(9), 1822–1835.

156. Han, P. Y., Tani, M., Pan, F., and Zhang, X. C., Use of the organic crystal DAST for terahertz beam applications. *Opt. Lett.* 2000, *25*(9), 675–677.

157. Hu, B. B., Zhang, X. C., and Auston, D. H., Terahertz radiation induced by subband-gap femtosecond optical excitation of GaAs. *Phys. Rev. Lett.* 1991, *67*(19), 2709–2712.

158. Vodopyanov, K. L., Fejer, M. M., Yu, X., Harris, J. S., Lee, Y. S., Hurlbut, W. C., Kozlov, V. G., Bliss, D., and Lynch, C., Terahertz-wave generation in quasi-phase-matched GaAs. *Appl. Phys. Lett.* 2006, *89*(14), 141119.

159. Löffler, T., Hahn, T., Thomson, M., Jacob, F., and Roskos, H., Large-area electro-optic ZnTe terahertz emitters. *Opt. Express* 2005, *13*(14), 5353–5362.

160. Wang, Z., Generation of terahertz radiation via nonlinear optical methods. *IEEE Trans. Geosci. Remote* 2001, *1*(1), 1–5.

161. Dexheimer, S. L., *Terahertz Spectroscopy: Principles and Applications.* Taylor & Francis, Boca Raton, FL, 2007.

162. D'Amico, C., Houard, A., Franco, M., Prade, B., Mysyrowicz, A., Couairon, A., and Tikhonchuk, V. T., Conical forward THz emission from femtosecond-laser-beam filamentation in air. *Phys. Rev. Lett.* 2007, *98*(23), 235002.

163. Nahata, A., Weling, A. S., and Heinz, T. F., A wideband coherent terahertz spectroscopy system using optical rectification and electro-optic sampling. *Appl. Phys. Lett.* 1996, *69*(16), 2321–2323.

164. Jepsen, P. U., Cooke, D. G., and Koch, M., Terahertz spectroscopy and imaging – Modern techniques and applications. *Laser Photon. Rev.* 2011, *5*(1), 124–166.

165. Rostami, A., Rasooli, H., and Baghban, H., *Terahertz Technology: Fundamentals and Applications.* Springer, Berlin, 2011.

166. Amico, C. D., Houard, A., Akturk, S., Liu, Y., Bloas, J. L., Franco, M., Prade, B., Couairon, A., Tikhonchuk, V. T., and Mysyrowicz, A., Forward THz radiation emission by femtosecond filamentation in gases: Theory and experiment. *New J. Phys.* 2008, *10*(1), 013015.

167. Zhang, Y., Chen, Y., Marceau, C., Liu, W., Sun, Z. D., Xu, S., Théberge, F., Châteauneuf, M., Dubois, J., and Chin, S. L., Non-radially polarized THz pulse emitted from femtosecond laser filament in air. *Opt. Express* 2008, *16*(20), 15483–15488.

168. Sprangle, P., Peñano, J. R., Hafizi, B., and Kapetanakos, C. A., Ultrashort laser pulses and electromagnetic pulse generation in air and on dielectric surfaces. *Phys. Rev. E* 2004, *69*(6), 066415.

169. Liu, Y., Houard, A., Prade, B., Akturk, S., Mysyrowicz, A., and Tikhonchuk, V. T., Terahertz radiation source in air based on bifilamentation of femtosecond laser pulses. *Phys. Rev. Lett.* 2007, *99*(13), 135002.

170. Minami, Y., Kurihara, T., Yamaguchi, K., Nakajima, M., and Suemoto, T., Longitudinal terahertz wave generation from an air plasma filament induced by a femtosecond laser. *Appl. Phys. Lett.* 2013, *102*(15), 151106-1–151106-3.

171. Chen, Y., Marceau, C., Liu, W., Sun, Z.-D., Zhang, Y., Theberge, F., Chateauneuf, M., Dubois, J., and Chin, S. L., Elliptically polarized terahertz emission in the forward direction of a femtosecond laser filament in air. *Appl. Phys. Lett.* 2008, *93*(23), 231116-1–231116-3.

Index

Printed and bound by CPI Group (UK) Ltd, Croydon, CR0 4YY

22/10/2024

01777613-0013